Mechanical Behavior of Polymeric Materials: Recent Study

Mechanical Behavior of Polymeric Materials: Recent Study

Emilia P. Collar
Jesús-María García-Martínez

MDPI

Basel • Beijing • Wuhan • Barcelona • Belgrade • Novi Sad • Cluj • Manchester

Editors

Emilia P. Collar
Polymer Engineering
Group (GIP)
Institute of Polymer Science
and Technology (ICTP)
Madrid
Spain

Jesús-María García-Martínez
Polymer Engineering
Group (GIP)
Institute of Polymer Science
and Technology (ICTP)
CSIC
Madrid
Spain

Editorial Office
MDPI AG
Grosspeteranlage 5
4052 Basel, Switzerland

This is a reprint of articles from the Special Issue published online in the open access journal *Polymers* (ISSN 2073-4360) (available at: www.mdpi.com/journal/polymers/special_issues/Mechanical_Behavior_Polymers).

For citation purposes, cite each article independently as indicated on the article page online and as indicated below:

Lastname, A.A.; Lastname, B.B. Article Title. *Journal Name* **Year**, *Volume Number*, Page Range.

ISBN 978-3-7258-2524-0 (Hbk)
ISBN 978-3-7258-2523-3 (PDF)
doi.org/10.3390/books978-3-7258-2523-3

Contents

About the Editors

Emilia P. Collar

Emilia P. Collar has a Ph.D. in Industrial Chemistry (U. Complutense, 1986). Since 1990, Dr. Collar has been a permanent staff member (Tenured Scientist) at the Consejo Superior de Investigaciones Cient´(CSIC), after two years (1986–88) as CSIC's postdoc fellow and one (1989) as a Chemical Engineering Assistant Teacher at the Universidad Complutense de Madrid. In the Polymer Science and Technology Institute (ICTP/CSIC), she works at the Polymer Engineering Group (GIP), founded in 1982 by Prof. O. Laguna, and has been the GIP's Head since 1999. Between 1990 and 2005, she supervised six doctoral theses and five post-graduate ones under 14 research publicly funded projects, as well as 18 research-funded contracts and 53 technical reports for different companies. Furthermore, from 2001 to 2005, she held the positions of Technical Director of the Physical and Mechanical Properties Laboratory and Deputy-Technical Director of the Thermal Properties Laboratory under the successful ISO 17025 Accreditation Project for the CSIC/ICTP's Laboratories, ACiTP, commanded by the ICTP's Head. She is also the author and co-author of more than 20 chapters on books, two currently active industrial patents on polymer recycling, and more than 150 papers, mainly in SCI Journals. Her research work lines deal with polymers and the environment under the general frame of heterogeneous materials based on polymers. From 2006 to the present, she has participated in four public Spanish-funded research projects and one EU project under its 7th Framework Program.

Jesús-María García-Martínez

Jesús-María García-Martínez holds a Ph.D. in Chemistry (Chemical Engineering) from Universidad Complutense de Madrid (1995), two M.Sc. level degrees (Industrial Chemistry and Polymer Science and Technology), and has received more than 70 highly specialized courses. He is a Tenured Scientist at the Institute of Polymer Science and Technology (ICTP) of the Spanish National Research Council (CSIC). Since 1992, within the Polymer Engineering Group (GIP), he has co-authored more than 200 scientific and/or technical works on topics related to polymer engineering, chemical modification of polymers, heterogeneous materials based on polymers, interphases and interfaces, polymer composites, polyblends and alloys, organic–inorganic hybrids materials, polymer recycling, quality, and so on. Furthermore, he has participated in 32 research and industrial projects (national and international programs) and is co-author of one currently active industrial patent on polymer recycling. From 2000 to 2005, he also assumed the position of Quality Director for the ICTP (CSIC) ISO 17025 Accreditation Project of the ICTP Laboratories. Additionally, Dr. García-Martínez is actively reviewing tasks for WOS and SCOPUS-indexed journals, with more than 450 reports in the last few years, and has been awarded twice with the Publons Reviewer Awards (2018, 2019) and once with the POLYMERS Outstanding Reviewer Award (2019). He is an editorial member of Polymer Science-related journals and Editor of more than 80 scientific articles. He has been a Guest Editor for seven Special Issues of Q1 journals. Since 2016, he has served as Head of the Department of Chemistry and Properties of Polymer Materials within the ICTP (CSIC).

Preface

This Special Issue is dedicated to the captivating and intricate field of polymer science and technology, particularly the myriad factors that shape the properties of polymer-based materials. In this context, it is crucial to equip our readers with a foundational understanding of what failure signifies in a polymer-based material. This understanding is not just a matter of academic interest but a pressing need in the field of polymer science and engineering.

From a fundamental perspective, fractures in a material involve surpassing the bond forces that hold its constituent atoms together. However, the diverse ways these atoms are bonded as molecules, leading to various supramolecular aggregates, give rise to intricate questions. This complexity, which underscores the heterogeneous nature of these materials, is a challenging and fascinating characteristic that may intrigue and captivate our readers.

In determining material strength capabilities, it has been undisputed that a fracture will always take place in the weakest region of a material; however, the search for empirical and semi-empirical approaches based on the results of conveniently designed mechanical testing procedures restricted to material or part performance still represents an open research question. These considerations have become especially relevant in the case of heterogeneous materials based entirely or partially on organic polymers. Also, it has been well established that the stress/strain relationship is not linear because it strongly depends on time and temperature effects on the applied external force level and morphological changes in the material bulk due to environmental conditions. This underscores the ongoing importance of research in this field, fulfilling essential research requirements and ensuring reproducible and reliable results.

Ultimately, the findings in the literature have made it clear that mechanical energy interacts with matter and materials in different ways in different environments and conditions. This, together with the correct analysis of material responses, will continue to be vital in the development of mathematical models to assist in material design optimization within a general framework of sustainability, with the end purpose of avoiding catastrophic failure situations such as structural collapses or material ruptures.

The works compiled in this work offer excellent examples of the current and future trends in the field of the mechanical properties of polymeric materials. The eleven novel articles in this Special Issue provide information and inspiration for future approaches in this fascinating field of science.

Emilia P. Collar and Jesús-María García-Martínez
Editors

polymers

MDPI

Editorial

Mechanical Behavior of Polymeric Materials: Recent Studies

Emilia P. Collar * and **Jesús-María García-Martínez** *

Polymer Engineering Group (GIP), Polymer Science and Technology Institute (ICTP), Spanish National Research Council (CSIC), C/Juan de la Cierva, 3, 28006 Madrid, Spain
* Correspondence: ecollar@ictp.csic.es (E.P.C.); jesus.maria@ictp.csic.es (J.-M.G.-M.)

This Special Issue is devoted to one of the most exciting fields in polymer science and technology: the many factors that influence the properties of polymer-based materials. Therefore, it is pertinent to provide readers with some fundamental information on what is meant by failure in a polymer-based material (and thus in its mechanical properties). From a fundamental approach, fracture in a given material means overpassing the bond forces that keep its constituent atoms together. However, the ways in which these atoms are bonded as molecules and appear as very different types of supramolecular aggregates lead to complex questions, implying a heterogeneous character in these materials. In determining material strength capabilities, it has been undisputed since early studies by Griffith [1] that a fracture will always take place in the weakest region of a material; however, the search for empirical and semi-empirical approaches based on the results of conveniently designed mechanical testing procedures restricted to material or part performance still represent an open research question.

These considerations have become especially relevant in the case of heterogeneous materials based entirely or partially on organic polymers. From early times in materials research [2–4], it has been well established that the stress/strain relationship is not linear, because it is strongly dependent on time and temperature effects on the applied external force level and morphological changes in the material bulk as a result of environmental conditions [5–7]. This underscores the ongoing importance of research in this field fulfilling key research requirements and ensuring reproducible and reliable results. The development of inter-laboratory protocols and international standardized procedures is testament to the reliability of our results, and these efforts are crucial in providing robust mechanical parameter values [8–13]. These values should be suitable for incorporation into solid databases to feed polymeric part and material design software [14] in order to forecast the mechanical behavior of a polymeric material once it reaches the solid state after the processing steps that determine the emerging morphologies of the organic fractions in this type of material [15,16].

Ultimately, findings have made it clear that the study of mechanical energy interacts with matter and materials in different ways in different environments and conditions. This, together with the correct analysis of material responses, will continue to be vital in the development of mathematical models to assist in material design optimization within a general framework of sustainability, with the end purpose of avoiding catastrophic failure situations. A significant number of manuscripts were submitted for consideration for this Special Issue, but only a limited number were published following the rigorous revision process of *Polymers*. This Special Issue includes eleven exciting works related to this hotspot in polymer R&D. The articles compiled in this volume fully align with the philosophies mentioned above. The aim of this Editorial is not to elaborate on each of the texts but to encourage the reader to browse them in depth.

Sun et al. [17] present a finite element numerical simulation calculation for pavement structure load responses, considering that there is still a lack of research on polyurethane (PU) mixture composite pavement load responses because of our lack of understanding

Citation: Collar, E.P.; García-Martínez, J.-M. Mechanical Behavior of Polymeric Materials: Recent Studies. *Polymers* **2024**, *16*, 2821. https://doi.org/10.3390/polym16192821

Received: 20 August 2024
Revised: 27 September 2024
Accepted: 27 September 2024
Published: 5 October 2024

of the mechanical characteristics of PU mixture composite pavement. To fill this gap, this article analyses the mechanical properties of PU mixtures using the dynamic modulus test, the uniaxial penetration test, fatigue tests, and the finite element theory calculation method, to perform a load response calculation for an orthogonal-design composite pavement structure. The results show that PU mixtures exhibit more obvious elastic characteristics with good shear resistance, fatigue, and temperature stability and can be employed as shear and anti-fatigue layers. Using these data, a comparison with conventional pavement is performed and fascinating conclusions are drawn.

The article by Shishkovsky et al. [18] presents a study on the mechanical and thermo-mechanical performance of shape memory PLA parts by employing many sets, considering five variable printing parameters, in an FDM printing method. The results show that the temperature of the extruder and the nozzle diameter were the most significant parameters when it came to achieving better mechanical properties in the final part. In other words, this comprehensive study demonstrated a complex operational relationship between the mechanical and thermomechanical properties determined, combining the characteristics of a thermoplastic material with the shape memory effect and FDM printing parameters.

Vozniak et al. [19] investigated the mass transfer process of binary esters of acetic acid in poly(ethylene terephthalate) (PET), poly(ethylene terephthalate) with a high degree of glycol modification (PETG), and glycol-modified poly(cyclohexanedimethylene terephthalate) (PCTG), finding that the desorption rate of the complex ether at the equilibrium point was significantly lower than the sorption rate, depending on the type of polyester and the temperature and permitting the accumulation of ester in the polyester bulk. This effect is beneficial as a physical blowing agent in the filament extrusion additive manufacturing (AM) process, and it represents the main advantage of the obtention of nonbrittle foams compared to conventional polyester foams.

The work by Ehrmann et al. [20] is focused on the use of textile fabrics in fabricating stab-resistant garments with the advantage of being lightweight, considering most materials on the market. These research attempts are currently focused on textile fabrics, mostly through impregnation with shear-thickening fluids (STFs) or ceramic coatings, as well as on lightweight composites. In this work, the authors discuss different measurement methods, including dynamic and quasistatic methods, and the correlations of stab resistance with other physical properties jointly with an overview of recent developments and research on stab-resistant polymers, by employing different materials and combinations of materials and structures.

The investigation by Wildemann et al. [21] is focused on the effect of various types of external loadings on composite structures, leading to a decrease in mechanical properties in a system as a whole. For this purpose, many experimental investigations into the mechanical behavior of composites under uniaxial cyclic loading had previously been carried out. In contrast, in this study, new data on reducing composite materials' mechanical characteristics under multiaxial cyclic loading were examined. Hence, these authors performed an experimental investigation into the mechanical behavior of fiberglass tubes under proportional cyclic loading and static and fatigue tests conducted under tension with torsion conditions. Further, the accumulation of structural damage was examined. Finally, the authors conclude that multiaxial cyclic loading significantly reduces the mechanical properties and must be considered in composite structure design.

The research article by Sebaey, Wagih et al. [22] demonstrates an enhancement in notch sensitivity in a hybrid carbon/epoxy (CFRP) composite with a Kevlar core sandwich compared to monotonic CFRP and Kevlar composites, bearing in mind that so-called fiber-reinforced plastic composites are sensitive to holes inducing out-of-plane stresses. The study is performed through open-hole tension (OHT) tests and further comparison of the open-hole tensile strength and strain and damage propagation. The results show that hybrid laminate has a lower notch sensitivity than CFRP and KFRP laminates because the strength reduction rate with the hole size is lower.

Belaadi et al. [23] present a comprehensive study of the elements that enhance the performance of biocomposites or sustainable ropes created from vegetable fibers, which need to have a high confidence interval (95% CI) for the mechanical characteristic data of performance materials in order to be useful for the design of parts due to the enormous variation in plant fibers properties. To achieve this, the authors implemented an experimental design involving four groups, varying the number of tests to determine the mechanical properties (modulus, strength, and strain at break) of sisal yarn in order to ensure that the CI was at least 95%.

The work by Peponi et al. [24] reports their study into the mechanical behavior of woven and non-woven PLA/OLA/MgO electrospun fibers in detail using the Box–Wilson surface response methodology as a follow-up to a previous work by the authors, where the diameters and thermal responses of these fibers were discussed in terms of the different amounts of magnesium oxide nanoparticles (MgO), as well as the oligomer (lactic acid, OLA) used as a plasticizer. The results suggest that these works can be strongly correlated. Critical points for both MgO and OLA were identified. Therefore, the approach presented by the authors permits the design of tailor-made electrospun nanocomposites with specific mechanical requirements.

The contribution by Koutsos et al. [25] addresses variability in the thermomechanical behavior of virgin and recycled polypropylene/high-density polyethylene blends without other components. The authors highlight the fact that understanding the performance variability in such blends is a crucial aspect in enabling recycled materials to re-enter the consumer market in terms of the circular economy, requiring further research to minimize the inhomogeneity of recycled materials. For these purposes, the authors performed a complete thermal and mechanical characterization of virgin and recycled polypropylene/high-density polyethylene blends, concluding that recycled and virgin blends are immiscible. This study was complemented by a dynamical mechanical analysis showing a slight variation in the storage modulus of recycled and virgin blends but lower alpha and beta relaxation temperatures in recycled blends. Additionally, the authors conclude that the tensile properties of recycled blends exhibit worse properties, which they explain.

The article authored by Cauich-Robriguez, Peponi et al. [26] explores the possibility of mimicking the architecture of human coronary artery native vessels for biomedical applications. The authors used roto-evaporation to engineer a three-layer polyurethane vascular graft (TVG). They synthesized two segmented polyurethanes using lysine (SPUUK) and ascorbic acid (SPUAA) to create the intima and adventitia layers, respectively. In contrast, the media layer was fabricated from a commercially available polyurethane and compared to single-layer vascular grafts (SVGs) from individual polyurethanes and a polyurethane blend (MVG). They found that the TVG exhibited the highest circumferential tensile strength and longitudinal forces compared to single-layer vascular grafts of lower thicknesses made from the same polyurethanes. The TVG also showed higher suture and burst strength values than native vessels. Additionally, the authors performed an indirect cytocompatibility test, showing 90 to 100% viability for all polyurethanes, surpassing the minimum 70% threshold needed for biomaterials deemed to have cytocompatibility. Except for SPUUK, all exhibited poor fibroblast adhesion and hemolysis values under the permissible limit of 5%, as well as longer coagulation times.

Finally, Minh et al. [27] present a study investigating the mechanical properties of coconut sawdust powder combined with polypropylene (PP) by observing the effect of the compatibilizer, wood powder (WP) content, and injection molding parameters on the properties of coconut wood powder composite (WPC). Based on a Taguchi experimental design, the authors performed an interesting investigation to determine the influential coordinates required to obtain optimized mechanical properties (tensile and flexural strength and hardness), aided by scanning electronic microscopy (SEM). They mainly concluded that the parameters influencing the ultimate properties follow a hierarchy, with the wood powder being the most influential and the melt temperature being the least influential.

To conclude, as the Guest Editors of this fascinating Special Issue, we can comfortably say that the topic "Mechanical Behavior of Polymeric Materials" represents an essential framework in the field of Polymer Science and Technology, now and in the near future. For this reason, a second Special Issue on this topic, to be published in 2025 in *Polymers*, is now in progress and open for submissions, and contributions are welcomed.

Conflicts of Interest: The authors declare no conflict of interest.

References

1. Griffith, A.A. The Phenomena of Rupture and Flow in Solids. *Philos. Trans. R. Soc. Lond.* **1921**, *221*, 163–198.
2. Flory, P.J. *Principles of Polymer Chemistry*; Cornell University Press: New York, NY, USA, 1953.
3. Morawetz, H. *Polymers: The Origins and Growth of a Science*; John Wiley & Sons: New York, NY, USA, 1985.
4. Peterlin, A. *Plastic Deformation of Polymers*; M. Dekker: New York, NY, USA, 1971.
5. Van Krevelen, D.W.; Nijenhuis, K.T. *Properties of Polymers: Their Correlation with Chemical Structure*; Their Numerical Estimation and Prediction from Additive Group Contributions; Elsevier: Amsterdam, The Netherlands, 2009.
6. Mascia, L. *The Role of Additives in Plastics*; Halsted Press: New York, NY, USA, 1975.
7. Gugumus, F. The Use of Accelerated Tests in the Evaluation of Antioxidants and Light Stabilizers. In *Developments in Polymer Stabilisation-8*; Scott, G., Ed.; Springer: Dordrecht, The Netherlands, 1987.
8. Al Malaika, S.; Golovoy, A.; Wilkie, C.A. *Chemistry and Technology of Polymer Additives*; Blackwell Science: Oxford, UK, 1999.
9. Mascia, L. *Thermoplastics: Materials Engineering*, 2nd ed.; Elsevier: New York, NY, USA, 1989; Chapter 4.
10. Katz, H.S.; Milewski, J.V. *Hand Book of Fillers and Reinforcements for Plastics*; Van Nostrand Reinhold: New York, NY, USA, 1978.
11. Cogswell, E.N. *Thermoplastic Aromatic Polymer Composites*; Buttenvorth-Heinemann: Oxford, UK, 1992.
12. Carlsson, L.A.; Adams, D.F.; Pipes, R.B. *Experimental Characterization of Advanced Composite Materials*, 4th ed.; CRC Press: Boca Raton, Fl, USA, 2014.
13. Roulin-Moloney, A.C. (Ed.) *Fractography and Failure Mechanisms of Polymers and Composites*; Kluwer Academic Publishers: Dordrecht, The Netherlands, 1989.
14. Japan Association for Chemical Innovation (Ed.) *Computer Simulation of Polymeric Materials: Applications of the OCTA System*; Springer: Singapore, 2016.
15. Han, C.D. *Rheology in Polymer Processing*; Academic Press: Cambridge, MA, USA, 1976.
16. Tadmor, Z.; Gogos, C.G. *Principles of Polymer Processing*, 2nd ed.; Wiley: New York, NY, USA, 1979.
17. Zhuang, W.; Bi, Y.; Liu, B.; Hou, D.; Jing, S.; Lu, X.; Sun, M. Mechanical Properties of Polyurethane Mixture and Load Response Behaviour of Polyurethane Composite Pavement. *Polymers* **2023**, *15*, 417. [CrossRef] [PubMed]
18. Issabayeva, Z.; Shishkovsky, I. Prediction of The Mechanical Behavior of Polylactic Acid Parts with Shape Memory Effect Fabricated by FDM. *Polymers* **2023**, *15*, 1162. [CrossRef] [PubMed]
19. Sova, N.; Savchenko, B.; Beloshenko, V.; Slieptsov, A.; Vozniak, I. Sorption Properties of PET Copolyesters and New Approach for Foaming with Filament Extrusion Additive Manufacturing. *Polymers* **2023**, *15*, 1138. [CrossRef] [PubMed]
20. Panneke, N.; Ehrmann, A. Stab-Resistant Polymers—Recent Developments in Materials and Structures. *Polymers* **2023**, *15*, 983. [CrossRef] [PubMed]
21. Wildemann, V.; Staroverov, O.; Strungar, E.; Mugatarov, A.; Kuchukov, A. Mechanical Properties Degradation of Fiberglass Tubes during Biaxial Proportional Cyclic Loading. *Polymers* **2023**, *15*, 2017. [CrossRef] [PubMed]
22. Khan, M.K.A.; Junaedi, H.; Alshahrani, H.; Wagih, A.; Lubineau, G.; Sebaey, T.A. Enhanced Open-Hole Strength and Toughness of Sandwich Carbon-Kevlar Woven Composite Laminates. *Polymers* **2023**, *15*, 2276. [CrossRef] [PubMed]
23. Gahgah, M.; Belaadi, A.; Boumaaza, M.; Alshahrani, H.; Khan, M.K.A. Effect of Number of Tests on the Mechanical Characteristics of Agave sisalana Yarns for Composites Structures: Statistical Approach. *Polymers* **2023**, *15*, 2885. [CrossRef] [PubMed]
24. Leonés, A.; Peponi, L.; García-Martínez, J.-M.; Collar, E.P. Study on the Tensile Behavior of Woven Non-Woven PLA/OLA/MgO Electrospun Fibers. *Polymers* **2023**, *15*, 3973. [CrossRef] [PubMed]
25. Jones, H.; McClements, J.; Ray, D.; Hindle, C.S.; Kalloudis, M.; Koutsos, V. Thermomechanical Properties of Virgin and Recycled Polypropylene—High-Density Polyethylene Blends. *Polymers* **2023**, *15*, 4200. [CrossRef] [PubMed]
26. Zumbardo-Bacelis, G.A.; Peponi, L.; Vargas-Coronado, R.F.; Rodríguez-Velázquez, E.; Alatorre-Meda, M.; Chevallier, P.; Copes, F.; Mantovani, D.; Abraham, G.A.; Cauich-Rodríguez, J.V. A Comparison of Three-Layer and Single-Layer Small Vascular Grafts Manufactured via the Roto-Evaporation Method. *Polymers* **2024**, *16*, 1314. [CrossRef] [PubMed]
27. Thiem, Q.V.; Nguyen, V.-T.; Phan, D.T.T.; Minh, P.S. Injection Molding Condition Effects on the Mechanical Properties of Coconut-Wood-Powder-Based Polymer Composite. *Polymers* **2024**, *16*, 1225. [CrossRef] [PubMed]

polymers

MDPI

Article

Mechanical Properties of Polyurethane Mixture and Load Response Behaviour of Polyurethane Composite Pavement

Wei Zhuang [1,2], Yufeng Bi [2,*], Baoju Liu [1], Derui Hou [3], Shuo Jing [3], Xiaojin Lu [3] and Min Sun [3,*]

1 School of Civil Engineering, Central South University, Changsha 410075, China
2 Shandong Provincial Communications Planning and Design Institute Co., Ltd., Jinan 250031, China
3 School of Transportation Engineering, Shandong Jianzhu University, Jinan 250101, China
* Correspondence: biyf@163.com or 18866130036@163.com (Y.B.); sunmin20@sdjzu.edu.cn or 15253170143@163.com (M.S.)

Abstract: Finite element numerical simulation calculation of pavement structure load response is widely applied; however, there is still a lack of research on the polyurethane (PU) mixture composite pavement load response. The mechanical characteristics of PU mixture composite pavement are not well understood, and there is a lack of research on typical pavement structures of PU mixtures, which limits their application in pavement structures. Therefore, herein, the mechanical properties of PU mixtures are analysed using the dynamic modulus test, uniaxial penetration test, and fatigue test. Further, the finite element theory calculation method is used to realize the load response calculation of orthogonal design composite pavement structure. The results show that PU mixtures exhibit more obvious elastic characteristics and have good shear resistance, fatigue stability, and temperature stability, and can be used as shear and anti-fatigue layers. The structure of '4 cm SMA-13 + 5 cm PUM-20 + 6 cm PUM-25 + semi-rigid base' is recommended for the PU mixture composite structure. In comparison to typical asphalt pavement, the analysis shows that except for shear stress, temperature has little effect on the load response of PU composite pavement structures, while high temperatures lead to a significant increase in the load response of typical asphalt pavement structures. The PU composite pavement can bear greater loads and has a reduced thickness of its surface layer by about 3 cm in comparison to conventional pavement. The results of this study provide theoretical support for the design of PU mixture pavement structures and promote the popularization and application of PU mixture pavement.

Keywords: mechanical properties; polyurethane mixture; load response; composite pavement; pavement structure

Citation: Zhuang, W.; Bi, Y.; Liu, B.; Hou, D.; Jing, S.; Lu, X.; Sun, M. Mechanical Properties of Polyurethane Mixture and Load Response Behaviour of Polyurethane Composite Pavement. *Polymers* **2023**, *15*, 417. https://doi.org/10.3390/polym15020417

Academic Editors: Jesús-María García-Martínez and Emilia P. Collar

Received: 12 December 2022
Revised: 3 January 2023
Accepted: 7 January 2023
Published: 12 January 2023

1. Introduction

Asphalt pavement has the properties of good driving comfort, low noise during driving, strong environmental adaptability, and technical and economic benefits, so it is widely used in all kinds of road pavement [1]. At present, the pavement structure of three-layer surfaces and cement-stabilized gravel bases is often adopted in China [2]. However, as a viscoelastic discontinuous material, the mechanical properties of asphalt mixtures are easily affected by external environmental factors. Early issues such as ruts, potholes, cracks, and uneven subsidence occur frequently during the use of asphalt mixtures [3]. Much fuel is consumed to heat mineral materials and asphalt; this produces large amounts of emissions [4]. The aforementioned defects of asphalt mixtures have become an important factor restricting the higher-quality development of pavement [5].

At present, there is an increasing body of research on polyurethane (PU) mixtures in pavement engineering [6–10]. Compared to traditional mixtures, PU mixtures have better durability, temperature stability, waterproof properties, and crack resistance, which can reduce maintenance frequency [11–15]. However, as a new type of road construction material, the conventional pavement structure design method is not applicable [16–20]. Therefore, it is necessary to study the mechanical properties of PU mixtures and their composite pavement load response behaviour, providing a basis for the structural design of PU mixture pavement.

At present, some research has been conducted into the load response of composite pavement structures. Angel Mateos et al. identified the two main distress mechanisms in concrete overlays on asphalt pavement based on the analysis of the structural response of test sections and finite element modelling [21]. A two-dimensional finite element model was developed using the Mich-Pave software to predict pavement responses, and the responses of flexible pavement were evaluated comprehensively [22]. A three-dimensional finite element (3D FE) model was used by Feng et al. to investigate the temperature field and thermal responses; it was found that the maximum principal tensile stresses of the contact interface between the pipe and the surrounding concrete were larger [23]. Haibin Wei used the ANSYS static structural analysis module to analyse three indicators of EPDM pavement [24]. Zhu et al. used a finite element method to analyse the mechanical response of hydronic asphalt pavement (HAP); the results indicated that the concrete between adjacent pipes was subjected to significant tensile stresses [25]. The mechanical response of hydronic asphalt pavement under temperature–vehicle coupled loading was evaluated by Zhu et al., who found that the concrete between adjacent pipes was subjected to significant tensile stresses [25]. Huang et al. established a numerical model to characterize the fracture process of a reinforced concrete (RC) beam strengthened with fiber-reinforced polymer (FRP) in detail [26]. Yu et al. carried out an investigation on marble, and an impact factor of weak disturbance was defined [27]. Li et al. proposed a transfer learning pipeline and enabled a distress detection model to be applied to other untrained scenarios [28]. Different models, such as finite element models [29], three-dimensional finite element models [30], and ANSYS static structure modules [31], have been adopted by scholars to assess and analyse pavement structural responses, composite pavement, and snow-melting heated bridge deck systems [32,33].

It is clear that the current calculation method of pavement structure load response is relatively mature; however, there is a lack of research on PU mixture composite pavement load responses. The mechanical characteristics of PU mixture composite pavement are not known, and there are no relevant suggestions for the typical pavement structure of PU mixtures, which limits the popularization of PU mixtures in pavement structures. Therefore, the load responses of orthogonal design composite pavement structures are analysed basing on mechanical properties of PU mixtures and numerical simulation calculations, and the PU mixture composite pavement structure is recommended. The load response of PU mixture composite pavement structures is comprehensively analysed and compared with that of typical asphalt pavement structures, which provides theoretical support for the design of PU mixture pavement structures and promotes the popularization and application of PU mixture green roads.

2. Material Composition and Typical Pavement Structure

2.1. Raw Materials

(1) Polyurethane binder

In this study, a one-component moisture-curable PU binder is used, produced by Wanhua Chemical Co., Ltd., Yantai, China. It is a modified isocyanate prepolymer containing a certain terminal isocyanate (NCO) group, which is polymerized by isocyanate, polyether polyol, vegetable oil, and catalysts. The specific technical indicators of 25 °C are shown in Table 1.

Table 1. Technical indexes of PU binder [34].

Technology Index	Unit	Technical Requirement	Typical Values
Viscosity	mPa·s	1200–2200	1600
Isocyanate group content	%	10.3–11.3	11.1
Density	g/cm^3	1.05–1.11	1.08
Surface dry time	h	5–10	6
Tensile strength	MPa	≥10	26
Elongation at break	%	≥100	200

The NCO group in the PU binder is extremely chemically active. The first step of the reaction is that the NCO group reacts with water to form diamines and releases carbon dioxide. The second step is that the diamine substance further reacts with NCO, undergoes chain extension reactions, and generates the urea group, which solidifies the material.

(2) Asphalt

The 70# base asphalt and SBS-modified asphalt are used, the technical indicators of which meet the provisions of JTG F40-2004 [35].

2.2. Mixture Composition

The composition design of SBS-modified asphalt with a maximum nominal particle size of 13.2 mm (SMA-13), PU mixture, and asphalt mixture with a maximum nominal particle size of 19.0 mm and 26.5 mm (PUM-20, AC-20, PUM-25, AC-25) are carried out respectively. The aggregate gradations of the mixtures are shown in Figure 1. The composition design results of the mixtures are shown in Table 2.

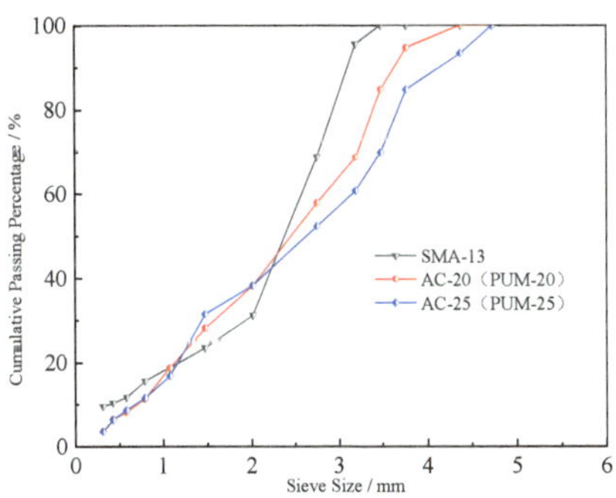

Figure 1. Mineral aggregate gradation of the mixture.

Table 2. The composition design results for various mixtures.

Mixture	Binder	Nominal Maximum Particle Sizes/mm	Binder Content/%	Void Ratio (VV)
SMA-13	SBS-modified asphalt	13.2	5.8	5.3
AC-20	70# based asphalt	19.0	5.0	4.5
AC-25	70# based asphalt	26.5	4.8	4.6
PUM-20	PU binder	19.0	4.8	4.9
PUM-25	PU binder	26.5	4.6	5.0

3. Test Scheme and Calculation Theory

3.1. Test Scheme

3.1.1. Dynamic Modulus Test

The dynamic modulus specimens are cylinders with a diameter of 100 mm and height of 150 mm, and one set of tests has three parallel specimens. According to the pavement working temperature, four temperatures of 10 °C, 20 °C, 35 °C, and 50 °C are selected, and nine loading frequencies of 0.1 Hz, 0.2 Hz, 0.5 Hz, 1.0 Hz, 2.0 Hz, 5.0 Hz, 10 Hz, 20 Hz, and 25 Hz are used. The specimens are kept at the test temperature for 5 h before tested. The dynamic modulus test is carried out in accordance with T 0738-2011 of JTG E20-2011 [36]. The five mixtures' dynamic modulus tests are carried out.

3.1.2. Uniaxial Penetration Test

According to JTG D50-2017, uniaxial penetration tests of the five mixtures are conducted [37]. The UTM-100 dynamic hydraulic test system is used for the test. A cylinder with a diameter of 42 mm is used. The loading rate is 1 mm/min and the pressure and displacement are recorded. When the stress value is reduced to 90% of the stress extreme point, the test is stopped, and the maximum shear failure load is taken at the inflexion point of the test curve, accurate to 0.001 KN.

The uniaxial penetration tests of the five mixtures are carried out at 15 °C, 20 °C, 30 °C, 40 °C, 50 °C, and 60 °C. The uniaxial penetration strength and modulus of the mixture are calculated as follows. The test apparatus and loading method are shown in Figure 2.

Figure 2. Uniaxial penetration test device.

3.1.3. Four-Point Bending Fatigue Test

Firstly, four-point bending static loading tests of five mixtures were carried out. The specimen size is 380 mm × 50 mm × 63.5 mm and the test temperature is 25 °C ± 0.5 °C. The displacement control mode is used to load at a speed of 0.01 mm/s until the specimen is broken, and the ultimate bearing capacity and flexural strength are obtained [32].

Secondly, according to the ultimate bearing capacity and stress level, the control stresses of the five mixtures are calculated. Previous studies have shown that the fatigue life of PU mixtures is long [16,17], so the stress levels of asphalt mixtures are 0.3, 0.4, and 0.5, and the stress levels of PU mixtures are 0.5, 0.6, and 0.7.

Finally, four-point bending fatigue tests are carried out under stress control mode. Continuous partial sinusoidal loading mode, loading frequency 10 Hz ± 0.1 Hz, loading waveform is a non-interrupted asymmetric constant amplitude sine wave. The cyclic characteristic value (the ratio of F_{min} to F_{max}) is R = 0.1, where the maximum cyclic loading is the product of the stress level S and the ultimate bearing capacity, and the minimum loading is the product of the maximum value and the cyclic characteristic value R [33]. Load is not stopped until fatigue fracture of the specimen occurs. The loading waveform curve is shown in Figure 3, and the loading method is shown in Figure 4.

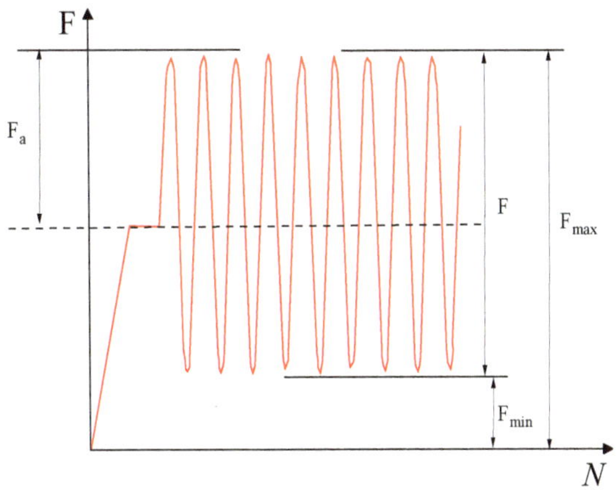

Figure 3. Loading waveform curve.

Figure 4. Specimen loading method.

3.2. Orthogonal Design of PU Mixture Composite Pavement Structure

Three-layer surfaces and cement-stabilized gravel bases are often adopted in China, and the uppermost layer is usually an SBS-modified asphalt mixture wear layer to maintain the pavement [13–15]. Therefore, the 4 cm SMA-13 mixture is used as the upper layer of the pavement, and this study focuses on the structural design of the middle and lower layers. Previous studies have shown that PU mixtures have significant elastic properties, and their dynamic modulus is obviously larger [16,17]. Therefore, the thickness of the PU mixture layers is appropriately reduced in the design of the pavement structure.

The four-factor mixed-level orthogonal design of composite pavement structures is shown in Table 3. The middle and lower layers are given different thicknesses: the middle layer thicknesses are 3 cm, 4 cm, and 5 cm; the lower layer thicknesses are 6 cm, 7 cm, and 8 cm. The AC-20 and PUM-20 mixtures are randomly matched to the middle layer, the AC-25 and PUM-25 mixtures are randomly matched to the lower layer, and the base layer is a double-layer cement stabilized gravel mixture.

Table 3. Orthogonal test design of PU and asphalt composite pavement.

Numbering	Factor A (Middle Layer Thickness mm)	Factor B (Lower Layer Thickness mm)	Factor C (Middle Layer Material)	Factor D (Lower Layer Material)
1	1 (30 mm)	1 (60 mm)	1 (AC-20)	2 (PUM-25)
2	2 (40 mm)	2 (70 mm)	1	2
3	3 (50 mm)	3 (80 mm)	1	2
4	1	2	1	2
5	2	3	1	2
6	3	1	1	2
7	1	2	2 (PUM-20)	1 (AC-25)
8	2	3	2	1
9	3	1	2	1
10	1	3	2	1
11	2	1	2	1
12	3	2	2	1

The deflection of the pavement surface, vertical compressive strain of the asphalt surface, shear stress between the asphalt layer and PU layer, tensile stress at the bottom of the base layer, and vertical compressive strain at the top of the subgrade are selected as key mechanical calculation indexes of the pavement load response. In addition, the material cost of the pavement structure is one of the key factors to be considered in pavement structure design. The pavement material cost calculation results of pavement with a width of 30 m and length of 100 m are used as an evaluation index for orthogonal design.

3.3. *Theoretical Calculation*

3.3.1. Finite Element

The load responses of PU mixture composite pavement are calculated using the ABAQUS finite element software. The model size is 10 m (length) × 6 m (width) × 6 m (height); the x-direction in the model is the driving direction, the y-direction is the road depth direction, and the z-direction is the lateral direction of the road. The standard axle load of BZZ-100 is selected for model loading, the tire grounding pressure P is 0.7 MPa, and the tire grounding size is 0.226 m × 0.156 m. The elastic continuous layered system theory is used for calculation [34]. The structural layers of the pavement are connected through binding. The materials of each layer are divided by eight-node linear hexahedral elements. The established model structure is shown in Figure 5.

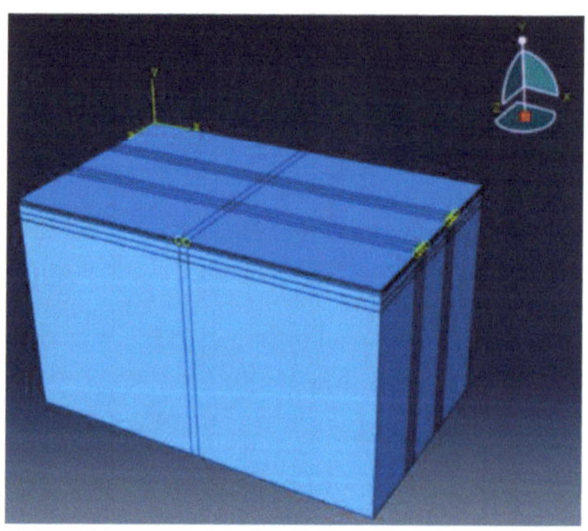

Figure 5. Finite element for the pavement structure.

3.3.2. Calculation Parameters and Conditions

The calculation models for the twelve pavement structures in the four-factor mixed-level orthogonal design table are established. Based on the analysis of the load responses of orthogonal design pavement structure, the PU mixture composite pavement structure is recommended. The load responses of the PU mixture composite pavement structure and typical asphalt pavement structure, with temperatures of 20 °C and 50 °C, are calculated and compared. The material parameters of different materials are shown in Table 4.

Table 4. Material parameters of each pavement structure layer.

Structure Layer Type	20 °C Dynamic Modulus/MPa	50 °C Dynamic Modulus/MPa	Poisson Ratio
SMA-13	10,184	720	0.25
AC-20	9383	642	0.25
AC-25	9483	692	0.25
PUM-20	11,310	7395	0.25
PUM-25	10,567	7642	0.25
Cement-stabilized gravel (base)	16,000	16,000	0.25
Cement-stabilized gravel (subbase)	12,000	12,000	0.25
subgrade	70	70	0.40

The vertical compressive strain of the asphalt surface layer, tensile stress of the base bottom, road surface deflection, maximum shear stress, and vertical compressive strain of the subgrade surface are taken as the orthogonal design analysis calculation indexes.

4. Results and Discussion

4.1. Dynamic Modulus Test Results

4.1.1. Dynamic Modulus

The dynamic modulus test results for the five mixtures are shown in Figures 6 and 7. The results show that the higher the test temperature, the lower the dynamic modulus of the five mixtures at the same loading frequency. The high temperature reduces the ability of the five mixtures to resist loading. With increases in the test temperature, the dynamic modulus of the SMA-13, AC-20, and AC-25 mixtures decrease greatly, while those of the PUM-20 and PUM-25 mixtures decrease slightly. This indicates that SMA-13, AC-20, and AC-25 mixtures are significantly affected by temperature, while the PUM-20 and PUM-25 mixtures have better temperature stability.

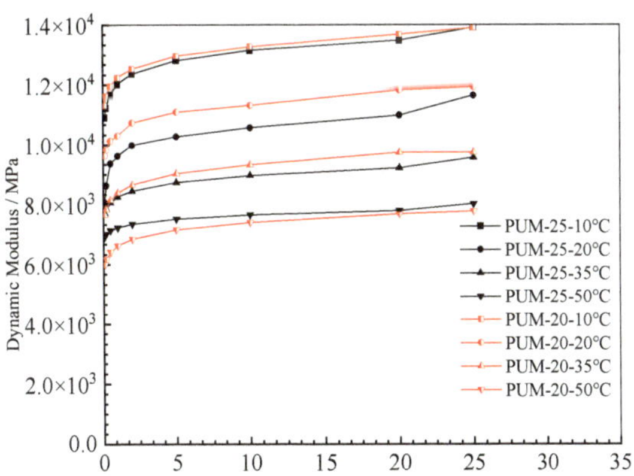

Figure 6. Dynamic modulus of PU mixtures.

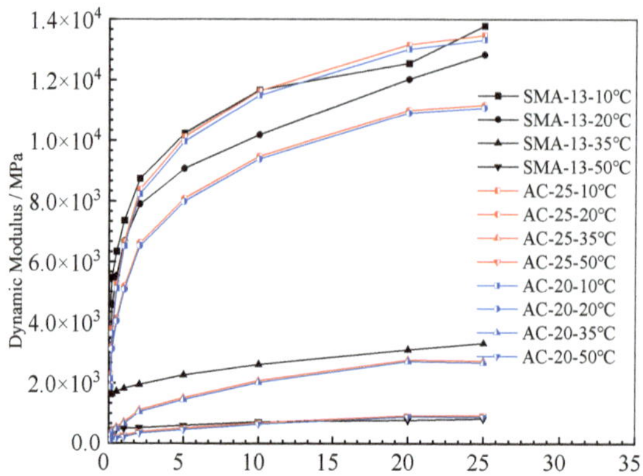

Figure 7. Dynamic modulus of asphalt mixtures.

The dynamic modulus of the two mixtures increased with increases in the loading frequency. The dynamic modulus of the SMA-13, AC-20, and AC-25 mixtures is obviously affected by the loading frequency. The dynamic modulus of PUM-20 and PUM-25 mixtures at different loading frequencies changes slightly, and the influence of different loading frequencies on PU mixtures is relatively small.

The reason for this is that asphalt is a temperature-sensitive binder. Increases in the temperature lead to a transition of asphalt from an elastic state to a viscous state, which affects its mechanical properties [16]. The strength of the PU mixture is formed by the curing reaction of the PU binder, which is irreversible. When the ambient temperature rises, the mechanical properties of PU mixtures do not change; thus, PU mixtures exhibit good temperature stability.

4.1.2. Phase Angle

The phase angle of 10 Hz is shown in Figure 8, and the phase angle of 20 °C is shown in Figure 9.

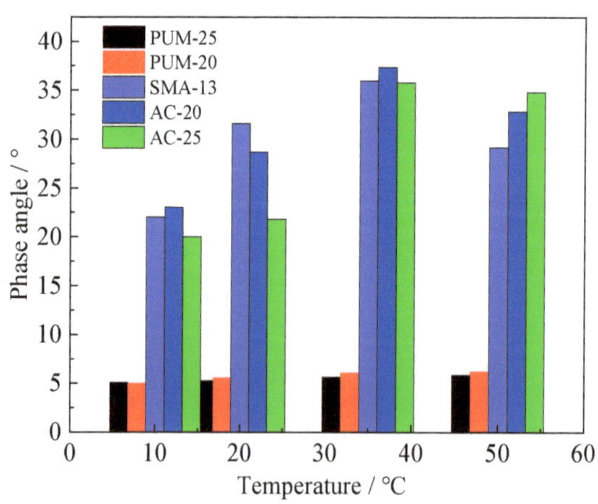

Figure 8. Phase angle at 10 Hz loading frequency.

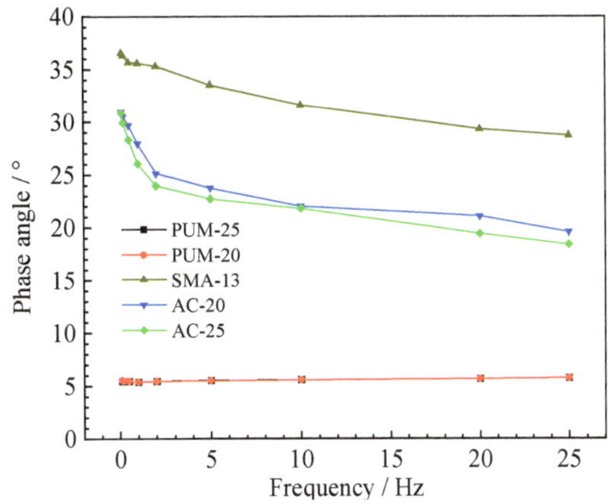

Figure 9. Phase angle at 20 °C.

Under different test temperatures and loading frequencies, the phase angles of PUM-20 and PUM-25 mixtures change between 4° and 6°; the overall change range is within 2°. The phase angles of SMA-13, AC-20, and AC-25 mixtures are between 15–40°, and the phase angle varies greatly with the loading frequency and test temperature. The phase angle is used as an index to analyse the elasticity and viscosity coefficient of the material. The larger the phase angle, the more obvious the viscosity characteristics of the material. The test results show that the elastic properties of PU mixtures are relatively obvious, while the viscous properties of the asphalt mixture are relatively obvious; further, the temperature has little effect on the viscoelastic properties of PU mixtures.

4.2. Uniaxial Penetration Test Results

The uniaxial penetration strength and shear modulus of the five mixtures are shown in Figures 10 and 11. The uniaxial penetration strength and shear modulus of PUM-20 and PUM-25 mixtures are far greater than those of SMA-13, AC-20, and AC-25 mixtures, and the temperature has little influence on the uniaxial penetration strength of PU mixtures [8]. The PU mixtures have high uniaxial penetration strengths and are less affected by temperature, indicating their high bearing capacity and good temperature stability. The shear moduli of PUM-20 and PUM-25 mixtures are much larger than those of SMA-13, AC-20, and AC-25 mixtures, and the temperature has little effect on the shear modulus of PU mixtures, which is consistent with the dynamic modulus test results. Therefore, PU mixtures can be applied to the shear layer of pavement structures.

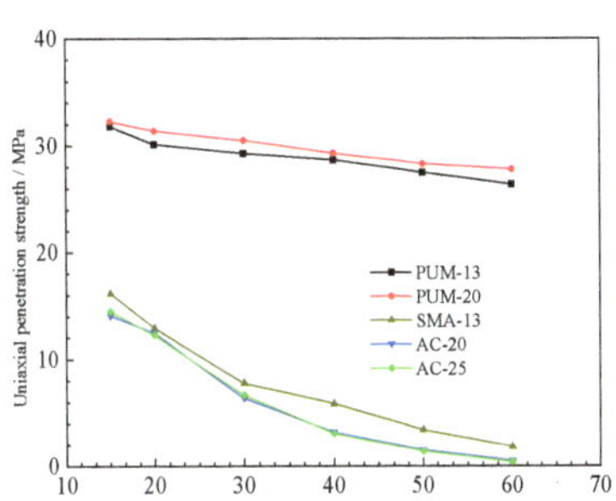

Figure 10. Uniaxial penetration strength results.

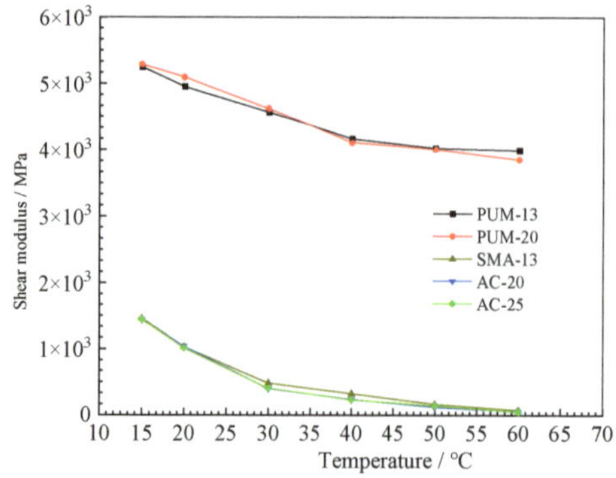

Figure 11. Shear modulus test results.

4.3. Four-Point Bending Fatigue Test Results

4.3.1. Four-Point Bending Static Loading Tests

The four-point bending load–displacement curves of the five mixtures are shown in Figure 12. The maximum deformation and bending failure stress are shown in Figure 13.

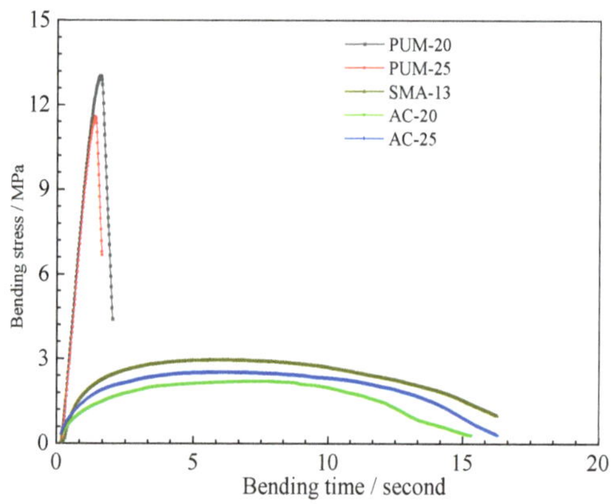

Figure 12. Time–stress curve.

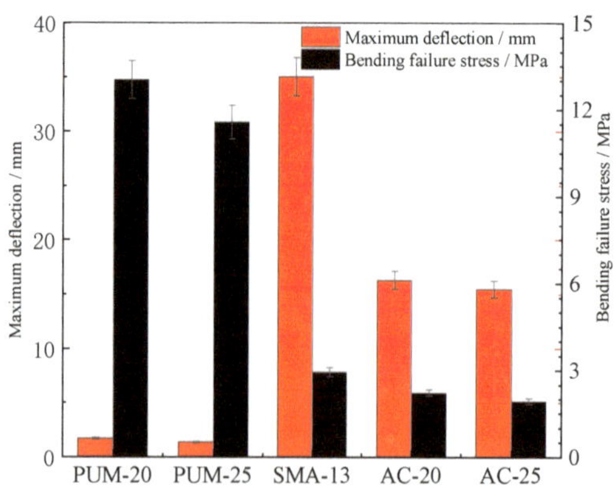

Figure 13. Maximum deformation and damage loads.

The load–displacement curves of the PUM-20 and PUM-25 mixtures differ from those of the SMA-13, AC-20, and AC-25 mixtures. The bending stresses of PUM-20 and PUM-25 mixtures increase rapidly under loading, and the mixture fails after bending for about 3 s, while the time–stress curves of PUM-20 and PUM-25 mixtures are relatively flat, and the mixture fails after loading for 15 s. In addition, the bending failure stress of PUM-20 and PUM-25 is about four times larger than that of SMA-13, AC-20, and AC-25, while the maximum deformation of SMA-13, AC-20, and AC-25 is about 15 times larger than that of PUM-20 and PUM-25. This indicates that the PU mixture has a strong ability to resist loading; however, its plastic deformation under loading is small, easily leading to brittle failure [12].

4.3.2. Four-Point Bending Fatigue Tests

The stress-controlled four-point bending fatigue life test results of the five mixtures are shown in Figure 14.

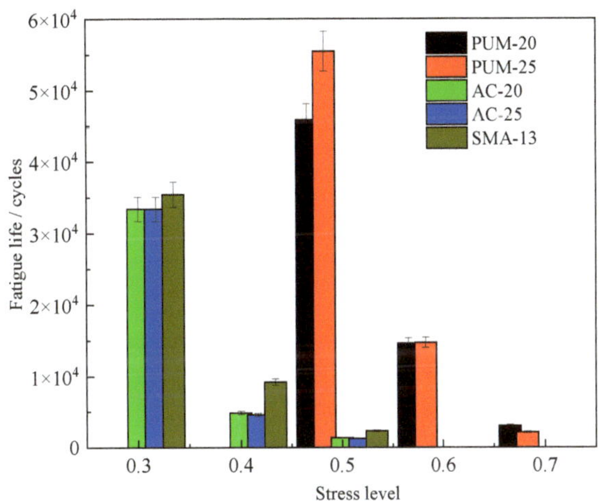

Figure 14. Fatigue life of mixtures.

With increases in the stress level, the fatigue life of the five mixtures gradually decreases. When the stress level of SMA-13, AC-20, and AC-25 mixtures increases to 0.5, the fatigue life is only 1000–2000 cycles, while the fatigue life of the PUM-20 and PUM-25 mixtures at the 0.7 stress level can still reach 2000–3000 cycles; that is, PU mixtures still have good fatigue resistance at high stress levels. At the 0.5 stress level, the fatigue life of PUM-20 and PUM-25 mixtures is more than 25 times that of the SMA-13, AC-20, and AC-25 mixtures, indicating that the PU mixture has a strong ability to resist fatigue load [12]. The four-point bending fatigue test shows that the PU mixture has strong fatigue resistance and can be applied at higher stress levels.

4.4. Analysis of Pavement Structure Orthogonal Design Results and Pavement Structure Recommendation

The calculation results for the twelve pavement structures of the orthogonal design table are shown in Table 5.

Table 5. Orthogonal test results for PU mixture composite pavement.

Number	Factor A	Factor B	Factor C	Factor D	Vertical Compressive Strain of Asphalt Surface/μ ε	Tensile Stress of Base Bottom/10^{-3} MPa	Road Surface Deflection/0.01 mm	Maximum Shear Stress/10^{-3} Mpa	Vertical Compressive Strain of Subgrade Top/μ ε	Material Cost/CNY Ten Thousand
1	1	1	1	2	52.86	137.5	17.10	49.10	39.6	89.5
2	2	2	1	2	48.49	138.0	26.24	49.80	39.42	97.9
3	3	3	1	2	40.13	126.4	13.00	38.00	36.70	106.4
4	1	2	1	2	53.91	134.6	16.85	48.93	38.75	94.8
5	2	3	1	2	40.24	128.7	13.11	36.40	37.32	103.3
6	3	1	1	2	39.95	132.0	13.37	39.80	38.31	95.7
7	1	2	2	1	52.38	135.1	16.60	49.35	39.05	94.8
8	2	3	2	1	39.60	128.8	13.07	43.60	37.48	103.3
9	3	1	2	1	39.65	130.8	13.18	39.90	38.00	95.7
10	1	3	2	1	39.62	131.8	13.26	43.10	38.37	100.2
11	2	1	2	1	39.62	133.8	13.77	43.40	38.89	92.6
12	3	2	2	1	39.65	128.3	13.03	43.60	37.31	101.0

The range (ΔK) under different factors and levels is calculated as shown in Table 6, and is then used to analyse the impacts of each factor on different indicators in the orthogonal test, and to determine the optimal factor combination corresponding to different indicators.

Table 6. Orthogonal data processing results.

Response	Range (ΔK)				Optimal Structure Combination
	Factor A	Factor B	Factor C	Factor D	
Vertical compressive strain of surface layer/με	39.39	34.84	25.06	25.06	A3B3C2D1
Tensile stress of base bottom/10^{-3} MPa	2.15	2.03	0.86	0.86	A3B3C2D1
Pavement surface deflection/0.01 mm	13.61	20.28	41.76	41.76	C2D2B3A3
Maximum shear stress/10^{-3} Mpa	29.18	30.58	0.92	0.92	B3A3C1D2
Vertical compressive strain of subgrade top/με	5.45	4.93	1	1	A3B3C1D2
Material cost/CNY ten thousand	19.5	39.7	0	0	B1A1CD

According to the results of the aforementioned method, the influence factors for the mechanical indexes of pavement structures are discussed as follows. The factors affecting the vertical compressive strain of the surface layer and the tensile stress of the base bottom are ranked as A > B > C = D. The pavement structure with A3B3C2D1 can obtain the minimum vertical compressive strain of the asphalt surface layer and the tensile stress at the bottom of the base layer. The influence order of the pavement surface deflection factors is C = D > B > A. The pavement structure of C2D2B3A3 can obtain the minimum pavement surface deflection and improve the pavement-bearing capacity. The influence of the maximum shear stress factors is B > A > C = D, and the B3A3C1D2 pavement structure can obtain the minimum interlayer shear stress and reduce the interlayer displacement. The influence of vertical compressive strain factors on the top surface of the subgrade is A > B > C = D, and the compressive strain of the top surface of the A3B3C1D2 pavement structure is the smallest, reducing the deformation of the subgrade. The influence of the material cost factors is B > A > C = D, and the pavement structure cost of B1A1 is the lowest.

Combining various load responses and pavement structure costs, the final preferred pavement structure is A3B1C2D2; that is, '4 cm SMA + 5 cm PUM-20 + 6 cm PUM-25' is proposed as the PU and asphalt mixture composite structure.

4.5. Load Response Behaviour of Composite Pavement Structures

To comprehensively characterize the load response of PU composite pavement, the load responses of typical asphalt pavement structure (typical structure) and PU and asphalt mixture composite pavement (composite structure) are also calculated and compared. The two pavement structures are shown in Figure 15.

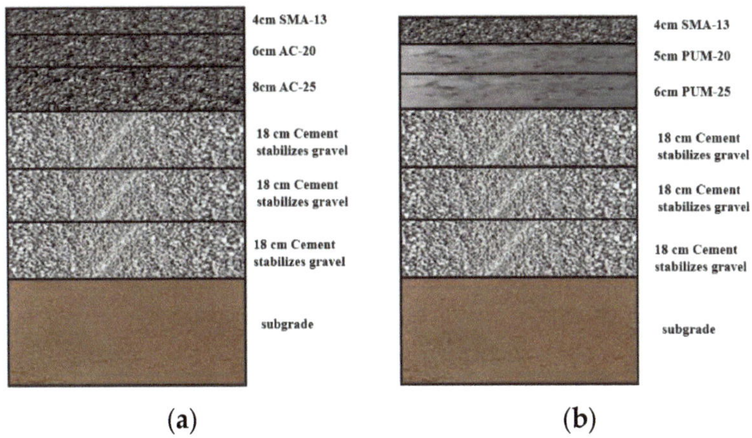

Figure 15. Schematic diagram of the two pavement structures. (**a**) Typical pavement structure. (**b**) Composite pavement structure.

4.5.1. Vertical Compressive Strain of the Subgrade Surface

The vertical compressive strain calculation results of the subgrade surface are shown in Figure 16.

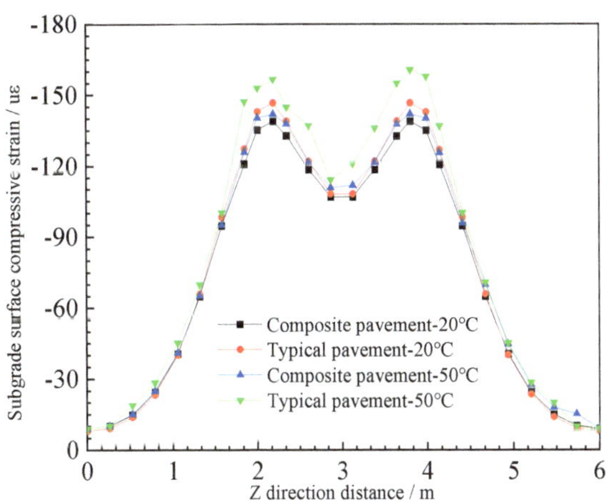

Figure 16. Subgrade surface compressive strain.

It can be seen that the load response curves of PU and asphalt composite pavement and typical asphalt mixture pavement are similar. The vertical compressive strain of the top surface of the subgrade decreases with increases in the load distance. With increases in the temperature, the subgrade surface compressive strain of the two pavement structures increases, and the temperature has a great influence on the subgrade surface compressive strain of typical asphalt mixture pavement. The maximum subgrade surface compressive strain changes from 142 με to 167 με, while the temperature has a relatively small influence on the subgrade surface compressive strain of the PU composite pavement; the maximum subgrade surface compressive strain changes from 139 με to 142 με. This may be because the PU mixture has good temperature stability, and the mechanical properties change slightly under high-temperature conditions, while the asphalt mixture changes from elastic to viscous under high-temperature conditions, resulting in large vertical compressive strains on the top surface of the subgrade of typical asphalt pavement [30,38].

4.5.2. Interlayer Shear Stress

The upper-layer bottom shear stress and the maximum shear stress in the depth direction of the two pavement structures are shown in Figures 17 and 18.

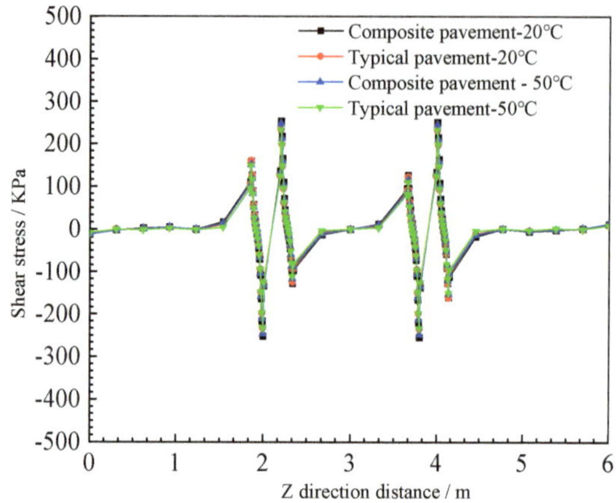

Figure 17. Upper-layer maximum shear stress.

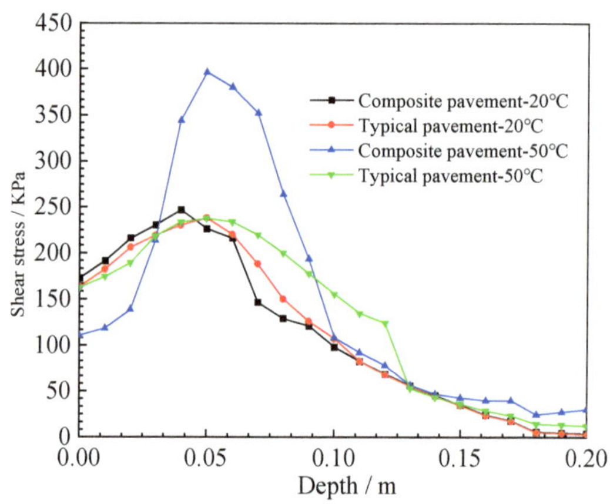

Figure 18. Shear stress in the depth.

The maximum shear stress at the bottom of the upper layer decreases with increases in the load distance, as shown in Figure 17. The maximum shear stress at the bottom of the upper layer of the PU composite pavement and the typical asphalt pavement structure at 10 °C and 50 °C is similar to the lateral distribution of the road. The temperature has little effect on the upper-layer maximum shear stress of the typical asphalt pavement structure, except the wheel load position. The interlayer shear stress of the composite pavement load position under high-temperature conditions reaches 254 KPa.

The shear stress distribution curve in the depth direction of the load position is shown in Figure 18. The shear stress value is larger at the depth of 4–10 cm of the two pavement structures. The shear stress distribution in the depth direction of typical asphalt pavement structures is only slightly affected by temperature changes; however, the shear stress of the PU mixture composite pavement structure increases at high temperatures, and the shear stress value reaches 400 KPa at a 4–6 cm depth of the pavement structure [31]. Therefore, it is necessary to carry out layer treatments between the PU mixture and asphalt mixture layer; on the basis of previous research results, a two-component PU adhesive layer of 0.4 L/m^2 should be sprayed.

4.5.3. Vertical Compressive Strain of the Composite Pavement Structure Surface

The vertical compressive strain curves of the surface layer in Figure 19 show that the two structures exhibit the same change trends. The vertical compressive strain of the PU mixture composite pavement is relatively small at different temperatures, and the values are close, indicating that the pavement structure has high-temperature stability, mainly due to the strength formed by the cross-linking curing reaction of the PU mixture, which does not easily deform with temperature changes [32,39]. However, the vertical compressive strain of typical asphalt pavement is the largest at 50 °C, primarily due to the high-temperature sensitivity of the asphalt binder.

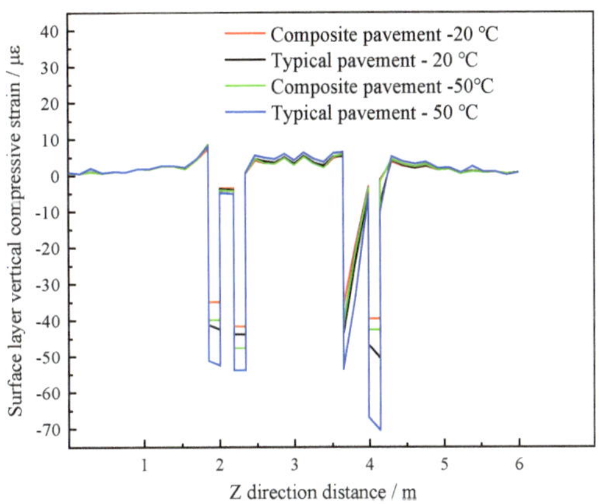

Figure 19. Vertical compressive strain of the surface layer.

4.5.4. Deflection

The deflection calculation results of the two pavement structures are shown in Figure 20. It can be seen that the maximum deflection appears at the position of load action, and the deflection of the composite pavement structure is smaller than that of typical asphalt pavement, which indicates that PU mixture composite pavement has a higher structural bearing capacity and can be adapted to higher-grade loads.

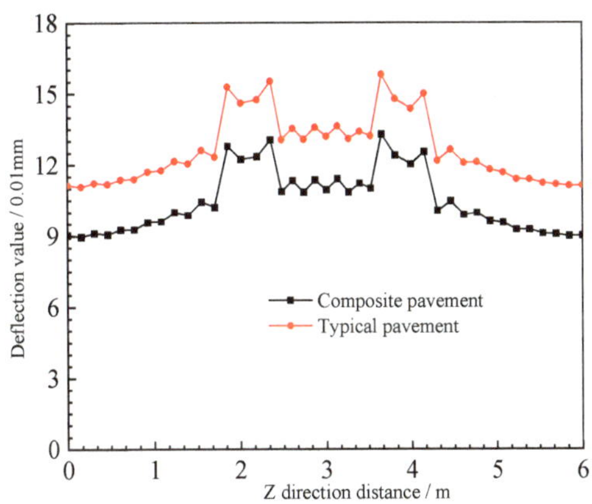

Figure 20. Deflections of the two pavement structures.

4.5.5. Composite Pavement Structure Base-Layer Bottom Tensile Stress

From the base-layer bottom tensile stress curves in Figure 21, it can be seen that the change trends of the two structures are the same; the stress decreases with increases in the load distance, and the base-layer bottom tensile stress at the same position increases with increases in the temperature. Further, the temperature has a great influence on the base-layer bottom tensile stress of typical asphalt pavement, and the base-layer bottom tensile stress of asphalt pavement under high-temperature conditions is the largest. Because of the lower base-layer bottom tensile stress, a relatively long service life of the PU composite pavement base is ensured [40].

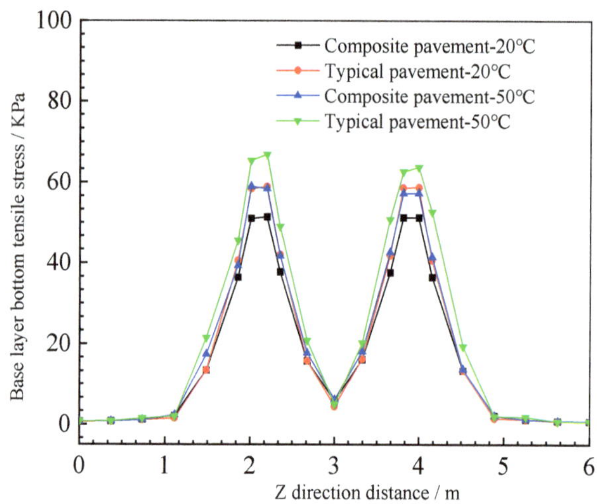

Figure 21. Bottom tensile stress of the base layer.

In conclusion, the responses of the two pavement structures near the edge of the pavement are relatively similar, but the mechanical response values of the PU and asphalt composite pavement structures near the load are relatively small; in addition to the shear stress, the temperature has little effect on the load responses of the PU composite pavement structure, and high-temperature effects lead to significant increases in the load response of the typical asphalt pavement. This shows that the PU and asphalt composite pavement can bear greater loads and resist higher temperatures, and the surface layer thickness of PU and asphalt composite pavement is about 3 cm smaller than that of typical asphalt pavement, which can save a significant amount of road construction material. Furthermore, the excellent durability of the PU mixture can effectively reduce the maintenance and reconstruction cost investment, reducing the interference of road traffic flow during operation.

5. Conclusions

The PU mixtures exhibit more obvious elastic characteristics and have good shear resistance, fatigue stability, and temperature stability, and can be used as shear and anti-fatigue layers; the conclusions are as follows:

(1) The influence of the temperature and loading frequency on the dynamic modulus and phase angle of PU mixtures is relatively small. The phase angle is between 4° and 6°, and the elastic characteristics of the PU mixture are relatively stable. The uniaxial penetration strength and shear modulus of the PU mixture are greater than those of asphalt mixtures, and are less affected by temperature. PU mixtures have good shear resistance and temperature stability, and can be used as shear layers.

(2) The bending failure stress of PU mixtures is much larger than that of asphalt mixtures, which have a strong ability to resist loads; however, its plastic deformation is small

under loading, and the failure speed is fast, which may be more prone to brittle failure. PU mixtures have excellent fatigue resistance and can be applied to anti-fatigue layers due to the need for use in layers with high stress levels.

(3) Based on the calculation of load responses of orthogonal design pavement structures, the typical PU mixture composite pavement structure is recommended, which is a '4 cm SMA-13 + 5 cm PUM-20 + 6 cm PUM-25 + semi-rigid base' pavement structure.

(4) In comparison to the load response calculation structure of typical asphalt pavement, the analysis shows that—other than for shear stress—temperature has little effect on the load response of PU composite pavement structure. The PU composite pavement can bear greater loads and has strong resistance to high temperatures. The load responses of PU composite can meet the requirements of high-grade highways, heavy load pavement, long longitudinal slope sections, and intersection sections.

(5) The theory of elastic layered system is adopted. The viscoelastic characteristics of asphalt mixture are not fully considered, which may affect the numerical analysis results to a certain extent. The recommended polyurethane mixture composite pavement structure is still more expensive than conventional asphalt pavement structure, and can potentially be applied to special road sections.

Author Contributions: Conceptualization, Y.B. and D.H.; methodology and data curation, W.Z. and M.S.; validation and investigation, B.L. and S.J.; formal analysis, X.L. and S.J.; resources, Y.B.; writing—original draft preparation, M.S.; writing—review and editing, B.L.; supervision, W.Z.; and funding acquisition, Y.B. All authors have read and agreed to the published version of the manuscript.

Funding: This work was supported by Key Scientific Research Projects in the Transportation Industry of the Ministry of Transport (2019MS2028), Shandong Expressway Group Project (HSB 2021-72).

Institutional Review Board Statement: Not applicable.

Informed Consent Statement: Not applicable.

Data Availability Statement: The data used to support the findings of this study are available from the corresponding author upon request.

Conflicts of Interest: The authors declare no conflict of interest.

References

1. Chem, Y.C.; Tseng, S.M.; Hsieh, K.H. Damping properties of interpenetrating polymer networks of polyurethane-modified epoxy and polyurethanes. *J. Appl. Polym. Sci.* **2015**, *74*, 328–335.
2. Wang, Q.; Chen, S.; Wang, T.; Zhang, X. Damping, thermal, and mechanical properties of polyurethane based on poly (tetramethylene glycol)/epoxy interpenetrating polymer networks: Effects of composition and isocyanate index. *Appl. Phys. A* **2011**, *104*, 375–382. [CrossRef]
3. Wang, J.; Yuan, J.; Xiao, F.; Li, Z.; Wang, J.; Xu, Z. Performance investigation and sustainability evaluation of multiple-polymer asphalt mixtures in airfield pavement. *J. Clean. Prod.* **2018**, *189*, 67–77. [CrossRef]
4. Chen, J.; Yin, X.; Wang, H.; Ding, Y. Evaluation of durability and functional performance of porous polyurethane mixture in porous pavement. *J. Clean. Prod.* **2018**, *188*, 12–19. [CrossRef]
5. Chen, J.; Ma, X.; Wang, H.; Xie, P.; Huang, W. Experimental study on anti-icing and deicing performance of polyurethane concrete as road surface layer. *Constr. Build. Mater.* **2018**, *161*, 598–605. [CrossRef]
6. Wang, D.; Schacht, A.; Leng, Z.; Leng, C.; Kollmann, J.; Oeser, M. Effects of Materials Composition on Mechanical and Acoustic Performance of Poroelastic Road Surface (PERS). *Constr. Build. Mater.* **2017**, *135*, 352–360. [CrossRef]
7. Wang, D.; Liu, P.; Leng, Z.; Leng, C.; Lu, G.; Buch, M.; Oeser, M. Suitability of Poroelatic Road Surface (PERS) for Urban Roads in Cold Regions: Mechanical and Functional Performance Assessment. *J. Clean. Prod.* **2017**, *165*, 1340–1350. [CrossRef]
8. Sun, M.; Bi, Y.; Zheng, M.; Wang, J.; Wang, L. Performance of Polyurethane Mixtures with Skeleton-Interlocking Structure. *J. Mater. Civ. Eng.* **2020**, *32*, 04019358. [CrossRef]
9. Lu, G.; Renken, L.; Li, T.; Wang, D.; Li, H.; Oeser, M. Experimental study on the polyurethane-bound pervious mixtures in the application of permeable pavement. *Constr. Build. Mater.* **2019**, *202*, 838–850. [CrossRef]
10. Li, T.; Lu, G.; Wang, D.; Hong, B.; Tan, Y.; Oeser, M. Key Properties of High-performance Polyurethane Bounded Pervious Mixture. *China J. Highw. Transp.* **2019**, *32*, 158–169. (In Chinese)

11. Lu, G.; Liu, P.; Wang, Y.; Faßbender, S.; Wang, D.; Oeser, M. Development of a sustainable pervious pavement material using recycled ceramic aggregate and bio-based polyurethane binder. *J. Clean. Prod.* **2019**, *220*, 1052–1060. [CrossRef]
12. Sun, M.; Qu, G.; Geng, L.; Hou, D.; Jing, S. Fatigue Properties and Damage Characteristics of Polyurethane Mixtures under a Stress Control Mode. *Sustainability* **2022**, *14*, 10966. [CrossRef]
13. Cong, L.; Wang, T.; Tan, L.; Yuan, J.; Shi, J. Laboratory evaluation on performance of porous polyurethane mixtures and OGFC. *Constr. Build. Mater.* **2018**, *169*, 436–442. [CrossRef]
14. Lu, G.; Liu, P.; Törzs, T.; Wang, D.; Oeser, M.; Grabe, J. Numerical analysis for the influence of saturation on the base course of permeable pavement with a novel polyurethane binder. *Constr. Build. Mater.* **2019**, *240*, 117930. [CrossRef]
15. Xu, S.; Zhang, Y.; Guo, Y.; Ma, C.; Gao, D.; Peng, G. Determination of polyurethane concrete compaction timing based on penetration resistance test system. *China, J. Highw. Transp.* **2021**, *34*, 226–235. (In Chinese)
16. Sun, M.; Bi, Y.; Zhuang, W.; Chen, S.; Zhao, P.; Pang, D.; Zhang, W. Mechanism of Polyurethane Binder Curing Reaction and Evaluation of Polyurethane Mixture Properties. *Coatings* **2021**, *11*, 1454. [CrossRef]
17. Gao, G.; Sun, M.; Xu, C.; Qu, G.; Yang, Y. Interlaminar Shear Characteristics of Typical Polyurethane Mixture Pavement. *Polymers* **2022**, *14*, 3827. [CrossRef]
18. Song-Tao, L.; Zhaohui, L.; Juan, X. Fatigue performance of aging asphalt mixtures. *Polimery* **2015**, *60*, 126–131. [CrossRef]
19. Lv, S.; Liu, C.; Yao, H.; Zheng, J. Comparisons of synchronous measurement methods on various modulus of asphalt mixtures. *Constr. Build. Mater.* **2018**, *158*, 1035–1045. [CrossRef]
20. Törzs, T.; Lu, G.; Monteiro, A.O.; Wang, D.; Grabe, J.; Oeser, M. Hydraulic properties of polyurethane-bound permeable pavement materials considering unsaturated flow. *Constr. Build. Mater.* **2019**, *212*, 422–430. [CrossRef]
21. Mateos, A.; Millan, M.A.; Harvey, J.T.; Paniagua, F.; Wu, R. Mechanisms of asphalt cracking and concrete-asphalt debonding in concrete overlay on asphalt pavements. *Constr. Build. Mater.* **2021**, *301*, 124086. [CrossRef]
22. Bayat, A.; Knight, M. Field evaluation and analysis of flexible pavement structural responses under dynamic loads. *Road Mater. Pavement Des.* **2012**, *13*, 26–37. [CrossRef]
23. Feng, J.; Yin, G. Thermal Analyses and Responses of Bridge Deck Hydronic Snow Melting System. *Adv. Civ. Eng.* **2019**, *2019*, 8172494. [CrossRef]
24. Wei, H.; Ma, Z.; He, X.; Han, S.; Jiang, B. Structural response analysis of conductive ethylene–propylene–diene monomer rubber composite pavement under validated temperature field. *Constr. Build. Mater.* **2022**, *328*, 127094. [CrossRef]
25. Zhu, X.; Zhang, Q.; Chen, L.; Du, Z. Mechanical response of hydronic asphalt pavement under temperature–vehicle coupled load: A finite element simulation and accelerated pavement testing study. *Constr. Build. Mater.* **2020**, *272*, 121884. [CrossRef]
26. Huang, Y.; Zhang, W.; Liu, X. Assessment of Diagonal Macrocrack-Induced Debonding Mechanisms in FRP-Strengthened RC Beams. *J. Compos. Constr.* **2022**, *26*, 04022056. [CrossRef]
27. Yu, J.; Zhu, Y.; Yao, W.; Liu, X.; Ren, C.; Cai, Y.; Tang, X. Stress relaxation behaviour of marble under cyclic weak disturbance and confining pressures. *Measurement* **2021**, *182*, 109777. [CrossRef]
28. Li, Y.; Che, P.; Liu, C.; Wu, D.; Du, Y. Cross-scene pavement distress detection by a novel transfer learning framework. *Comput. Civ. Infrastruct. Eng.* **2021**, *36*, 1398–1415. [CrossRef]
29. Liu, K.; Dai, D.; Fu, C.; Li, W.; Li, S. Structural investigation of the snow-melting heated bridge deck based on the thermal field distribution. *Appl. Therm. Eng.* **2019**, *161*, 114–132. [CrossRef]
30. Liu, K.; Xie, H.; Xu, P.; Wang, Z.; Bai, H.; Wang, F. The thermal and damage characteristics of an insulated-conductive composite structure for the heated bridge deck for snow-melting. *Constr. Build. Mater.* **2019**, *216*, 176–187. [CrossRef]
31. Zhou, B.; Pei, J.; Hughes, B.R.; Nasir, D.S.; Vital, B.; Pantua, C.A.J.; Calautit, J.; Zhang, J. Structural response analysis of road pavement solar collector (RPSC) with serpentine heat pipes under validated temperature field. *Constr. Build. Mater.* **2020**, *268*, 121110. [CrossRef]
32. Liu, K.; Xu, P.; Wang, F.; Jin, C.; Liu, Q.; Pang, H.; Xie, H. The accumulated stress damage and residual life prediction of unreinforced concrete pavement with electric heating pipes. *Constr. Build. Mater.* **2021**, *278*, 122258. [CrossRef]
33. Xu, H.; Tan, Y. Modeling and operation strategy of pavement snow melting systems utilizing low-temperature heating fluids. *Energy* **2015**, *80*, 666–676. [CrossRef]
34. *GB/T 16777-2008*; Test Methods for Building Waterproofing Coatings. General Administration of Quality Supervision, Inspection and Quarantine of the People's Republic of China: Beijing, China, 2008.
35. *JTG F40-2004*; Technical Specifications for Construction of Highway Asphalt Pavements. Ministry of Transport of the People's Republic of China: Beijing, China, 2004.
36. *JTG E20-2011*; Standard Test Methods of Bitumen and Bituminous Mixtures for Highway Engineering. Ministry of Transport of the People's Republic of China: Beijing, China, 2011.
37. *JTG D50-2017*; Specifications for Design of Highway Asphalt Pavement. Ministry of Transport of the People's Republic of China: Beijing, China, 2017.
38. Maalej, M.; Quek, S.T.; Zhang, J. Behavior of Hybrid-Fiber Engineered Cementitious Composites Subjected to Dynamic Tensile Loading and Project Impact. *J. Mater. Civ. Eng.* **2005**, *17*, 143–152. [CrossRef]

39. Zhang, J.; Gong, C.; Guo, Z.; Zhang, M. Engineered Cementitious Composites with Characteristic of Low Drying Shrinkage. *Cem. Concr. Res.* **2009**, *39*, 303–312. [CrossRef]
40. Guo, Z.; Wang, L.; Feng, L.; Guo, Y. Research on fatigue performance of composite crumb rubber modified asphalt mixture under freeze thaw cycles. *Constr. Build. Mater.* **2022**, *323*, 126603. [CrossRef]

polymers

MDPI

Article

Prediction of The Mechanical Behavior of Polylactic Acid Parts with Shape Memory Effect Fabricated by FDM

Zhamila Issabayeva and Igor Shishkovsky *

Center for Materials Technologies, Skolkovo Institute of Science and Technology, 121205 Moscow, Russia
* Correspondence: i.shishkovsky@skoltech.ru; Tel.: +7-495-2801481

Abstract: In this study, the mechanical as well as thermomechanical behaviors of shape memory PLA parts are presented. A total of 120 sets with five variable printing parameters were printed by the FDM method. The impact of the printing parameters on the tensile strength, viscoelastic performance, shape fixity, and recovery coefficients were studied. The results show that two printing parameters, the temperature of the extruder and the nozzle diameter, were more significant for the mechanical properties. The values of tensile strength varied from 32 MPa to 50 MPa. The use of a suitable Mooney–Rivlin model to describe the hyperelastic behavior of the material allowed us to gain a good fit for the experimental and simulation curves. For the first time, using this material and method of 3D printing, the thermomechanical analysis (TMA) allowed us to evaluate the thermal deformation of the sample and obtain values of the coefficient of thermal expansion (CTE) at different temperatures, directions, and running curves from 71.37 ppm/K to 276.53 ppm/K. Dynamic mechanical analysis (DMA) showed a similar characteristic of curves and similar values with a deviation of 1–2% despite different printing parameters. The glass transition temperature for all samples with different measurement curves ranged from 63–69 °C. A material crystallinity of 2.2%, considered by differential scanning calorimetry (DSC), confirmed its amorphous nature. From the SMP cycle test, we observed that the stronger the sample, the lower the fatigue from cycle to cycle observed when restoring the initial shape after deformation, while the fixation of the shape did not almost decrease with each SMP cycle and was close to 100%. Comprehensive study demonstrated a complex operational relationship between determined mechanical and thermomechanical properties, combining the characteristics of a thermoplastic material with the shape memory effect and FDM printing parameters.

Keywords: shape memory polymer (SMP); fused deposition modeling (FDM); viscoelasticity; thermomechanical testing; biodegradable polymer; polylactic acid; 4D printing; mechanical strength

check for updates

Citation: Issabayeva, Z.; Shishkovsky, I. Prediction of The Mechanical Behavior of Polylactic Acid Parts with Shape Memory Effect Fabricated by FDM. *Polymers* **2023**, *15*, 1162. https://doi.org/10.3390/polym15051162

Academic Editors: Emilia P. Collar and Jesús-María García-Martínez

Received: 25 January 2023
Revised: 19 February 2023
Accepted: 22 February 2023
Published: 25 February 2023

1. Introduction

The widespread use of additive manufacturing (AM) techniques has been linked to the fact that AM technologies save product development costs and time. The mechanical performance of the most widely used polymers is often the factor limiting possible applications because the AM process and parameters influence the resulting mechanical properties, which could be radically different from the unprocessed material. The most common additive manufacturing technique is fused deposition modeling (FDM), where a material is layered from a molten thermoplastic filament extruded by a heated nozzle. By adjusting several variables of the printing regime including the infill, density, filament orientation, and print speed, the FDM technique allows for the mechanical qualities to be customized. According to the literature, the parameters, which include printer settings and part characteristics, have a direct impact on the strength properties of the part [1,2]. Customizing printing parameters is a delicate procedure with numerous aspects that influence the quality and characteristics of the components. Printing and process parameters influence the bonding process between adjacent filaments that, in turn, influence the void density and bond strength, which means directly defining the mechanical properties of the FDM components. Some parameters have a negligible effect, while others are decisive. For example, the height of the layer affects the interlayer bonding, and the nozzle temperature affects the detail and strength. Furthermore, the rougher the surface of the previous layer of the printout, the better its adhesion with the subsequent layer will be. Better adhesion can be achieved by adjusting the bed temperature and even the print speed, as examined by Vannamei, El Magri et al. [3]. They noticed that the variation of 40 °C in the platform temperature resulted in an 11% increase in tensile strength and the increase of 20 mm/s in print speed resulted in a 14% increase.

Rankouhi et al. [4] examined the effect of the thickness and orientation of the layer on the tensile properties. It was found that the acrylonitrile butadiene styrene (ABS) part with a layer thickness of 0.2 mm showed higher tensile strength properties than with a layer thickness of 0.4 mm. A similar study presented by Wang [5] concluded that a smaller layer thickness increased the tensile strength due to smaller interlayer gaps and fewer air pores in the cross-section. Chacon et al. [6], in contrast, argued that in the vertical samples, the tensile and bending strength increased as the layer thickness increased. A more reliable approach would be to estimate the ratio of the nozzle diameter and the layer height and its effect on the strength of the part. This ratio is characterized by the area of the interlayer contact surface [7]. An increase in the ratio of the nozzle diameter to the height of the layer results in a cross-section of the thread similar to a circle; with a decrease in the layer's height, the cross-section of the thread resembles a rectangle. Most of the air remains at the junction of the perimeters, which reduces the strength of the printout.

Regarding the filling pattern, the published study by Khan et al. [8] showed greater strength for ABS parts with a rectilinear infill pattern, while a concentric infill pattern yielded the best elongation, and the honeycomb pattern was recognized as irrational due to a waste of material. Yeoh et al. showed that different filling patterns influenced the strength and flexibility of the part to varying degrees [9], and PLA with a zigzag infill pattern showed the highest tensile strength in comparison with grid and concentric patterns.

The 3D printing of smart materials such as shape memory polymers (SMP) has led to the development of 4D printing technology [10]. The novelty of the topic is that the shape of 3D printed products can be transformed over time, achieving 4D printing. The difference is that the 3D object is static, but the 4D object is already dynamic. In this case, the fourth dimension is the response time to the stimulus and the subsequent change in shape [11]. Shape memory materials have the capability to change their shape upon the application of an external stimulus [12]. The ability to physically change the size and shape, particularly compression, expansion, and twisting, has allowed for the use of SMPs for artificial muscles, parts for aerospace ships, transforming medical devices, and intelligent textiles. In our recent review [13], it was shown that the 4D printing of SMP opens up the bio-fabrication possibilities of hierarchical auxetic metamaterials. There are artificial devices such as coronary stents, filtration and drug delivery systems, implants and many others where the ability for high-precision tuning are highlighted.

Typically, the materials used for 4D printing have a SME. However, a group of scientists has proposed a new approach to 4D printing without SMP and extra operations such as synthesis and blending [14,15]. They created encapsulated and bilayer macroscopic SMP structures (PCL-TU) using multimaterial printing of nonshape memory polymers. This solved the problem of a limited selection of printable SMP materials, making it possible to use all commercial filaments.

The SME in a material can be estimated by the recovery coefficient from cycle to cycle. Therefore, it is appropriate to emphasize that FDM parameters can impact the shape recoverability. In [16], Liu et al. showed that the thickness of the sample layer printed on a 3D printer had a significant impact on the recovery efficiency. The lower the layer height, the higher the elastic modulus and yield strength corresponding to it. The choice of nozzle temperature also plays an important role. Therefore, it was revealed that lower printing temperatures and higher printing speeds improve the characteristics of self-bending and shape recovery [17,18]. In another study, the authors summarized that the combination of a lower nozzle temperature with a higher printing speed had a negative effect on the recovery time and percentage, since high-viscosity materials have a limited time for extrusion as a result of the fast printing process [19]. Soleyman et al. [20] investigated the 4D printing capability of shape memory thermoplastic PET as well as the influence of printing temperature and speed on the emergent curved third shape during recovery. Despite the different values of the printing parameters and the degree of the polymer chain extension, the shape recovery for the two samples exceeded 96%.

As above-mentioned, the ability to keep and restore shape characterizes SMPs, which is due to the glass transition of thermally activated amorphous SMPs. The amorphous shape memory effect (SME) is dependent on molecular interactions [21]. Structurally, polymers look like long flexible macromolecules capable of changing the spatial arrangement of atoms under the influence of mechanical and thermal action, while the mobility and speed of conformations depend on the temperature regime [22]. Temperature has a considerable influence on the mechanical characteristics of polymeric materials [23]. At higher temperatures, an amorphous polymer that is stiff at ambient temperature becomes bendable and stretchy. The mechanical characteristics of the SMPs change with temperature, particularly around their glass transition temperatures. The glass transition temperature is the temperature at which the polymer chains become more mobile and there is a transition from a glassy rigid state to a rubbery deformable state [24]. The glass transition temperature is an essential result of dynamic mechanical analysis (DMA) as well as differential scanning calorimetry (DSC) and thermomechanical analysis (TMA). In fact, the value is not strictly defined, as depending on the way that it is determined, the value will vary [25]. There are three ways to determine the glass transition temperature from DMA data, and all three curves can show their own value of a given temperature. The T_g can be determined by the drop in the storage module, the peak of the loss module, or the peak of the $\tan(\delta)$. Viscoelastic properties are manifested at temperatures exceeding T_g.

The mechanical properties of specimens from the same PLA material can vary significantly. In our study for the above-mentioned material, we not only obtained the mechanical characteristics for different printing parameters, but also investigated the thermal behavior of the FDM printed samples providing TMA + DSC as well as DMA analyses. Aside from everything else, we conducted cyclic thermomechanical tests to study the SME. The results were brought together to create a more detailed portrait of the PLA material, its parts, and its behavior under different conditions. Furthermore, we were able to compare the experimental results of the tensile test and DMA with the FEM calculations.

2. Materials and Methods

2.1. Materials

In all stages of the research, the material used was a SMP from polylactic acid—PLA (Createbo, Ningbo, China). Despite the variety of printing materials, PLA is the most usable and widespread as a building material in FDM [26]. PLA is a synthetic aliphatic polyester derived from agricultural resources. PLA is favored due to its biodegradability, renewability, and its low toxicity, so is strongly recommended for medical applications. The material has a lower shrinkage and melting temperature than other 3D printing polymers, making polylactic filament easier to extrude through the FDM nozzle. The PLA filament is an amorphous polymer in which the glass transition temperature is a critical point of phase transition, and hence the critical temperature to study the SME.

There are three types of the polylactic acid structure due to the different optical rotation, L-polylactic acid, DL-polylactic acid, and D-polylactic acid [27]. The degree of amorphous branching inside the polymer chains differs. Both PLLA and PDLLA polymers are naturally crystalline with a glass transition temperature of 60 °C, which means that they take on ordered molecular structures. Poly(DL-Lactide) PDLLA is an amorphous polymer with a glass transition temperature of about 55 °C. According to the supplier, the type of material used in this project was PDLLA.

2.2. Uniaxial Tensile Tests

2.2.1. Sample Fabrication

A dog-bone shape was chosen as the specimen geometry with dimensions of the rectangular part of 60 mm in length, 3 mm in width, and 3 mm in thickness according to ASTM D638-10. Five printing process parameters were selected. For each parameter, the following values were defined based on the values reported in the literature and the capabilities of the FDM systems used in this research:

- Infill density: 20%, 25%, 50%, 70% and 80%;
- Infill pattern: grid, line, gyroid;
- Layer thickness: 0.2 mm, 0.3 mm;
- Diameter of the nozzle: 0.3 mm, 0.5 mm;
- Temperature: 210 °C, 240 °C.

Other printer parameters remained constant throughout the study. The bed temperature was considered to be equal to 60 °C. The printing speed was set to 60 mm/s. Printing parameters were chosen based on the preliminary literary analysis and our own experience in working with a 3D printer. The literature states that the most important printing parameters are layer thickness, infill, and temperature [28,29]. The values for the infill density were selected based on preliminary attempts to evaluate the quality of the printed parts. Therefore, details with a density of less than 20% were unacceptable, and a huge difference in strength was not observed between 80% and 100%, where only time consumption and waste material were highlighted.

According to the parameters and values selected, 120 combinations were fabricated using a Picasso Designer X Pro 3D printer (see Supplementary Materials). Every regime comprised 4 of the same specimens, which means that 480 specimens were printed and evaluated for the tensile test.

2.2.2. Uniaxial Tensile Experiments

The uniaxial tensile tests were performed following the ASTM D638-02 standard. Uniaxial tensile tests were performed on an Instron 5969 universal material testing machine (Instron, Norwood, MA, USA). Displacements of the center section were measured with an extensometer with a gauge length of 25 mm. The allowable travel of the extensometer was 2.5 mm, which corresponded to the limit of measuring strain of 10%. The basic experimental conditions were a room temperature of 22 ± 0.5 °C, and a relative humidity of around 35%.

The initial sample dimensions (width and thickness) were measured. The specimens were settled into the grips of the testing machine. A load at a constant speed of 1 mm/min was applied. A gradual load of 1.0 kN was applied to the samples until the moment of failure, and the resulting loads and displacement were recorded by the testing machine for each individual sample. The tests ended automatically when the specimens fractured.

2.3. Differential Scanning Calorimetry

Differential scanning calorimetry (DSC) is the most commonly used method of thermal analysis, in which the difference in the amount of heat required to increase the temperature of the sample and the reference is measured as a function of temperature. The value of measuring the energy flow is that it allows us to determine the range of different transitions that can occur in a sample when it is heated or cooled. To avoid temperature gradients inside the sample, a small sample size is preferable. A DSC test was conducted on the ASTM D3418 standard. The dimensions of the studied sample were 3 mm × 3 mm × 3 mm. A sample was heated or cooled and the changes in its heat capacity were tracked as changes in the heat flow. This allows for the detection of transitions such as melts, glass transitions, phase changes, and curing. The experiment was carried out on the DSC3+ Excellence system (Mettler Toledo, Greifensee, Switzerland). To conduct the measurements, the sample was placed in a calorimetric crucible and sealed. The DSC temperature program was determined in the range of 0–180 °C with a heating rate of 10 K/min in a nitrogen medium with a flow rate of 50 mL/min in an Al crucible at 40 µL.

2.4. Thermomechanical Analysis

The TMA experiment was carried out on the TMA/SDTA2+ system (Mettler Toledo, Greifensee, Switzerland) under static load using various sensors to measure the changes in the length of the sample depending on temperature. The most important measurements of TMA include the determination of the coefficient of linear thermal expansion—α and the glass transition temperature T_g (ASTM E831 and ASTM E1545). The size of the test sample experiment should not exceed 4–5 mm. The dimensions of the tested sample were 3 mm × 3 mm × 3 mm. It is important to note that the initial dog-bone specimen was cut in the central part into cube form. The surface was pre-polished to achieve smoothness. The experiment program was based on the results of the DSC measurements and consists of several consecutive segments:

- Isotherm 0 °C during 10 min;
- First heating 0–90 °C;
- Cooling 90–0 °C;
- Second heating 0–150 °C.

The heating rate was 3 K/min in a nitrogen medium with a flow rate of 30 mL/min, and a probing constant load of 0.02 N in the dilatometric mode. Purge gas, N_2, provided a continuous laminar gas flow to prevent the formation of air turbulence when the temperature rose, prevent the deposition of decomposition products inside various parts of the device, increase heat transfer to the sample, and prevent oxidation at high temperatures. To measure the internal properties of a material, the thermal history of the samples must first be erased. The treatment was performed in a thermal analysis device by heating the sample to a temperature exceeding its glass transition temperature.

2.5. Dynamic Mechanical Analysis

The DMA tests were conducted using a DMA Q800 machine (TA Instruments, Houston, TX, USA) to determine the glass transition temperature T_g of the SMP specimen as described in ASTM D4065. Four specimens with different printing configurations were selected for analysis. A rectangular sample with dimensions of 60 mm × 5 mm × 2 mm was placed in the DMA device using a dual cantilever clamp configuration. A small dynamic load with an amplitude of 10 μm with a frequency of 1 Hz was applied to the roller. The sample was examined in the range from room temperature to 150 °C. The heating rate of 3 K/min was used to avoid the effects of a fast temperature rise on the experimental results. The data were collected every 3 s during thermomechanical analysis.

2.6. Thermomechanical Shape Memory Cycle

Uniaxial tensile tests were carried out on a servo-hydraulic testing machine Instron 8801 (Instron, Norwood, MA, USA) equipped with a temperature chamber and an optical window to conduct observations inside the chamber. The overall length was set to 210 mm, the gauge length of sample was 60 mm, and the cross-section was 50 mm × 7 mm. Deformation control was carried out using the Digital Image Correlation method (DIC) [30,31]. Vic-3D (Correlated Solutions, Irmo, SC, USA) software was used to keep track of the shooting and process the photos that were captured. Figure 1 shows sample no. 17 during the first and fifth cycles with a field of measurement in the DIC system before and after mechanical testing.

A typical cycle to study the SME is a sequence of heating–loading–cooling–unloading–heating steps [32]. In this study, this heating–cooling process was carried out on up to five shape memory cycles. A specimen was heated up to 70 °C with rate of 3 °C/min and subsequently deformed with 60% strain, then allowed to cool down, keeping the deformation. The cooling rate was set to 3 °C/min. The length of the sample during stretching was set to 60 mm, and the stretching speed was 50 mm/min.

After heating, the recovery ratio could be calculated from the recovered shape of the samples. Tandon et al. proposed the following definition of the shape memory ratios [24]:

$$R_{fix} = \frac{\varepsilon_u}{\varepsilon_p} \cdot 100\%, \tag{1}$$

where R_{fix} measures the ability of the specimens to fix the shape; ε_u is a strain obtained after unloading; and ε_p represents the maximum strain. The shape recovery ratio is expressed as:

$$R_{rec} = \left(1 - \varepsilon_f\right) \cdot 100\%, \tag{2}$$

where ε_f is the final strain expressed through the relation between the change in length after recovery and the initial specimen length.

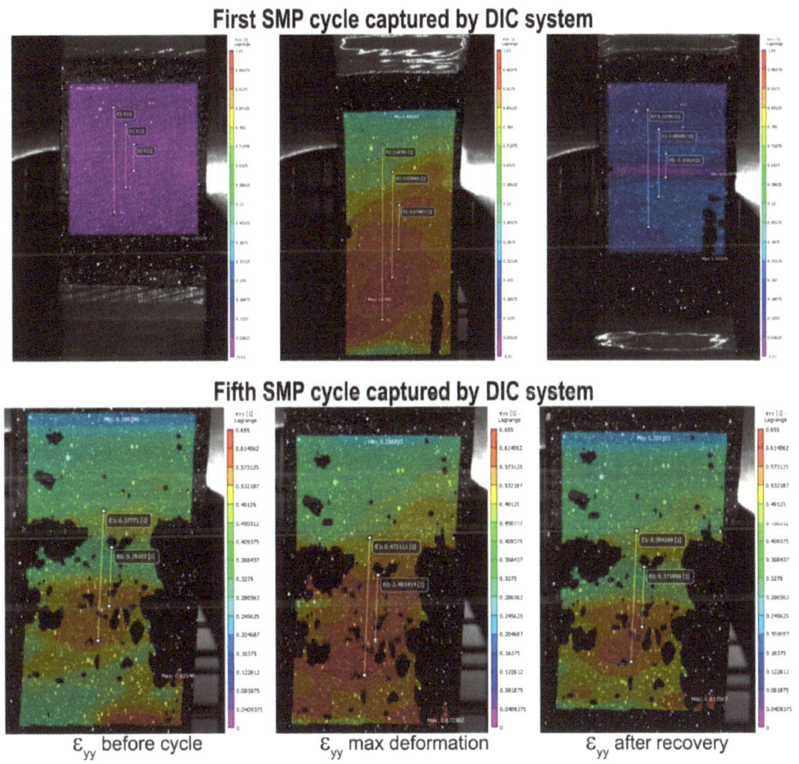

Figure 1. Specimen no. 17 at the maximum ε_{yy} displacement levels in the first and fifth cycles.

2.7. Numerical Simulations

2.7.1. Uniaxial Tensile Test

The finite element method to solve basic control equations and related boundary conditions specific to a particular task was implemented in Ansys Workbench 2019 (Ansys Inc, Canonsburg, PA, USA). Finite element analysis (FEA) was used to simulate the tensile load on the sample up to the moment before destruction. The built-in library contains a large selection of various materials including PLA. However, for nonlinear analysis, only having embedded data is not sufficient. To process and apply experimental data as input, it is necessary to define a model. Here, we considered the Mooney–Rivlin model with 3, 5, and 9 parameters [33]. The model with 3 parameters showed the worst fit, while 5 and 9 parameters fitted the curve equally well. Therefore, we chose the Mooney–Rivlin model with 5 parameters as the determining one. The coefficients of the five-parameter Mooney–Rivlin model were found by approximating the experimental data using Ansys Workbench. The specimens were meshed at the resolution of 5 mm/cell with HEX20 elements. The load was applied to the upper plane of the sample and the boundary conditions were applied to the lower part.

2.7.2. Viscoelastic Behavior

PLA material showed viscoelastic behavior, as studied by DMA. This viscoelastic behavior is typical for amorphous polymers at temperatures close to the glass transition. Linear models are most often used to describe viscoelastic behavior. The linear viscoelastic Prony model is relatively simple, and at the same time, quite informative [34]. The Prony model contains a time relaxation series for each component, so relaxation occurs not at once, but over a certain time. This model, or rather the components characterizing it, can be specified in the finite element calculations of structures using the Ansys Workbench software. Based on the limited amount of data obtained experimentally, it is possible to determine the rheological parameters of the viscoelastic model.

Here, we considered the method to calculate the Prony series from DMA data. The determining relation is represented by the viscoelastic Prony model in the case of the absence of volumetric relaxation [35]:

$$G(t) = G_0\left[\alpha_0^G + \sum_{i=1}^{N}\alpha_i^G \exp\left(-\frac{t}{\tau_i^G}\right)\right], \tag{3}$$

where $G(t)$ is the shear relaxation modulus; α_i^G are the relative shear modulus for the shear relaxation times τ_i^G; N is the number of shear relaxation times.

In the equation described above, it is assumed that the material only experiences shear relaxation and is characterized by a constant volume compression modulus. Often, to obtain experimental parameters, test data are used not for shear, but for uniaxial tension, as was also used in our study. Therefore, the relaxation module for uniaxial stretching will be represented as [35]:

$$E(t) = E_0\left[c_0 + \sum_{i=1}^{N} c_i \exp\left(-\frac{t}{\beta_i}\right)\right], \tag{4}$$

where c_i is the relative module of tension–compression for relaxation times β_i; N is the number of the shear relaxation times of tension–compression relaxation; E_0 is a modulus when $t = 0$. All c_i must be positive and in sum <1 [36].

Using the calculation part with related equations in the approach by Smetannikov et al., we assumed [35]:

$$E' = E_\infty + \omega^2 E_0 \sum_{i=1}^{N} \frac{\beta'^2_i c_i}{1+\beta'^2_i\omega^2}, \tag{5}$$

$$E'' = \omega E_0 \sum_{i=1}^{N} \frac{\beta'_i c_i}{1+\beta'^2_i\omega^2}, \tag{6}$$

where E' and E'' are the real and imaginary parts of the complex modulus of elasticity; E_∞ is a modulus when $t = \infty$. The relaxation times given are in accordance with the principle of the temperature–time analogy:

$$\beta'_i = \frac{\beta_i}{A(t)}, \tag{7}$$

where $A(t)$ is the shift function. Time–temperature superposition is achieved by the William–Landel–Ferry (WLF) shift factor as given in the next equation [35,36]:

$$\lg(A(T)) = \frac{C_1(T - T_r)}{C_2 + (T - T_r)}, \tag{8}$$

where $\lg(A(T))$ is the shift factor in logarithmic form; T_r represents the glass transition temperature obtained from the experiment; the constants C_1 and C_2 are empirical constants of the material and approximated as 17.44 and 51.6, respectively [36].

The storage and damping properties of the generalized Maxwell model (GMM) [36] were used to estimate the master curve for viscoelastic behavior. The GMM model with a number of Maxwell elements using Prony series coefficients of relaxation time β_i and weights α_i^G fit the experimental data. For practical use, curve fitting can be implemented using software, however, it is necessary to determine the optimal number of Prony components. Each commercial CAM software package may have a different definition of the Prony series. We used the Origin 2021 software (Origin Lab, Northampton, MA, USA) to set the user-defined function, which contained a viscoelastic equation (Equation (5)) for the storage modulus, to fit a nonlinear curve.

2.7.3. Finite Element Modeling of DMA

The mechanical properties of the proposed model, which depend on temperature, were evaluated via step-by-step modeling of the DMA experiment with a dual cantilever. Along with determining the Prony coefficients, the thermomechanical characteristics should also be considered. The PLA material specification in different temperatures were into engineering data of Ansys. The defining relations corresponded to the viscoelastic Prony model using the Williams–Landell–Ferry shift function (Equation (8)) in the calculation.

The preliminary convergence study with refined meshing allowed the appropriate mesh size to be set to 1 mm. The mesh was created using TET10 elements. During the simulation, a rectangular sample was subjected to cyclic loading in accordance with a sinusoidal waveform at a frequency of 1 Hz with an amplitude of 10 microns. The displacement of 10×10^{-3} mm was applied to the small cross-section of the specimen. Boundary conditions were applied both to the upper surface and to the lower one. The value of the dynamic elastic modulus was calculated as [37]:

$$E_{dyn}(\omega) = |E^*(\omega)| = \frac{\sigma_0}{\varepsilon_0(\omega)}, \tag{9}$$

where σ_0 and ε_0 are the stress and strain amplitude, respectively.

3. Results

3.1. Thermomechanical Analysis and Differential Scanning Calorimetry

The thermogram of the PLA sample contained several thermal events, as shown in Figure 2.

The printed material was examined in three directions relative to the original shape of the bar:

1. Normal to the sample is the thickness (Z coordinate of the 3D printer);
2. Across the sample is the width (X coordinate of the 3D printer);
3. Along the sample is the length (Y coordinate of the 3D printer).

The coefficient of thermal expansion (CTE) was determined from the TMA measurements. From the upper part of Figure 2, we observed how significant different CTE was not only found in different directions, but also in the first heating, second heating, and cooling run. The cyclical heating–cooling–heating method is commonly applied in TMA studies to eliminate the thermal history and reduce the internal stresses in the sample. The transition temperatures are determined by the sample's internal stresses, and hence its thermal history. This explains why the curves observed in the first and second heating runs had distinct forms.

As it turned out, the most dramatic effects were observed in the first heating, while geometry changes not related to thermal expansion began to take place already at temperatures higher than 41 °C. The nature of the change in the measured deformation effect differed significantly for the three directions of the coordinate grid of the sample, which indicated a serious CTE anisotropy of the studied material. Without considering the thermal expansion, the thickness of the sample (Z coordinate black) decreased (maximum) by 1.3%, the width (X coordinate red) decreased by 0.07%, and the length (Y coordinate blue) increased by 0.7%. Furthermore, the sample sought to compensate for the deformation effect, which may be associated with the beginning of the crystallization process.

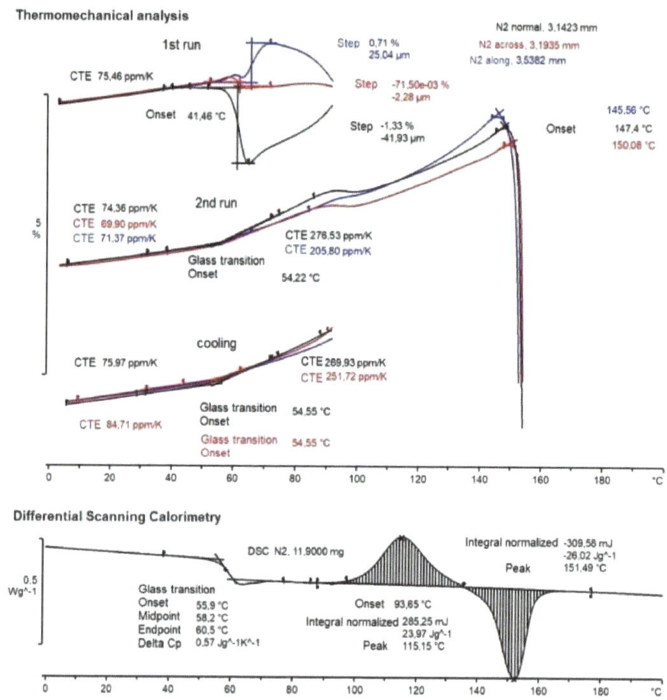

Figure 2. The TMA (top) and DSC (bottom) experimental results.

The cooling and second heating curves looked more predictable. While cooling, the CTE changed in three directions and was less visible: the normal direction at a glassy state had a CTE equal to 75.97 ppm/K and above the T_g, the CTE was equal to 269.93 ppm/K and 251.72 ppm/K for the normal and across directions, respectively. The second heating run curve was examined in the same way. In the glassy state, the CTE was 74.36 ppm/K for the normal direction, 69.9 ppm/K for the across direction, and 71.37 ppm/K for the along direction. Above the glass transition temperature, the CTE showed 276.53 ppm/K for the normal direction, and 205.8 ppm/K for both the along and across directions. The CTE values are necessary in the process of modeling the behavior of the material in order to take into account the change in stress with temperature changes. From [38], the value of CTE is 98 ppm/K in the glassy state. This is more than the value we received of 14–20 ppm/K. The changes in the CTE values are shown in Table 1.

The glass transition temperature itself was well fixed on the cooling (54.55 °C) and heating (54.22 °C) curves and located as the intersection of tangents corresponding to the change in the CTE before and after the phase transition. Residual anomalies of geometry change continued to be observed in the second heating curve in the range of 85–105 °C. After 147 °C, the sample melted and began to deform under the influence of probing stress.

Table 1. Coefficients of thermal expansion for different cases.

Direction	Glassy State	Above T_g
	First heating curve	
Normal (Z) Across (X) Along (Y)	75.46 ppm/K	-
	Cooling curve	
Normal (Z)	75.97 ppm/K	269.93 ppm/K
Across (X) Along (Y)	84.71 ppm/K	251.72 ppm/K
	Second heating curve	
Normal (Z)	74.36 ppm/K	276.53 ppm/K
Across (X)	69.9 ppm/K	205.8 ppm/K
Along (Y)	71.37 ppm/K	

The DSC curve is shown at the bottom of Figure 2. In the range of 55–60 °C, a jump in the heat capacity (Delta C_p) of 0.57 J/(g K) was observed, corresponding to the glass transition of the polymer with onset T_g = 55.9 °C, midpoint T_g = 58.2 °C, and endpoint T_g = 60.5 °C. Then, in the range of 93–135 °C, there was an exothermic peak of polymer crystallization at 115.15 °C, followed by the latent heat of crystallization of about 23.97 J/g. After 135 °C, melting occurred with an endothermic peak at 151.4 °C and the latent heat of the melting was 26.02 J/g.

The percent crystallinity was determined using the following equation [39]:

$$\%Crystallinity = \frac{\Delta H_m - \Delta H_c}{\Delta H_m{}^{\circ}} \cdot 100\% \tag{10}$$

where ΔH_c is the latent heat of crystallization and ΔH_m is the latent heat of melting; $\Delta H_m{}^{\circ}$ is a reference heat of melting for a 100% crystalline polymer (for a 100% crystalline PLA $\Delta H_m = 93$ J/g) [39,40].

Based on the ratio from [40] of the energy effects of the exo- and endo-peaks, it can be assumed that $\%Crystalinity = 2.2\%$ and the polymer is predominantly amorphous, which confirmed the information from the supplier.

3.2. Uniaxial Tensile Test

A total of 480 samples were tested for tension. For each sample, a strain–stress curve was obtained, which is the maximum value of the stress that the sample experiences until the moment of fracture. The test results were processed by the methods of statistical analysis (see Section 3.3) using Statistica (ver. 13). Table 2 consists of the tensile strength values for the top ten sample printing modes that were of greatest interest.

Conditionally, they can be divided into three groups: modes with the best mechanical properties, modes with the worst mechanical properties, and the two modes that were recorded as outliers on graphs when performing the ANOVA analysis. Moreover, the nature of the behavior of the properties in these two modes did not correspond to the general logic of the behavior of the other samples. These specimens were selected to be tested with the shape memory cycle.

Table 2. Specimens selected for the shape memory cycle test.

Specimen	D of the Nozzle	Layer Height	Infill Pattern	Infill Density	Extrusion Temperature	Tensile Strength
Specimens with the highest tensile strength values						
10	0.5 mm	0.2 mm	Line	80%	240 °C	50.1 MPa
16	0.5 mm	0.3 mm	Grid	20%	240 °C	49.9 MPa
17	0.5 mm	0.3 mm	Grid	25%	240 °C	49.8 MPa
15	0.5 mm	0.2 mm	Gyroid	80%	240 °C	49.1 MPa
Specimens with the lowest tensile strength values						
113	0.3 mm	0.3 mm	Line	50%	210 °C	32.9 MPa
117	0.3 mm	0.3 mm	Gyroid	25%	210 °C	34.2 MPa
115	0.3 mm	0.3 mm	Line	80%	210 °C	34.9 MPa
107	0.3 mm	0.3 mm	Grid	25%	210 °C	35.2 MPa
Samples with ambiguous behavior						
2	0.5 mm	0.2 mm	Grid	25%	240 °C	43.2 MPa
45	0.5 mm	0.2 mm	Gyroid	80%	210 °C	44.5 MPa

According to Table 2, the highest values of the tensile stress were shown by samples with a print temperature of 240 °C and a nozzle diameter of 0.5 mm while the lowest values were for samples with a print temperature of 210 °C and a nozzle diameter of 0.3 mm. A sample with 80% filling and line pattern provided the highest ultimate tensile strength. This sample had a maximum tensile strength of 50.1 MPa. As can be seen from the table, the same line drawing with 50% filling, but already with a nozzle temperature of 210 °C, nozzle diameter of 0.3, and a layer height of 0.3 mm, showed the lowest ultimate tensile strength of 32.9 MPa. The study in [41] showed that the value of the tensile strength for a linear pattern was greater than for grid filling, which, by and large, was implicit in our study regarding the presence of more significant parameters. In a study by Corapi et al. [42], the tensile strength of the PLA sample ranged from 28 to 58 MPa, depending on the orientation of the print. Rao et al. [43] estimated the effect of the layer height in the PLA sample on the tensile strength, which was in the range from 21 to 26 MPa. The strength values obtained by us during the experiment corresponded to the reality and were in the upper limits for this material.

Below, we demonstrate the outcomes of the Ansys numerical approach on a single specimen out of 480. Figure 3 shows the simulation of the dog-bone specimen for a tensile test based on the experimental data for printing regimen no. 17, which presented the highest tensile strength. The difference between the experimental and simulation data is presented in Figure 4.

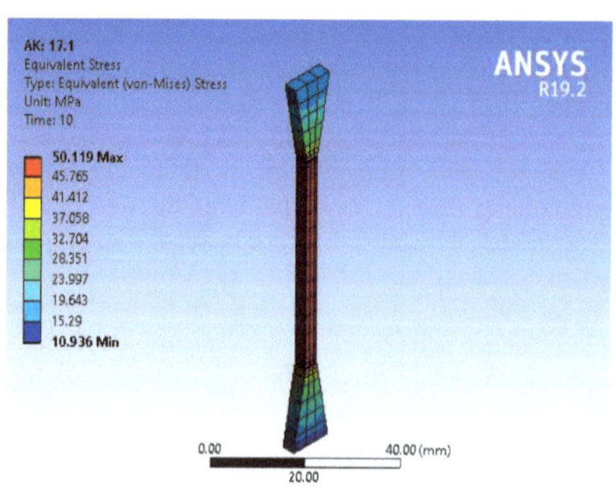

Figure 3. Tensile test simulation.

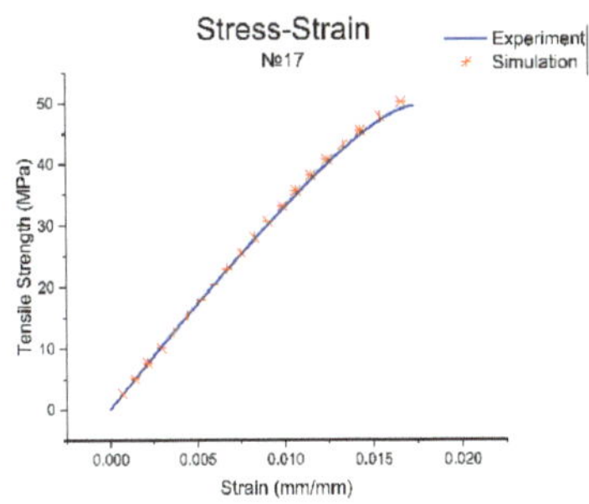

Figure 4. Experimental and simulation curves.

Thus, the results of the tensile test simulation were in good agreement with the experimental data and confirmed the adequacy of the selected Mooney–Rivlin model.

3.3. Analysis of Variance

The ANOVA analysis was performed in this study to determine the degree of influence of one or more parameters on the sample's tensile strength. This allowed us to determine the most notable printing regimes for the further thermomechanical study after examining the dependencies. Statistica software (ver. 13) was used to conduct the analysis. The tensile strength was chosen as the output parameter, while all five printing parameters were chosen as inputs. As a result, the multivariate analysis of variance was used to determine the impact of numerous factors on the dependent variable.

First, each parameter's independent impact on the stress value was analyzed. The dependence of strength on parameters such as nozzle diameter and extrusion temperature, for which the p value was less than 0.05, is shown in Tables 3 and 4. The same data in terms of plots are presented in Figures S1 and S2 (Supplementary Materials).

Table 3. Descriptive statistics for the nozzle diameter factor.

Nozzle Diameter	Mean (MPa)	Standard Deviation
0.5 mm	46.87	0.45
0.3 mm	38.31	0.45

Table 4. Descriptive statistics for the extrusion temperature factor.

Extrusion Temperature	Mean (MPa)	Standard Deviation
240 °C	43.18	0.45
210 °C	41.99	0.45

The difference between the average values of the factor is significant if the p value is small ($p < 0.05$). As a result, these two factors had a large influence on the dependent variable. The primary assumption was that a greater value of the nozzle diameter corresponded to a greater stress value and the higher the temperature of the extruder, the greater the strength of the part we observed. Whether everything was so unambiguous was found out during our evaluations. Considering other printing parameters such as the layer height, infill pattern, and density, in our research, their single impacts on the tensile strength were too insignificant with a p-value >> 0.5.

The next step was to consider the combined influence of factors. First, we studied the combined effects of two factors. Among the 10 combinations, only one showed statistically significance—extrusion temperature and layer height. Next, we continued to increase the number of factors for the total influence assessment. The mutual influence of two and three parameters on the tensile strength is presented in Tables 5 and 6. The same data in in terms of plots are presented in Figures S1 and S2 (Supplementary Materials).

Table 5. Descriptive statistics for the combined influence of the extrusion temperature and layer height.

Extrusion Temperature	Layer Height	Mean (MPa)	Standard Deviation
240 °C	0.3 mm	43.69	0.64
240 °C	0.2 mm	42.67	0.64
210 °C	0.2 mm	42.51	0.64
210 °C	0.3 mm	41.47	0.64

Table 6. Descriptive statistics for the combined influence of the extrusion temperature, nozzle diameter, and layer height.

Nozzle Diameter	Extrusion Temperature	Layer Height	Mean (MPa)	Standard Deviation
0.5 mm	240 °C	0.3 mm	47.47	0.9
0.5 mm	240 °C	0.2 mm	47.06	0.9
0.5 mm	210 °C	0.2 mm	46.53	0.9
0.5 mm	210 °C	0.3 mm	46.39	0.9
0.3 mm	240 °C	0.3 mm	39.92	0.9
0.3 mm	210 °C	0.2 mm	38.48	0.9
0.3 mm	240 °C	0.2 mm	38.27	0.9
0.3 mm	210 °C	0.3 mm	36.55	0.9

Here, we focused on the parameters that had previously been highlighted as being significant. However, since there are so many parameters and determining how they interact is a complex challenge, we summarized the results of our analysis for a further selection and study of the samples. It can be assumed that the nozzle diameter of 0.5 mm corresponded to higher values of maximum stress, and the temperature of the extruder significantly affected the result. The smaller the ratio of the layer height and diameter, the better for the strength properties in the case when the temperature was 210 °C. The reverse situation was typical for a temperature of 240 °C.

3.4. Dynamic Mechanical Analysis

The printing modes shown in Table 7 were selected and studied for DMA based on the ANOVA results. The choice of the tested samples can be explained by the versatile combinations of the printing parameters and different values of mechanical strengths. We selected two samples, no. 10 and no. 17, with the highest strength values, and two samples no. 113 and no. 117, with the worst. Sample no. 2 showed average stress values under uniaxial tension.

Table 7. Specimens and printing modes selected for DMA.

No.	Printing Regime	D of the Nozzle	Layer Height	Infill Pattern	Infill Density	Extrusion Temperature	Tensile Strength
1	10	0.5 mm	0.2 mm	Line	80%	240 °C	50.1 MPa
2	17	0.5 mm	0.3 mm	Grid	25%	240 °C	49.8 MPa
3	2	0.5 mm	0.2 mm	Grid	25%	240 °C	43.2 MPa
4	117	0.3 mm	0.3 mm	Gyroid	25%	210 °C	34.2 MPa
5	113	0.3 mm	0.3 mm	Line	50%	210 °C	32.9 MPa

We tried to consider critical cases to expect different data, since there was the assumption that the DMA results for one material would be similar despite the print parameters. One essential result of DMA was the glass transition temperature, which corresponded to a rapid fall in the storage modulus or a peak in the tan delta and loss modulus curve. The behavior of the storage modulus, loss modulus, and tan delta as a function of temperature, obtained as a result of the measurements, is depicted in Figure 5.

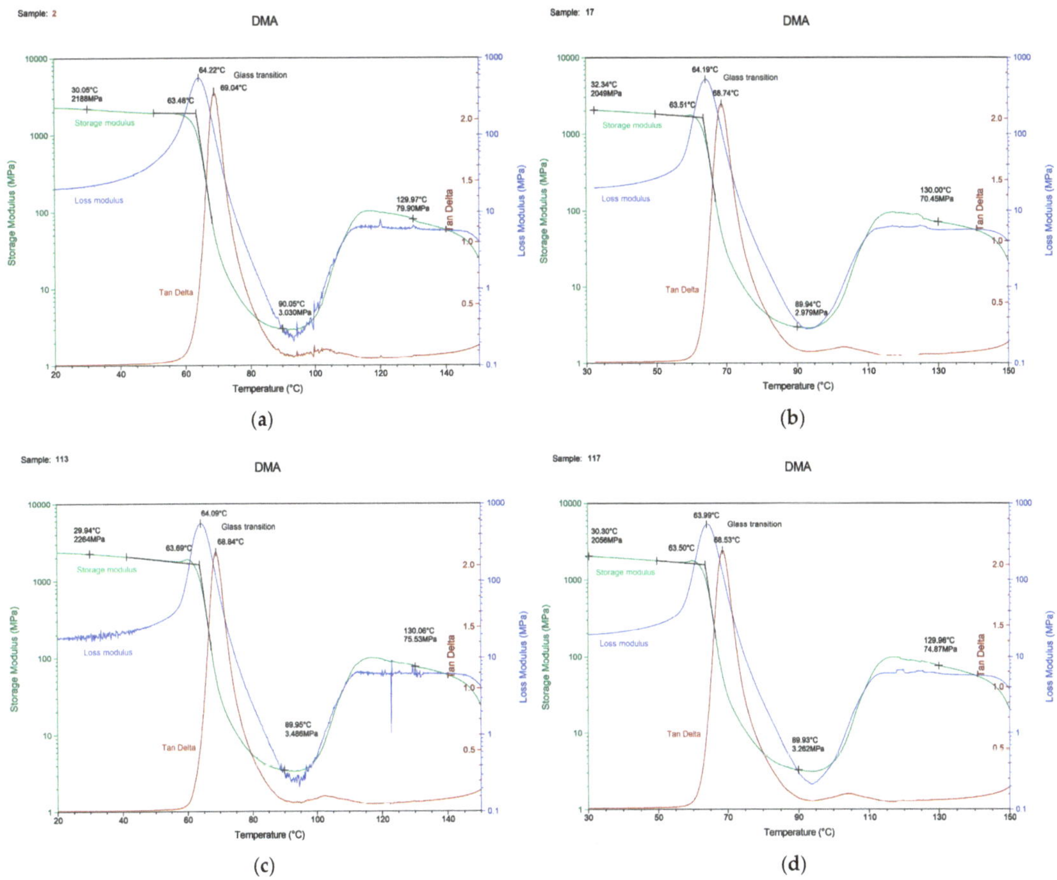

Figure 5. The DMA results for the (**a**) specimen with printing regime no. 2, (**b**) no. 17, (**c**) no. 113, (**d**) no. 117.

The glass transition temperatures obtained by these three curves varied by up to 5 °C. In addition, we observed a 10 °C deviation from the DSC defined glass transition temperature (Figure 2). It is crucial to describe how the T_g should be obtained and in which experimental conditions. The experimental results are shown in Table 8.

Table 8. The DMA results.

Specimen	T_g (Storage Modulus)	T_g (Loss Modulus)	T_g (Tan Delta)	Storage Modulus at T = 30 °C	Storage Modulus at T_g	Loss Modulus at T = 30 °C	Loss Modulus at T_g
2	64.22 °C	64.09 °C	69.04 °C	2188 MPa	859.6 MPa	20.37 MPa	555 MPa
17	63.51 °C	64.22 °C	68.74 °C	2049 MPa	1051 MPa	20.04 MPa	543.8 MPa
113	63.69 °C	64.19 °C	68.84 °C	2264 MPa	1064 MPa	19.02 MPa	553.8 MPa
117	63.50 °C	63.99 °C	68.53 °C	2056 MPa	1078 MPa	19.36 MPa	556.7 MPa

The glass transition temperature was in the range of 63–64 °C from the storage or loss modulus and 69 °C determined from the tan delta. The T_g value for each sample varied within 1%, which means that this is an acceptable measurement error, and its value is unique for samples from the same material.

Experimental data were used to fit the curve specified in Equation (5), which allowed us to determine the components of the Prony series, which are necessary to use the viscoelastic model in Ansys. The number of Prony terms (relaxation time and modulus) can be any; the only condition is the convergence of the curves. In this study, we tried to use only two terms, but the convergence was far from ideal, after which we experimentally increased the number of terms to $N = 8$. In our case, Ansys requests a table with the c_i terms representing the normalized Prony coefficients for the shear behavior, and the β_i values represent the relaxation times of the Prony series. Figure 6 contains graphs showing the result of fitting the curve to the experimental points.

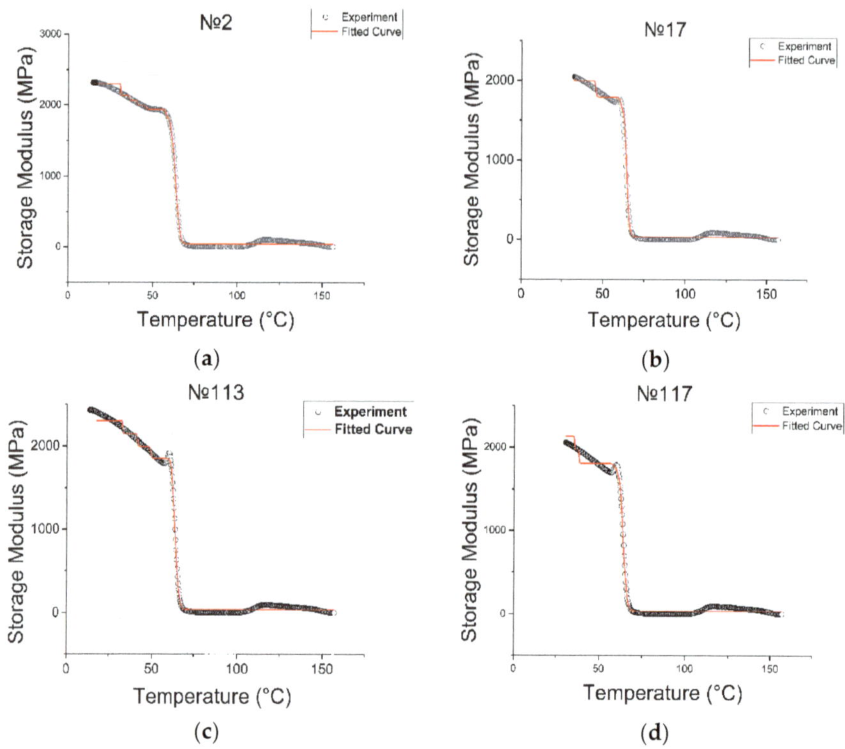

Figure 6. Convergence of the experimental and fitted curves of the specimen: (**a**) specimen with printing regime no. 2, (**b**) no. 17, (**c**) no. 113, (**d**) no. 117.

In a section up to 50 °C, the fitted curve had a non-uniform representation in the form of steps. Increasing the Prony series to eight arguments led to a decrease in the move of these steps, but it was not possible to achieve a fully linear representation of this section. Increasing the number of arguments up to nine, convergence was not observed. The obtained results were evaluated as optimal. Each curve corresponded to its own set of Prony series.

The experimental results of DMA were used to extract the viscoelastic characteristics, which were subsequently used as defining parameters of the model in the simulation. The modulus of elasticity was calculated from the curve of the storage modulus according to Equation (5). In addition, the values of the coefficient of thermal expansion at room temperature and at the glass transition temperature were obtained during thermomechanical analysis. Here, we used the CTE values for the normal direction and the second heating curve from Table 1. Table 9 contains the data used in Ansys simulation to define the PLA material model.

Table 9. Data of the temperature dependent material from the TMA and DMA data.

Temperature	Tensile Moduli	Poisson's Ratio	Thermal Expansion Coefficient [Table 1]
22 °C	2300 MPa	0.36	74.36 ppm/K
64 °C	929 MPa	0.42	276.53 ppm/K

As output values from the Ansys simulation, the amplitude of strain and stress were successfully obtained. Every temperature point starting from 30 °C to 80 °C with an increment of 5–10 °C (near T_g 1 °C) was simulated and had its own value of strain and stress. The phase shift δ was taken from the experimental $\tan(\delta)$ data. For each temperature point, these values were substituted into the corresponding formula (Equation (9)) to calculate the storage module (Equation (5)). All calculations were performed and save in an Excel file and then exported to Origin 2021 software. To compare the simulation and the experiment, we plotted two curves of each sample on the same graph. The temperature dependence of the real part of the complex module was determined during the experiment and as a result of the numerical solution, is shown in Figure 7.

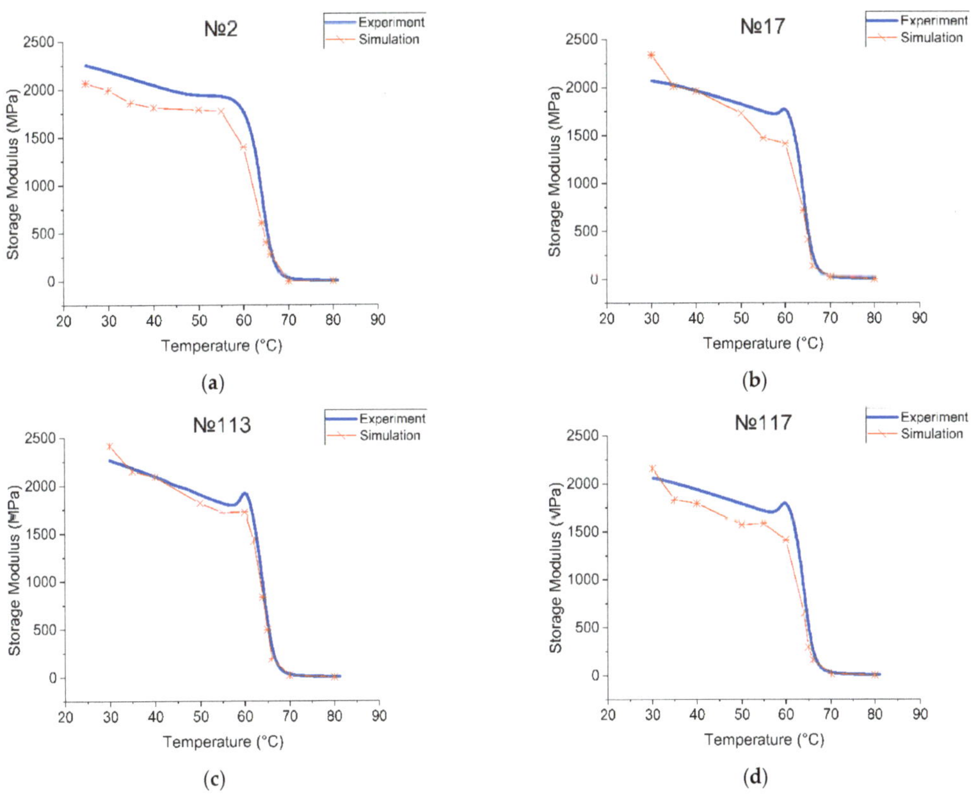

Figure 7. Comparison between the experimental results and simulation for the (**a**) specimen with printing regime no. 2, (**b**) no. 17, (**c**) no. 113, (**d**) no. 117.

Consider the printing regimes no. 2 and no. 17, where the difference in parameters was the value of the layer height (0.2 mm and 0.3 mm, respectively). At the starting point, the storage modulus experimental value was 2188 MPa for sample no. 2 (2070 MPa for sample no. 17) and the modeling value was 2064 MPa for sample no. 2 (2337 MPa for sample no. 17). As the free volume continued to increase with the increasing temperature, the glass transition occurred. The onset of the storage modulus curve drop was used to detect this transition point. The experimental curve had a step at T_g = 64.22 °C for sample no. 2 (63.51 °C for sample no. 17) and the simulation had a step at T_g = 63 °C for sample no. 2 (62.8 °C for sample no. 17). The experimental storage modulus value at T_g was 859.6 MPa and the simulation value was 1029 MPa for sample no. 2. The sample with printing mode no. 17 had an experimental storage modulus equal to 1211 MPa, and the simulation curve showed 1314 MPa. Thus, we received the overestimated storage module values per 103–170 MPa. The curve also showed the softening process at the glass transition, where the modulus fell to less than 10 MPa. The storage modulus changed by about two orders during the glass transition. The area after the T_g had a rubbery plateau. The modulus in the plateau region was 1 MPa for sample no. 2 (6 MPa for sample no. 17).

Printing regimes no. 113 and no. 117 had common parameter values (nozzle diameter 0.3 mm, layer height 0.3 mm, extruder temperature 210 °C) and differed in filling (no. 113—lines 50%, no. 117—gyroid 25%). These samples showed the worst mechanical tensile properties. In the very beginning (T = 30 °C), the experimental storage modulus was 2264 MPa for sample no. 113 (2056 MPa for sample no. 117) and the modeling value was 2414 MPa for sample no. 113 (2155 MPa for sample no. 117). The glass transition temperature was T_g = 63.69 °C for sample no. 113 (63.50 °C for sample no. 117) and from the modeling curve, the T_g = 63 °C for sample no. 113 (63 °C for sample no. 117). The storage modulus at the T_g point in the experiment was 1064 MPa and the simulation value was 1071 MPa for sample no. 113. For sample no. 117, the experimental storage modulus was equal to 1078 MPa and the simulation curve was 1085 MPa. The modulus in the plateau region was 5 MPa for both. In the cases discussed, we connected the difference between the experimental and numerical values of the storage modulus with the need to further improve our numerical model.

The apparent difference between the numerical and experimental curves can be explained by the insufficient correctness of the temperature–time analogy used in the calculation when passing through the glass transition interval. It is important to note that the values of the physical characteristics that were obtained experimentally are not truly accurate; various factors of influence must be taken into account, for example, the interaction of the sample with the grips of the testing machine. The causes and elimination of inaccuracies require further research.

3.5. Thermomechanical Shape Memory Cycle

SMPs showed excellent performance at temperatures below and above the glass transition temperature T_g. The shape recovery ratio indicates the ability of the SMP to recover its original shape from a temporary or deformed shape. The shape recovery and shape fixity ratios for different printing regimes and five cycles per each are presented in Tables 10 and 11, respectively.

Table 10. The experimental shape recovery ratio for different specimens.

Specimen	R_{rec} 1 Cycle	R_{rec} 2 Cycles	R_{rec} 3 Cycles	R_{rec} 4 Cycles	R_{rec} 5 Cycles
2	86.1%	81.2%	77.9%	64.5%	59.9%
10	91.4%	86.5%	84.0%	76.2%	71.6%
15	75.0%	60.5%	52.4%	48.0%	42.1%
16	84.4%	73.5%	50.9%	47.4%	45.5%
17	93.1%	78.7%	68.3%	63.1%	61.3%
45	74.6%	60.8%	57.5%	47.7%	41.8%
107	73.0%	59.9%	56.7%	52.7%	51.2%
113	78.2%	67.8%	63.1%	62.6%	55.3%
115	76.0%	56.0%	35.0%	32.0%	29.0%
117	79.2%	91.0%	96.4%	97.1%	96.5%

Table 11. The experimental shape fixity ratio for different specimens.

Specimen	R_{fix} 1 Cycle	R_{fix} 2 Cycles	R_{fix} 3 Cycles	R_{fix} 4 Cycles	R_{fix} 5 Cycles
2	98.46%	99.14%	97.94%	97.22%	96.96%
10	99.17%	99.47%	99.33%	99.36%	99.58%
15	99.73%	99.05%	98.14%	98.63%	98.86%
16	97.48%	99.48%	99.21%	98.98%	98.66%
17	99.26%	99.05%	98.34%	98.33%	98.11%
45	99.74%	99.46%	99.31%	99.18%	97.21%
107	99.72%	98.06%	98.64%	98.46%	98.19%
113	99.74%	99.06%	99.55%	99.41%	97.22%
115	99.77%	99.66%	99.56%	99.10%	98.86%
117	99.56%	97.96%	97.22%	98.14%	97.80%

The shape memory cycle was repeated up to five times for each sample in order to assess the effect of fatigue on the characteristics of the SMP. The deformation remained constant at 60% relative to the original gauge length. Figure 8 illustrates the specimen with printing mode no. 17 before and after the shape memory cycles.

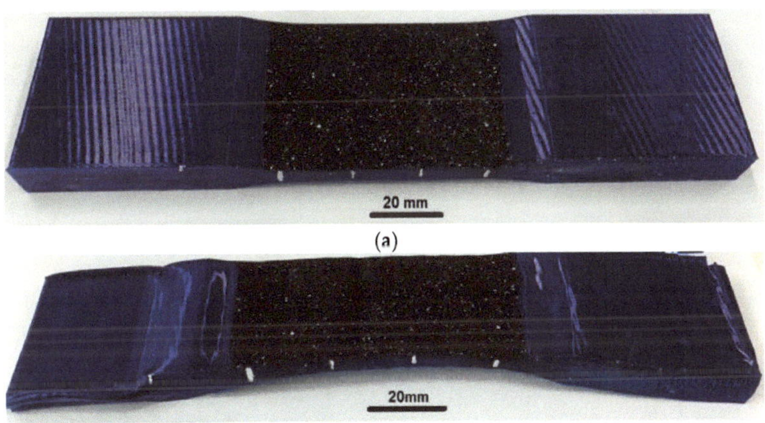

(**a**)

(**b**)

Figure 8. The specimen no. 17: (**a**) before the shape memory test; (**b**) after 5 SMP cycles.

The fixation of the shape practically did not change during the entire experiment and was close to 100%. Another situation arose when assessing the ability to recover the shape, which, even after the first cycle, was not comparable to the maximum 100%, and each cycle deteriorated significantly. Printing modes no. 10, no. 15, no. 16, and no. 17 showed the best mechanical tensile properties (tensile strength 49–50 MPa). All samples were printed at an extruder temperature of 240 °C and a nozzle diameter of 0.5 mm. When the density changed from 20% to 25% (printing modes no. 16 and no. 17, respectively), the coefficient of shape recovery for the first cycle increased from 84.4% to 93.1%, and for the last cycle decreased from 45.5% to 61.3%, accordingly. The average value of form fixation for all cycles was 98.76% for no. 16 and 98.62% for no. 17. With different line and gyroid filling patterns (printing modes no. 10 and no. 15, respectively) and the same other parameters, the coefficient of shape recovery for the first cycle was 91.4% for the line pattern (no. 10) and only 75% for the gyroid (no. 15). After five cycles, the recovery ratios changed to 71.6% and 42.1%, respectively. The average fixation coefficient for all cycles of sample no. 10 turned out to be 99.38%, and for sample no. 15, it was 98.88%. SME in the PLA after five cycles of stress–strain training in comparison with the first cycle is shown in Figure 1. We can conclude that the density and method of filling the sample volumes during printing significantly affected the appearance of the SME in the polymer. However, these manifestations were multidirectional.

The shape fixity and recovery for the specimens with the highest tensile strength are shown in Figure 9.

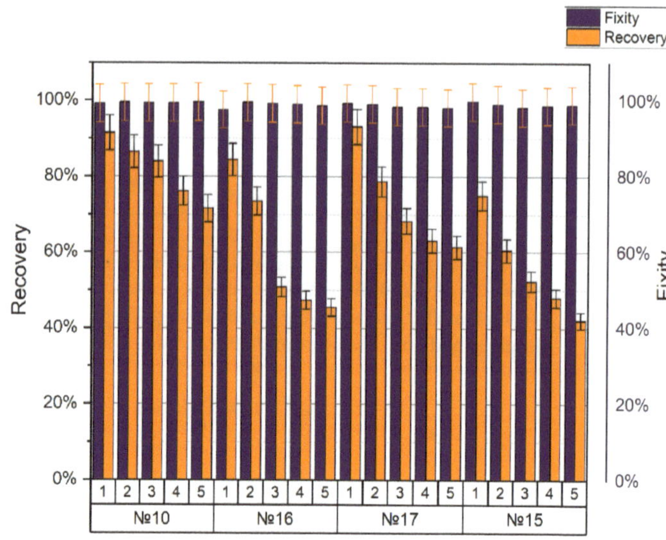

Figure 9. Shape recovery and fixity for the specimens with the highest tensile strength.

Information about the regimes that showed the worst mechanical properties (tensile strength 32–35 MPa) is presented in Figure 10.

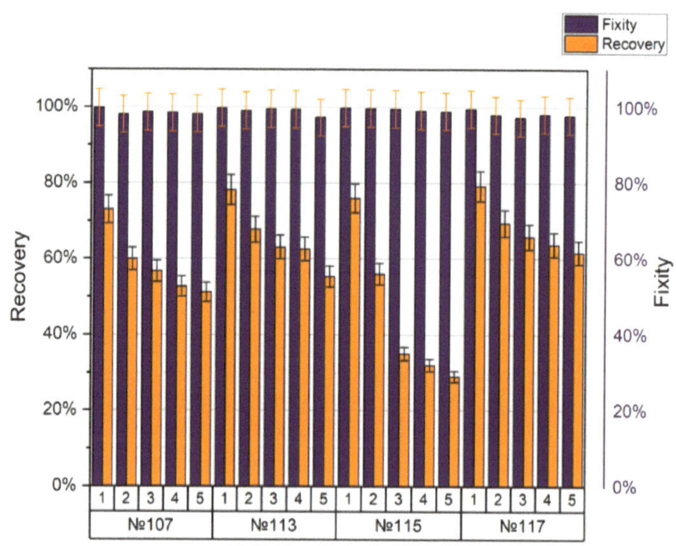

Figure 10. Shape recovery and fixity for the specimens with the lowest tensile strength.

The following printing values were common for all modes (no. 107, no. 113, no. 115, no. 117): nozzle diameter 0.3 mm, layer height 0.3 mm, extruder temperature 210 °C. Comparing the two modes, no. 107 and no. 117, one with a grid filling of 25%, and the other with a gyroid of 25%, respectively, we can say that the coefficient of shape recovery for the first cycle was 73% and 79.2% and after five cycles, their values changed by 51.2% and 61.7%, respectively. For both modes, the average fixation coefficient was 98%. Samples no. 113 and no. 115 differed only in density (lines 50% for no. 113, lines 80% for no. 115). The coefficient of form recovery after the first cycle was 78.2% and 76%, and after the fifth cycle, it was 55.3% and 29%, respectively. For both modes, the average fixation coefficient was around 99%.

The shape memory and shape fixity coefficients of printing modes no. 2 and no. 45 are presented in Figure 11.

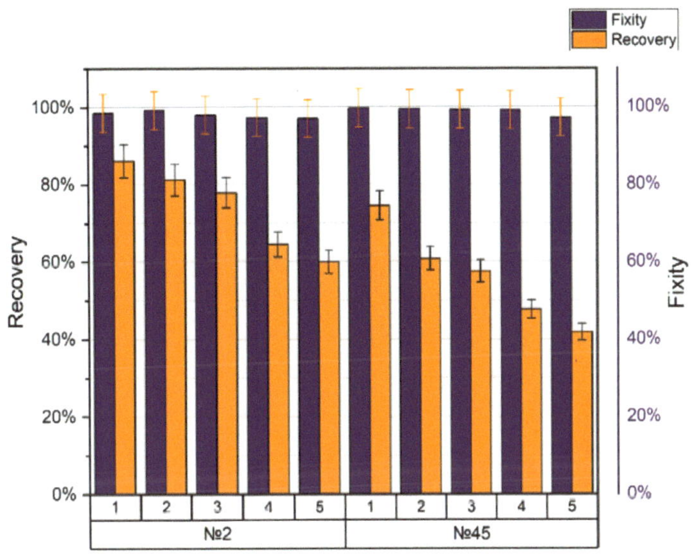

Figure 11. Shape recovery and fixity for specimens with ambiguous behavior.

Samples with printing modes no. 2 and no. 45 (tensile strength 43–44 MPa) were not in the categories with the best or worst mechanical characteristics. They aroused interest at the ANOVA analysis stage because they had an atypical behavior among other modes (see Section 3.3). Sample no. 2 (nozzle diameter 0.5 mm, layer height 0.2 mm, with a mesh filling of 25% and an extruder temperature of 240 °C) showed recovery after the first cycle by 86.1%, and after the fifth cycle by 59.9%. The average fixation of the form was 97.4%. Sample no. 45 (nozzle diameter 0.5 mm, layer height 0.2 mm, with gyroid filling 80%, and extruder temperature 210 °C) showed recovery after the first cycle by 74.6%, and after the fifth, it was 41.8%. The average fixation of the form was 98.9%.

The results of the stress–strain–temperature curve in 3D space for four different samples after the first cycle are shown in Figure 12. As expected, the stress–strain curve followed a linear trend for the first step (1), with a maximum strain value of 60%. As the temperature decreased, residual stresses accumulated due to the interaction of thermal and mechanical energy (2). The third stage showed stress relaxation and the ability of the material to memorize the shape, since the restriction is removed at low temperature (3). The shape was restored only after the temperature exceeded 64 °C (4). The deformation after the shape recovery stage was fixed at a level of more than 20%. The maximum stress value for mode no. 2 was 1.68 MPa, and the value of the final deformation was 0.23 mm/mm.

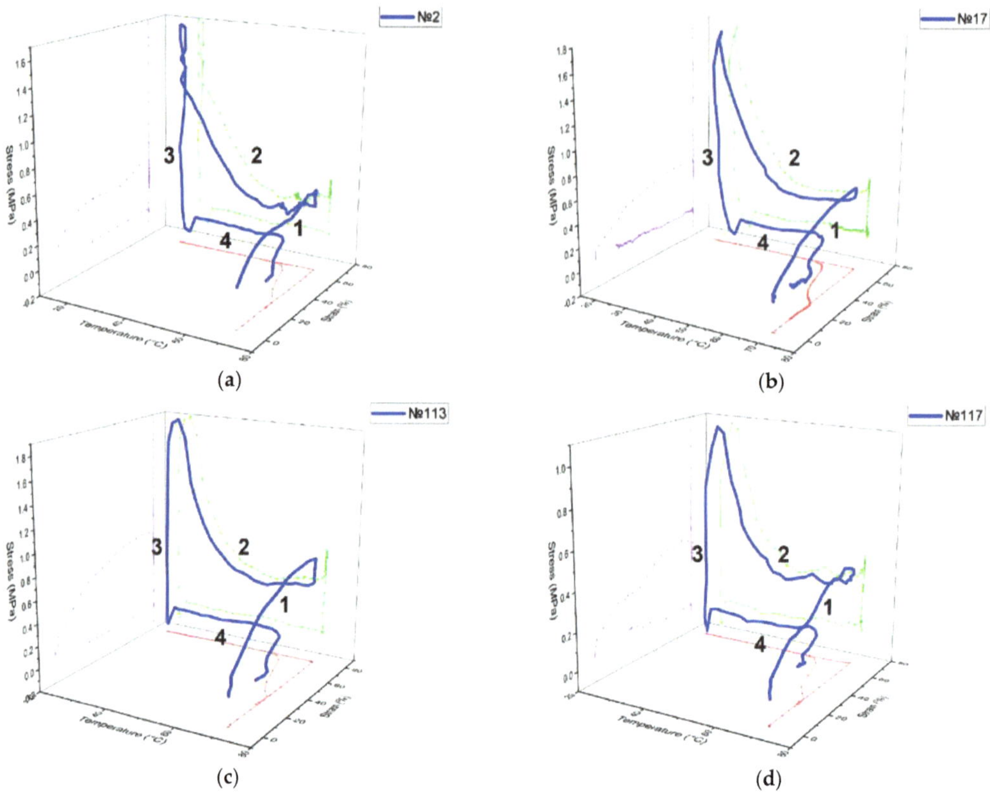

Figure 12. Shape memory baseline for the (**a**) specimen with printing regime no. 2, (**b**) no. 17, (**c**) no. 113, (**d**) no. 117.

Printing regimes no. 2 and no. 17 only showed a difference in the layer height (0.2 mm and 0.3 mm, respectively). Other parameters such as nozzle diameter 0.5 mm, infill line 80%, and an extruder temperature of 240 °C were the same for both. The coefficients of shape recovery were 86.1% and 93.1% due to the differing residual stresses of 1.68 MPa and 1.72 MPa, respectively. Two printing regimes, no. 113 and no. 117, with a difference in the infill and density (line 50% and gyroid 25%) had the same nozzle diameter of 0.3 mm, layer height of 0.3 mm, and an extruder temperature of 210 °C. The values of their shape recovery ratios were equal to 78.2% and 79.2%, respectively. The residual stresses rose to 1.87 MPa and 1.06 MPa, and the values of the final strain were 0.218 and 0.208 mm/mm, respectively.

4. Discussion

The results reported here confirm that the initial parameters of the printing mode directly influence the mechanical performance. From the ANOVA, we observed that the contribution of the extrusion temperature and nozzle diameter was the maximum compared to the other process parameters. However, it is important to emphasize that when talking about the diameter of the nozzle, it is necessary to consider the height of the layer. The experimental maximum in tensile strength of 50.1 MPa was observed for the 0.5 mm nozzle diameter, 0.2 mm layer thickness, line infill with 80% density raster, and temperature of 240 °C. The lowest tensile strength values were found in samples with printing modes in which the layer height and nozzle diameter were equal, and the extruder temperature was 210 °C.

We suppose that the number and size of the voids formed between the strands had a significant impact on the total strand adhesion. While the expected finding was presented in [44], this study showed that when the ratio between the layer height and the nozzle diameter decreased from 1 to 0.125, the porosity decreased from about 20% to 3%. In our study, this ratio took values from 1 to 0.4, where 1 corresponds to modes with a nozzle diameter D = 0.3 mm and a layer height H = 0.3 mm, and a ratio of 0.4 corresponds to modes with D = 0.5 mm and H = 0.2 mm. Therefore, for modes with a layer height of 0.2 mm and 0.3 mm with a nozzle diameter of 0.3 mm, the porosity was much higher than for modes with a nozzle diameter of 0.5 mm. This means that an influence of the ratio of these two parameters on mechanical performance was observed.

The experimental strain–stress curve and the identical simulation curve exhibited a similar trend with low inaccuracy when they were compared. On this basis, it can be assumed that the Mooney–Rivlin model with five parameters used to represent the PLA material in the simulation corresponded well with the actual data, implying that the model can predict and recreate the material's real mechanical response to the applied uniaxial tensile load. The same conclusion was reached by Meng et al. in his study of the tensile properties of polymers [45].

The obtained DSC results show the energy effects of the exo- and endo-peaks, the ratio of which it can be concluded that the samples under study were predominantly amorphous. This means that the transition point between the phases was characterized by the glass transition temperature, the value of which had also been estimated as T_g = 55 °C. Based on the obtained TMA information, it was interesting to see how the thermal expansion of the sample changed in the temperature range before the transition to the rubber state. In the glassy state, the CTE was about 70–80 ppm/K only, and above the glass transition temperature, it was 250–270 ppm/K. Moreover, the significantly anisotropic nature of the CTE behavior was discovered and its changes at the stages of heating and cooling were not obvious.

The DMA experiment showed that, regardless of the configuration of the printing process, the values of the glass transition temperature of the elasticity modulus at the initial stage of the curve were in the same range, and did not differ from each other by a value greater than 1–2%. The glass transition temperature for all samples with different measurement curves ranged from 63 to 69 °C. This is quite acceptable, given that the values of the glass transition temperature obtained within the same curve (for example, the curve of the storage module) differed within 1 °C. The output values confirmed our assumption that the DMA was not sensitive to the internal parameters of the FDM part, and that only the material under study is important. Similar conclusions were drawn by Bopp and Behrendt in [46], who demonstrated that frequency dependence in DMA is constant across a wide range of parameter settings.

When comparing the simulation curve with the experimental curve, a noticeable difference in the values of the elastic modulus could be observed. The reason for this could be an inaccurate reconstruction of the PLA material model in the simulation environment. Thus, the model requires further clarification. We assume that in order to obtain a more accurate curve, we could use a Prony series from a possible relaxation experiment, which was not conducted in our research.

Thermocycle tests showed that the samples had a SME. For this study, the samples were divided into groups with the best mechanical and worst mechanical properties as well as those modes that showed average strength values, but did not fall into the typical trend of the ANOVA curve. Sample no. 10 (nozzle diameter 0.5 mm, layer height 0.2 mm, with a line filling of 80%, and an extruder temperature of 240 °C) showed one of the highest tensile strengths (50.1 MPa) and the best values of the shape recovery coefficient for five cycles. There was a slight deterioration from cycle to cycle, but its drop was significantly less than in other samples (2–6%). The worst values were demonstrated by sample no. 115 (nozzle diameter 0.3 mm, layer height 0.3 mm, with a line filling of 80%, and an extruder temperature of 210 °C). Since these two samples (no. 10 and no. 115) showed high and low values in tensile strength, respectively, we can believe that the strength featured had an effect on the shape recoverability. Speaking of the fixity ratio, with each cycle, it decreased by 0.1–1%; for all samples, the values were close to 100%.

5. Conclusions

This study presented mechanical and thermomechanical tests aimed to predict the SMP behavior in PLA printed samples, especially in the range of glass transition temperature. In addition to evaluating the mechanical tensile strength, DSC, TMA, DMA, and the shape memory tests for five cycles were investigated.

We obtained results showing that the use of different printing parameters in sample production affects the tensile strength. It has been shown that samples with an extrusion temperature of 240 °C and a nozzle diameter of 0.5 mm had the highest values of tensile strength, while the lowest values were observed with an extrusion temperature of 210 °C and a nozzle diameter of 0.3 mm. The Mooney–Rivlin hyperelastic model with five parameters resulted in a good match between the simulation and the experimental data.

For the first time for the PLA material and FDM approach, the TMA experiment allowed us to evaluate the features of the thermal deformation of the samples to obtain the CTE values at different temperatures, space directions, and running curves. Experimentally obtained CTEs showed strong anisotropy and were successfully used in the material description during DMA simulation in Ansys. Based on the results of the DSC experiment, we were convinced of the amorphous structure of the sample, and determined the glass transition temperature as the phase transition point for the SME.

The FEM was implemented to simulate the dynamic response of the samples in a temperature regime. The viscoelastic characteristics of the PLA material were obtained by fitting the experimental curve to the Prony series, the time–temperature superposition method was applied. A comparison between the experiment and simulation curves showed that the values of the storage modulus in the range of glass transition temperature varied, but the curves showed a similar nature.

Our study highlighted that PLA really has a SME. Moreover, we noticed the influence of the mechanical characteristics of the samples on its ability to recover the shape. Thus, the stronger the sample, less fatigue from cycle to cycle was observed when restoring the initial shape after deformation. A significant decrease in the recovery ratio from cycle to cycle can be explained by the too large value of the applied deformation, despite the shape fixity, which did not decrease with each SMP cycle.

Future efforts should aim to create a polymer model with the SME, which will include elastic, viscoplastic, and thermal deformation, and is likely to contribute to the correct prediction of the SMP behavior. The final goal of such endeavors is to combine all of the experimental and simulation data into a neural network as well as to build a digital twin of the FDM process for SMP materials.

Supplementary Materials: The following supporting information can be downloaded at: https://www.mdpi.com/article/10.3390/polym15051162/s1, supplementary file (S1) presents a dataset with all of the printing regimes (Table S1 All printing regimes.) and ANOVA results with Figure S1 The effects of a single parameter on strength: (a) the effect of nozzle diameter; (b) the effect of extrusion temperature. and Figure S2 The combined influence of two and three factors: (a) extrusion temperature and layer height; (b) extrusion temperature, nozzle diameter and layer height.

Author Contributions: Conceptualization, Z.I. and I.S.; Methodology, Z.I. and I.S.; Software, Z.I.; Validation, Z.I.; Formal analysis, Z.I. and I.S.; Investigation, Z.I.; Resources, I.S.; Data curation, Z.I.; Writing—original draft preparation, Z.I. and I.S.; Writing—review and editing, Z.I. and I.S.; Visualization, Z.I.; Supervision, I.S.; Project administration, I.S.; Funding acquisition, I.S. All authors have read and agreed to the published version of the manuscript.

Funding: This study was partially supported by RFBR, project number 20-51-56011.

Institutional Review Board Statement: Not applicable.

Informed Consent Statement: Not applicable.

Data Availability Statement: The data presented in this study are available on request from the corresponding author after obtaining the permission of an authorized person.

Acknowledgments: The authors are thankful to O. Dubinin for his help with the sample 3D printing and S. Konev for their mechanical testing (both are engineers at the Center for Materials Technology, Skolkovo Institute of Science and Technology).

Conflicts of Interest: The authors declare no conflict of interest. The funders had no role in the design of the study, in the collection, analyses, or interpretation of data, in the writing of the manuscript, or in the decision to publish the results.

References

1. Yao, T.; Deng, Z.; Zhang, K.; Li, S. A method to predict the ultimate tensile strength of 3D printing polylactic acid (PLA) materials with different printing orientations. *Compos. Part B Eng.* **2019**, *163*, 393–402. [CrossRef]
2. Valvez, S.; Silva, A.; Reis, P. Optimization of Printing Parameters to Maximize the Mechanical Properties of 3D-Printed PETG-Based Parts. *Polymers* **2022**, *14*, 2564. [CrossRef]
3. Vanaei, H.; El Magri, A.; Rastak, M.; Vanaei, S.; Vaudreuil, S.; Tcharkhtchi, A. Numerical–Experimental Analysis toward the Strain Rate Sensitivity of 3D-Printed Nylon Reinforced by Short Carbon Fiber. *Materials* **2022**, *15*, 8722. [CrossRef]
4. Rankouhi, B.; Javadpour, S.; Delfanian, F.; Letcher, T. Failure Analysis and Mechanical Characterization of 3D Printed ABS With Respect to Layer Thickness and Orientation. *J. Fail. Anal. Prev.* **2016**, *16*, 467–481. [CrossRef]
5. Wang, P.; Pan, A.; Xia, L.; Cao, Y.; Zhang, H.; Wu, W. Effect of process parameters of fused deposition modeling on mechanical properties of poly-ether-ether-ketone and carbon fiber/poly-ether-ether-ketone. *High Perform. Polym.* **2022**, *34*, 337–351. [CrossRef]
6. Chacón, J.M.; Caminero, M.A.; García-Plaza, E.; Núñez, P.J. Additive manufacturing of PLA structures using fused deposition modelling: Effect of process parameters on mechanical properties and their optimal selection. *Mater. Des.* **2017**, *124*, 143–157. [CrossRef]
7. Kuznetsov, V.E.; Solonin, A.N.; Urzhumtsev, O.D.; Schilling, R.; Tavitov, A.G. Strength of PLA components fabricated with fused deposition technology using a desktop 3D printer as a function of geometrical parameters of the process. *Polymers* **2018**, *10*, 313. [CrossRef]
8. Khan, S.A.; Siddiqui, B.A.; Fahad, M.; Khan, M.A. Evaluation of the effect of infill pattern on mechanical stregnth of additively manufactured specimen. *Mater. Sci. Forum* **2017**, *887*, 128–132. [CrossRef]
9. Yeoh, C.K.; Cheah, C.S.; Pushpanathan, R.; Song, C.C.; Tan, M.A.; Teh, P.L. Effect of infill pattern on mechanical properties of 3D printed PLA and cPLA. *IOP Conf. Ser. Mater. Sci. Eng.* **2020**, *957*, 012064. [CrossRef]
10. Carrell, J.; Gruss, G.; Gomez, E. Four-dimensional printing using fused-deposition modeling: A review. *Rapid Prototyp. J.* **2020**, *26*, 855–869. [CrossRef]
11. Biswas, M.C.; Chakraborty, S.; Bhattacharjee, A.; Mohammed, Z. 4D Printing of Shape Memory Materials for Textiles: Mechanism, Mathematical Modeling, and Challenges. *Adv. Funct. Mater.* **2021**, *31*, 2100257. [CrossRef]
12. Lendlein, A.; Kelch, S. Shape-memory polymers as stimuli-sensitive implant materials. *Clin. Hemorheol. Microcirc.* **2005**, *32*, 105–116.
13. Mazur, E.; Shishkovsky, I. Additively Manufactured Hierarchical Auxetic Mechanical Metamaterials. *Materials* **2022**, *15*, 5600. [CrossRef]
14. Rahmatabadi, D.; Aberoumand, M.; Soltanmohammadi, K.; Soleyman, E.; Ghasemi, I.; Baniassadi, M.; Abrinia, K.; Zolfagharian, A.; Bodaghi, M.; Baghani, M. A New Strategy for Achieving Shape Memory Effects in 4D Printed Two-Layer Composite Structures. *Polymers* **2022**, *14*, 5446. [CrossRef]
15. Rahmatabadi, D.; Aberoumand, M.; Soltan Mohammadi, K.; Soleyman, E.; Ghasemi, I.; Baniassadi, M.; Abrinia, K.; Bodaghi, M.; Baghani, M. 4D Printing Encapsulated Polycaprolactone-Thermoplastic Polyurethane with High Shape Memory Performances. *Adv. Eng. Mater.* **2022**, 2201309. [CrossRef]
16. Liu, H.; He, H.; Huang, B. Favorable Thermoresponsive Shape Memory Effects of 3D Printed Poly(Lactic Acid)/Poly(ε-Caprolactone) Blends Fabricated by Fused Deposition Modeling. *Macromol. Mater. Eng.* **2020**, *305*, 2000295. [CrossRef]
17. Aberoumand, M.; Soltanmohammadi, K.; Soleyman, E.; Rahmatabadi, D.; Ghasemi, I.; Baniassadi, M.; Abrinia, K.; Baghani, M. A comprehensive experimental investigation on 4D printing of PET-G under bending. *J. Mater. Res. Technol.* **2022**, *18*, 2552–2569. [CrossRef]
18. Mehrpouya, M.; Gisario, A.; Azizi, A.; Barletta, M. Investigation on shape recovery of 3D printed honeycomb sandwich structure. *Polym. Adv. Technol.* **2020**, *31*, 3361–3365. [CrossRef]
19. Barletta, M.; Gisario, A.; Mehrpouya, M. 4D printing of shape memory polylactic acid (PLA) components: Investigating the role of the operational parameters in fused deposition modelling (FDM). *J. Manuf. Process.* **2021**, *61*, 473–480. [CrossRef]
20. Soleyman, E.; Aberoumand, M.; Soltan Mohammadi, K.; Rahmatabadi, D.; Ghasemi, I.; Baniassadi, M.; Abrinia, K.; Baghani, M. 4D printing of PET-G via FDM including tailormade excess third shape. *Manuf. Lett.* **2022**, *33*, 1–4. [CrossRef]

21. Safranski, D.L. Chapter One—Introduction to Shape-Memory Polymers. In *Shape-Memory Polymer Device Design*; Safranski, D.L., Griffis, J.C., Eds.; Plastics Design Library; William Andrew Publishing: Cambridge, MA, USA, 2017; pp. 1–22. ISBN 978-0-323-37797-3. [CrossRef]
22. Matveenko, V.P.; Smetannikov, O.; Trufanov, N.A.; Shardakov, I. Models of thermomechanical behavior of polymeric materials undergoing glass transition. *Acta Mech.* **2012**, *223*, 1261–1284. [CrossRef]
23. Melly, S.K.; Liu, L.; Liu, Y.; Leng, J. Active composites based on shape memory polymers: Overview, fabrication methods, applications, and future prospects. *J. Mater. Sci.* **2020**, *55*, 10975–11051. [CrossRef]
24. Tandon, G.P.; Goecke, K.E.; Cable, K.M.; Baur, J.W. Durability Assessment of Fabric-Reinforced Shape-Memory Polymer Composites. In Proceedings of the ASME 2009 Conference on Smart Materials, Adaptive Structures and Intelligent Systems, Oxnard, CA, USA, 21–23 September 2009.
25. Hutchinson, J. Determination of the glass transition temperature. *J. Therm. Anal. Calorim.* **2009**, *98*, 579–589. [CrossRef]
26. Choi, Y.-H.; Kim, C.-M.; Jeong, H.-S.; Youn, J.-H. Influence of Bed Temperature on Heat Shrinkage Shape Error in FDM Additive Manufacturing of the ABS-Engineering Plastic. *World J. Eng. Technol.* **2016**, *4*, 186–192. [CrossRef]
27. Sun, Q.; Sheng, J.; Yang, R. Controllable biodegradation and drug release behavior of chitosan-graft-poly(D, L-lactic acid) synthesized by an efficient method. *Polym. Degrad. Stab.* **2021**, *186*, 109458. [CrossRef]
28. Mohamed, O.A.; Masood, S.H.; Bhowmik, J.L. Optimization of fused deposition modeling process parameters: A review of current research and future prospects. *Adv. Manuf.* **2015**, *3*, 42–53. [CrossRef]
29. Tontowi, A.E.; Ramdani, L.; Erdizon, R.V.; Baroroh, D.K. Optimization of 3D-Printer Process Parameters for Improving Quality of Polylactic Acid Printed Part. *Int. J. Eng. Technol.* **2017**, *9*, 589–600. [CrossRef]
30. Quanjin, M.; Rejab, M.R.M.; Halim, Q.; Merzuki, M.N.M.; Darus, M.A.H. Experimental investigation of the tensile test using digital image correlation (DIC) method. *Mater. Today Proc.* **2020**, *27*, 757–763. [CrossRef]
31. Le, V.T.; Goo, N.S. Thermomechanical Performance of Bio-Inspired Corrugated-Core Sandwich Structure for a Thermal Protection System Panel. *Appl. Sci.* **2019**, *9*, 5541. [CrossRef]
32. McClung, A.J.W.; Tandon, G.P.; Baur Air, J.W. Fatigue cycling of shape memory polymer resin. In *Mechanics of Time-Dependent Materials and Processes in Conventional and Multifunctional Materials*; Conference Proceedings of the Society for Experimental Mechanics Series; Springer: New York, NY, USA, 2011; Volume 3. [CrossRef]
33. Mooney, M. A Theory of Large Elastic Deformation. *J. Appl. Phys.* **1940**, *11*, 582–592. [CrossRef]
34. Tapia Romero, M.A.; Dehonor Gomez, M.; Lugo Uribe, L.E. Prony series calculation for viscoelastic behavior modeling of structural adhesives from DMA data. *Ing. Investig. y Tecnol.* **2020**, *21*, e1668. [CrossRef]
35. Smetannikov, O.Y.; Faskhutdinova, Y.B.; Subbotin, E.V. Ansys research of shape memory effects in cross-linked polyethylene products. *Comput. Contin. Mech.* **2020**, *13*, 134–149. [CrossRef]
36. Anoop, M.S.; Senthil, P.; Sooraj, V.S. An investigation on viscoelastic characteristics of 3D-printed FDM components using RVE numerical analysis. *J. Brazilian Soc. Mech. Sci. Eng.* **2021**, *43*, 38. [CrossRef]
37. Panchenko, S.P. Numerical simulation of viscoelastic materials. *Sci. Transp. Progress. Bull. Dnipropetr. Natl. Univ. Railw. Transp.* **2014**, *53*, 157–165. [CrossRef]
38. Jayanth, N.; Jaswanthraj, K.; Sandeep, S.; Mallaya, N.H.; Siddharth, S.R. Effect of heat treatment on mechanical properties of 3D printed PLA. *J. Mech. Behav. Biomed. Mater.* **2021**, *123*, 104764. [CrossRef]
39. Setiawan, A.H. Determination of Crystallization and Melting Behaviour of Poly-lactic Acid and Polypropyleneblends as a Food Packaging Materials by Differential Scanning Calorimeter. *Procedia Chem.* **2015**, *16*, 489–494. [CrossRef]
40. Coppola, B.; Cappetti, N.; Di Maio, L.; Scarfato, P.; Incarnato, L. 3D printing of PLA/clay nanocomposites: Influence of printing temperature on printed samples properties. *Materials* **2018**, *11*, 1947. [CrossRef]
41. Selvamani, S.K.; Rajan, K.; Samykano, M.; Kumar, R.R.; Kadirgama, K.; Mohan, R.V. Investigation of tensile properties of PLA–brass composite using FDM. *Prog. Addit. Manuf.* **2022**, *7*, 839–851. [CrossRef]
42. Corapi, D.; Morettini, G.; Pascoletti, G.; Zitelli, C. Characterization of a polylactic acid (PLA) produced by Fused Deposition Modeling (FDM) technology. *Procedia Struct. Integr.* **2019**, *24*, 289–295. [CrossRef]
43. Durga Prasada Rao, V.; Rajiv, P.; Navya Geethika, V. Effect of fused deposition modelling (FDM) process parameters on tensile strength of carbon fibre PLA. *Mater. Today Proc.* **2019**, *18*, 2012–2018. [CrossRef]
44. Fischer, D.; Eßbach, C.; Schönherr, R.; Dietrich, D.; Nickel, D. Improving inner structure and properties of additive manufactured amorphous plastic parts: The effects of extrusion nozzle diameter and layer height. *Addit. Manuf.* **2022**, *51*, 102596. [CrossRef]
45. Meng, L.; Yang, X.; Salcedo, E.; Baek, D.C.; Ryu, J.E.; Lu, Z.; Zhang, J. A Combined Modeling and Experimental Study of Tensile Properties of Additively Manufactured Polymeric Composite Materials. *J. Mater. Eng. Perform.* **2020**, *29*, 2597–2604. [CrossRef]
46. Bopp, M.; Behrendt, M. Dynamic mechanical analysis of FFF printed parts in ABSplus. In Proceedings of the 2020 International Congress on Noise Control Engineering, INTER-NOISE 2020, Seoul, Republic of Korea, 23–26 August 2020.

polymers

MDPI

Article

Sorption Properties of PET Copolyesters and New Approach for Foaming with Filament Extrusion Additive Manufacturing

Nadiya Sova [1], Bohdan Savchenko [1], Victor Beloshenko [2], Aleksander Slieptsov [1] and Iurii Vozniak [3,*]

1 Department of Applied Ecology, Technology of Polymers and Chemical Fibers, Kyiv National University of Technologies and Design, Nemirovicha Danchenko Street, 2, 01011 Kyiv, Ukraine
2 Donetsk Institute for Physics and Engineering named after O.O. Galkin, National Academy of Sciences of Ukraine, pr. Nauki, 46, 03028 Kyiv, Ukraine
3 Centre of Molecular and Macromolecular Studies, Polish Academy of Sciences, Sienkiewicza Street, 112, 90-363 Lodz, Poland
* Correspondence: wozniak@cbmm.lodz.pl; Tel.: +48-42-6803237

Abstract: The mass transfer process of binary esters of acetic acid in polyethylene terephthalate (PET), polyethylene terephthalate with a high degree of glycol modification (PETG), and glycol-modified polycyclohexanedimethylene terephthalate (PCTG) was studied. It was found that the desorption rate of the complex ether at the equilibrium point is significantly lower than the sorption rate. The difference between these rates depends on the type of polyester and temperature and allows the accumulation of ester in the volume of the polyester. For example, the stable content of acetic ester in PETG at 20 °C is 5 wt.%. The remaining ester, which has the properties of a physical blowing agent, was used in the filament extrusion additive manufacturing (AM) process. By varying the technological parameters of the AM process, foams of PETG with densities ranging from 150 to 1000 g/cm^3 were produced. Unlike conventional polyester foams, the resulting foams are not brittle.

Keywords: polyethylene terephthalate copolyesters; sorption properties; mechanical properties; foaming; additive manufacturing

check for updates

Citation: Sova, N.; Savchenko, B.; Beloshenko, V.; Slieptsov, A.; Vozniak, I. Sorption Properties of PET Copolyesters and New Approach for Foaming with Filament Extrusion Additive Manufacturing. *Polymers* **2023**, *15*, 1138. https://doi.org/10.3390/polym15051138

Academic Editor: Emilia P. Collar

Received: 30 January 2023
Revised: 21 February 2023
Accepted: 21 February 2023
Published: 24 February 2023

1. Introduction

Low-molecular-weight organics are commonly used as solvents in a variety of industrial applications. Their ability to dissolve organic material is very valuable, but solubility is not always desirable, especially when material resistance or permeation properties are involved [1]. The interaction of polymeric materials with solvents may have some important applications [2]. Solubilization of low-molecular-weight polymers is a common basis for coatings [3]. For high-molecular-weight polymers, interaction with solvents is usually a problem of chemical resistance and stability or permeation under the influence of certain media [4].

For high-polymer solvents, the extent of compatibility may vary. Depending on the chemical and molecular structure, solvent diffusion may or may not be limited. When diffusion is not limited, the polymer and solvent form a solution; in contrast, limited diffusion can produce a wide range of structures, such as gel-like structures and plasticized systems. Polymeric materials consist of long molecules organized in some form under the molecular structure, which is a result of the manufacturing process of the thermochemical history of the material. Low-molecular-weight solvents can penetrate the molecular structure by diffusion processes and affect the sub-molecular structure of the polymer [5]. The influence of the solvent on the polymer can be summarized in terms of solubility parameters, thermodynamic compatibility, and intermolecular bonds that refer to specific chemical groups. In general, diffusion without influence on material properties can be described as permeability properties [6], diffusion with influence on intermolecular distance—as plasticization, diffusion with influence on the sub-molecular structure—as solvent crystallization and solvent cracking. The sorption of low-molecular-weight molecules affects the dimensional stability of the material [7].

Polyesters are a widely used and industrially important class of high-molecular-weight polymers. Members of the polyester family include polyethylene terephthalate (PET), polybutylene terephthalate, and polylactide, which are the most important polymers nowadays [8]. From a historical perspective, PET was one of the first industrial polymers and is now one of the most important standard polymers [9]. During the decades of industrial use of PET, polyester has been modified and improved for various end uses [10]. Containers made of PET polyester represent a revolution in the field of liquid foodstuffs, as they have exceptional properties and are easy to manufacture [11]. Moreover, the importance of packaging applications has changed the manufacturing technology of polyester from homopolymers to copolymers, which is now the common name for PET polyester. Further development of PET polyester is aimed at further influencing the crystallization behavior through more complex chemical modifications, leading to the creation of polymers of the PET copolymer family. PET polyester with a high degree of glycol modification, or PETG, is the second member of the family to exist as a stand-alone polymer [12]. PETG polymers with low crystallization rates are the ideal choice for thick-walled transparent packaging. The low crystallization rate is the result of the inhomogeneity of the molecular structure, which affects the folding of long molecules into thermodynamically stable structures. Low crystallization ability leads to amorphous structures with low glass transmission temperature, which is responsible for the thermal stability of the material during the application [13]. Another important member of the PET polyester family is glycol-modified polycyclohexanedimethylene terephthalate (PCTG) [14].

The interaction with volatile organic compounds (VOC) can be a key factor in the final application of the article made from a particular polymer material [15]. Contact sorption and vapor-phase sorption are common types of VOC interaction in real industrial applications [16]. Mass transfer analysis is the most reliable and rapid method to investigate possible interactions between polymer and VOC [17].

Additive manufacturing (AM) is a modern manufacturing approach and an important element of the fourth industrial revolution that is changing traditional manufacturing principles [18]. Adding or patterning materials layer by layer in a programmable manner is an evolutionary approach in many manufacturing industries. Polymeric materials were one of the first applications for additive manufacturing technology, which is now spreading to all industries [19–21]. Additive manufacturing is a modern application of PETG that takes advantage of certain material properties. The low melting point and amorphous structure create ideal conditions for cohesion between layers and low shrinkage of parts. Polymer AM uses different materials in different physical forms—liquids, filaments, powders, and granulates [22]. The most widely used process is material extrusion, specifically, melt extrusion and its most widely used type—fused filament fabrication. The main advantages of fused filament fabrication are its wide availability and the simplicity of equipment [23].

AM consists of a physical and programmable technology layer. A unique future of AM is programmable shape transformation into a core-shell structure, which enables the creation of new structures and materials with new property combinations [24]. Programmable transformation offers myriad possibilities for material distribution in the space of the final product and provides a background for new materials and technologies such as metamaterials, shape-memory materials, and smart materials. New AM technologies are emerging and gaining the attention of researchers, such as 4D printing, multi-material printing, controllable shape transformation materials, and many others [25–29]. At the current stage of development of AM technology, there are already so many ways to influence material properties that it is difficult to imagine how to investigate and evaluate all of them. One of the most important possibilities for a programmable part in an AM process is the infill structure or programmable distribution of the material in the volume. The volume of the part can be completely or partially filled with material, allowing control of the direction and distribution of the material flow and the creation of a cellular structure similar to conventional foam.

Foaming is a technology for the production and application of polymer composites with the goal of density reduction, insulation, etc. Generally, foaming technologies are classified according to the method of pore formation. There are two main types of foaming agents to form bubbles: physical and chemical. The most common is a chemical foaming agent, in which the blowing gas is formed as a result of the reaction, e.g., heat-induced decomposition of the chemical substance at the process temperature. Physical foaming agents, classified as either atmospheric gases (such as argon and helium) or as volatile liquids (e.g., propanes or heptanes), are metered directly under pressure into the polymer melt to generate bubbles [30].

In additive manufacturing, foaming technology can be applied as follows: (1) in a programmable way as a filling pattern, and (2) by using materials that can foam during the AM process. Foaming technology for extrusion AM is available as commercial filamentary material with a premixed chemical blowing agent [31–33]. The decomposition temperature of the blowing agent is in the same temperature range as material extrusion.

For PET polyester, industrial foaming technology is available as physical foaming with carbon dioxide or fluoro-organic compounds injected under pressure into the polymer melt. The application of supercritical carbon dioxide to PET polyester with extended sorption is used to foam materials [34]. Currently, physical foaming has not been developed for the AM process. In the current study, a special type of foaming is proposed for a series of PET polyester polymers. The application of polymer material with equilibrium VOC can also be used for other polymers. Polymer additive manufacturing uses interaction with solvents and VOC to smooth and bond the surface of parts [35–40].

The aim of the present study is to investigate the influence of binary acetic acid esters on the long- and short-term properties of common and modern PET polyesters. The investigation of the long-term mass transfer properties of polyester materials leads to the identification of a specific material property and its perspective application in a common additive manufacturing process.

2. Materials and Methods

2.1. Materials

Acetic esters—ethyl acetate and butyl acetate—were supplied by Telco LLC, Kyiv, Ukraine as commercial purity solvents. Glycol-modified PET polyester Skygreen® KN100—PETG, PCTG Skygreen® JN100, and PET copolymer Skypet® BR were supplied by Biesterfeld Special Chem, Kyiv, Ukraine.

2.2. Experimental Setup

Sorption of acetic acid esters in the polyester polymers was carried out in saturated vapor and in direct liquid contact under stabilized temperature conditions in a temperature-controlled chamber. Vapor sorption was carried out in a closed glass exicator in a saturated vapor medium directly under the liquid phase.

The polyester polymers were converted into films and monofilaments using laboratory extrusion equipment. Prior to extrusion, the material was dried in a Memmert ULP 500 convection oven.

PET copolymer—drying time 10 h, drying temperature −160 °C. PETG and PCTG polyesters—drying time 8 h, drying temperature 60 °C.

The tape samples were extruded through a flat slit die into a hot water bath in the vertical direction. The samples were cooled in the hot water bath to achieve the fastest possible cooling. The water bath temperature was 70 °C. The extrusion rate of the material was 1.8 kg/h. Before further experiments, the tapes were measured with a contact micrometer to check the uniformity of the thickness with reasonable tolerances in the range of ±1.5%. The spinneret draw ratio of 200% was kept constant for all extruded materials. The temperature profile of the extrusion line was constant for all materials and was

240–275–260–265 °C from the first heating zone to the die head. A single screw extruder with a diameter of 27 mm and a ratio of 30 L/D was used with a speed of 22 to 26 rpm depending on the bulk density of the material and constant mass productivity.

Monofilament extrusion was carried out in the horizontal direction with the same extruder setup and a round spinneret head. The melt was extruded through a 2.5 mm round die with an L/D ratio of 10. The draw ratio of the spinneret was kept in the range of 200 to 250% to obtain the desired filament diameter. The melt was cooled in a horizontal water bath with forced hot water circulation. Water bath temperature 70 °C. The extrusion rate of the material was 1.2 kg/h. The extruded samples were wound onto a plastic spool with as little tension as possible.

The film specimens were used for punching out standard specimens for tensile testing. Samples and cutting blade were preheated to 60 °C before the cutting procedure in order to minimize edge cracking.

2.3. Characterization

The samples of polyester material were prepared by melt extrusion and cooling. The sample was cooled at a high cooling rate to obtain an amorphous structure. The material density was measured by hydrostatic weighing.

Sorption and desorption of acetic acid esters were studied by weighing the samples over a time scale. Samples for sorption were cut to a size of 20 × 20 mm with a thickness of 0.5–0.6 mm.

Liquid-phase sorption was performed by immersing the polyester sample in a glass containing ester, which was placed in a temperature-controlled chamber. The jar containing the ester and the sample was sealed with a lid to prevent evaporation of the ester and heat absorption. To observe the sorption kinetics, the samples were accurately weighed after exclusion from the liquid phase.

The desorption was carried out under ambient conditions, elevated temperature, and reduced pressure in the vacuum-drying chamber LMM LP40412. The desorption of acetic ester was performed under a normal atmosphere in a temperature-controlled chamber. Forced desorption was carried out in a forced-air oven at an elevated temperature and in the vacuum-drying chamber at reduced pressure. Polyester samples were studied at different stages of sorption and desorption. Mechanical properties—tensile strength and elongation at the break—were measured on standardized specimens according to ISO 527-2:2012 [41]. Shore D hardness was measured on the same specimens with a thickness of 4 mm ISO 7619-1:2010 [42]. Material density was measured by hydrostatic weighing in water ISO 1183-1:2019 [43]. The relative error in the determination of density was not more than 2%. The melt flow rate was measured according to ISO 1133 [44] using a capillary of 2.095 mm diameter and weight of 2.16 kg.

The material samples were analyzed by surface FTIR using Perkin Elmer Spectrum 3 infrared spectrometer according to ASTM D5477—18 [45]. Thermal analysis was conducted using DSC TA devices, such as DSC 2920 according to ISO 11357-1:2009 [46]; the heating rate was 5 °C per minute. The material structure was evaluated by SEM JEOL JSM −5500 LV.

3. Results and Discussions

3.1. Liquid-Phase and Vapor-Phase Sorption of Ethyl Acetate for Polyesters

Acetic ester is a common ingredient in industrial solvents and some detergent formulations. When developing product packaging, the chemical formulation of the product must be tested for short- and long-term compatibility with the polymer material of the packaging. In routine compatibility testing of PETG polyester, the authors observed that limited sorption of ethyl acetate occurs in liquid phase contact sorption after a prolonged period of time. This limited sorption leads to swelling of the material and changes in hardness and flexibility. Acetic acid esters of different molecular weight—ethyl acetate and butyl acetate—were used as sorption media in the liquid and vapor phases.

Liquid-phase sorption of ethyl acetate is concentration-limited and has a stable maximum extent that is temperature dependent. At 20 °C sorption extent reaches 19.2% for PETG, 18.04 wt.% for PCTG, and 16.0 wt.% for PET. Temperature increase from 20 °C to 40 °C results in increasing the sorption extent to 25%. The PETG exhibits the highest value of consumed ester for both temperature set points, probably due to the higher content of amorphous phase among the polyesters studied.

An important observation in liquid-phase sorption is an intense cohesive bonding of the material surfaces, which is probably due to the self-healing effect of the surface and the plasticization effect. This effect is known for most semi-crystalline polyesters when reaching or exceeding the glass transmission temperature [47]. The main reason for cohesive bonding is molecular mobility at the surface, which may be favored by temperature or sorption of low-molecular-weight substances.

Another observed effect is a change in sample color and light transmission, with transparent samples changing their base color to opalescent white due to possible propagation of the solvent-induced crystallization process (Figure 1).

Figure 1. The appearance of PETG polyester sample before and during sorption of acetic ester.

The sorption of ethyl acetate in the vapor phase at 20 °C is shown in Figure 2a. The sorption maximum in the vapor phase is lower by 12–14% than in the liquid phase for all the polyesters studied. An increase in temperature from 20 °C to 40 °C leads to an increase in the maximum extent sorption for PETG by 14%, for PET by 1%, and for PCTG by 3%.

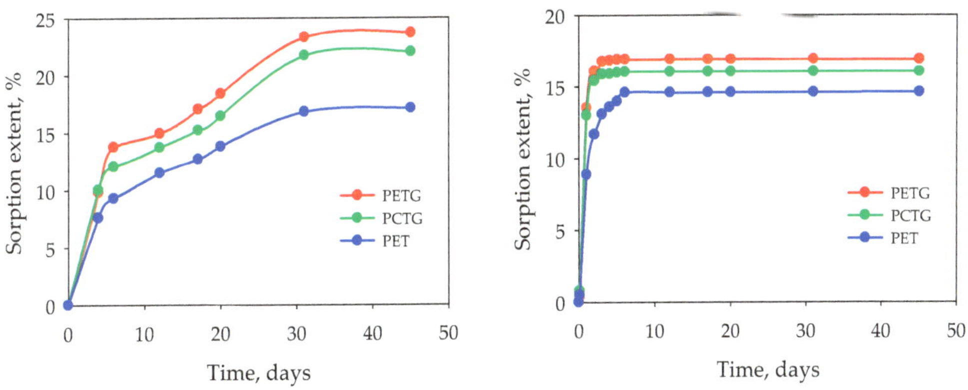

Figure 2. Sorption in the vapor phase: (**a**)—ethyl acetate, (**b**)—butyl acetate.

To describe the influence of the molecular weight of the acetic acid ester, the acetic acid ester with a higher molecular weight such as butyl acetate was used for comparison (Figure 2b). Butyl acetate exhibits different kinetics and extent. The sorption of butyl acetate has a two-step equilibrium behavior, the sorption rate is lower compared with ethyl acetate, but the maximum sorption extent at equilibrium is much higher. Butyl acetate has a higher molecular weight than ethyl acetate; however, both have a high maximum sorption degree. This observation may be related to molecular conformation and compatibility.

Due to the condensation production process, polyesters are known for their affinity for water vapor sorption [48]. The influence of initial water content in polyester was studied in the vapor sorption of ethyl acetate and polyester samples in the ambient and dried conditioned state. The water content was determined by weighing the samples before and after the vacuum-drying process at 40 °C and 10 mbar pressure, and it was 0% and 0.32%.

Dried samples show higher maximum sorption and sorption rate: for PETG by 15%, for PET by 1.6%, and for PCTG by 3.5% higher than for standard samples. These results can be explained by the higher affinity of the solid molecular structure of the polyester for the smaller water molecules and the same type of polar intermolecular interaction. The increase in the maximum sorption range for dried samples is significantly higher than the water content in conditioned samples. Thus, water has some simultaneous sorption with polyesters and influences the extent of sorption of acetic acid esters.

The rate and extent of sorption are the highest in the first 24 h of the sorption process (Figure 3). There is little change with respect to sample weight during the first two hours, and the rapid mass transfer begins after this induction period. This behavior can possibly be explained by the surface penetration of acetic ester.

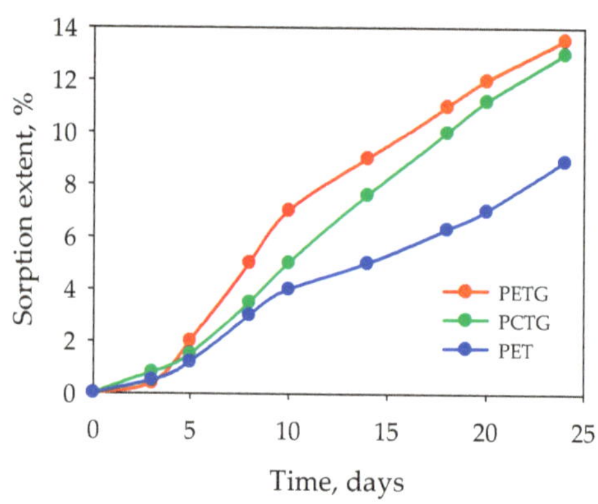

Figure 3. Sorption in the vapor phase of ethyl acetate. T = 20 °C in the first 24 h.

3.2. Desorption of Acetic Esters from Polyesters

Limited sorption of acetic esters in the saturated vapor is reversed by evaporation when the sample is stored at ambient conditions. Desorption of acetic esters was carried out under ambient conditions and at elevated temperatures in a temperature-controlled chamber with natural air circulation.

It was found that the desorption of acetic esters from polyester under ambient conditions and at elevated temperature is partial or incomplete, which is called incomplete desorption (Figure 4a). The desorption rate is significantly lower than the sorption rate for all polymers and esters studied. Butyl acetate shows a twofold desorption curve (Figure 4b). The desorption does not reach its full value for all the polyesters studied and shows a stable equilibrium value over time. The extent of residual equilibrium desorption is temperature-dependent for all polyesters and esters studied.

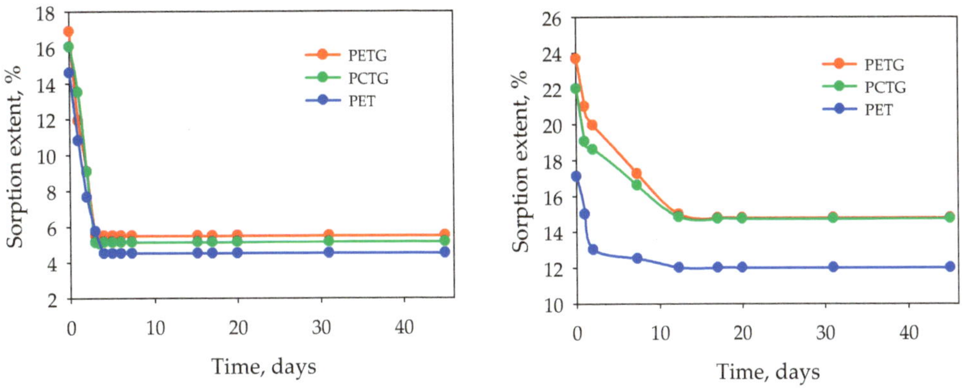

(a) (b)

Figure 4. Desorption of acetic acid esters from polyesters: (**a**)—ethyl acetate, (**b**)—butyl acetate.

A long-term study of the residual ester content shows a very slow decrease in ester content over time. At 20 °C and ambient pressure, the ester content can decrease to 0.9% by weight after one year of desorption. The slow rate of desorption in the equilibrium zone is significantly affected by temperature. At 60 °C, complete desorption for PETG and ethyl acetate can be achieved after 27 days. At 60 °C and vacuum (pressure 10 kPa), complete desorption for the same system is achieved in 36 h. For other polyesters, complete desorption can be reached in a similar way but within a different time period. The desorption of esters from polyesters is relatively slow, which offers the possibility to control the residual ester content by choosing specific conditions for the desorption process.

3.3. Influence of Sorption of Acetic Acid Esters on Material Properties

The process of sorption of selected acetic acid esters is accompanied by dramatic changes in the material physical properties (Figures 5 and 6). Sorption is found to be accompanied by swelling of the polyesters and a decrease in their hardness, which is due to the plasticizing effect of the esters on the polyesters. The behavior of density correlates with the behavior of hardness.

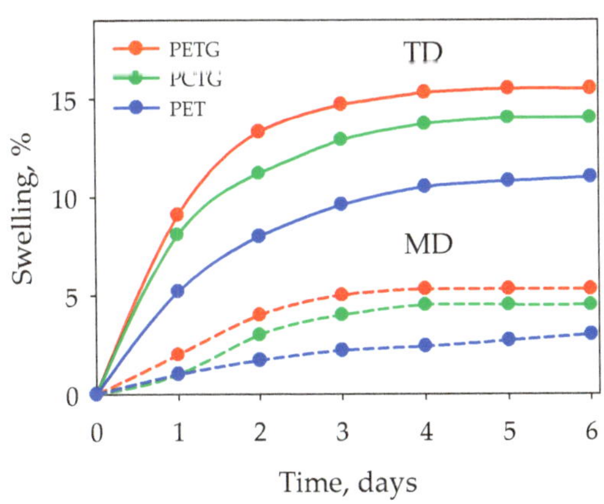

Figure 5. Material swelling during the ester sorption. MD—machine direction, TD—transverse direction.

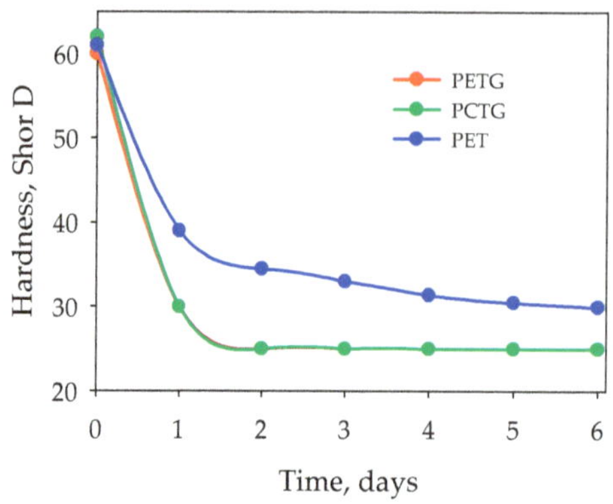

Figure 6. Material hardness during the sorption of ethyl acetate.

The tensile properties of polyesters at different stages of sorption and desorption of acetic ester were studied on tape and filament samples. The following measurements were carried out on the samples: under steady-state conditions—initial condition, conditions of maximum sorption extent; and under stable conditions—incomplete desorption and conditions of complete desorption. The results are listed in Table 1.

Table 1. Mechanical properties of polyesters under stable sorption conditions (the table shows average values).

Polyesters		Tensile Strength, MPa	Tensile Elongation, %	Tensile Modulus, MPa	Density, kg/m^3	Hardness, Shore D Scale
Initial	PETG	68 ± 2	32 ± 2	2160 ± 108	1264	60 ± 3
	PCTG	70 ± 2	48 ± 2	2060 ± 103	1245	62 ± 2
	PET	72 ± 2	56 ± 1	2240 ± 112	1340	61 ± 3
Maximum sorption	PETG	38 ± 2	158 ± 8	240 ± 22	1234	25 ± 1
	PCTG	22 ± 1	370 ± 18	150 ± 13	1220	25 ± 2
	PET	34 ± 2	320 ± 16	230 ± 21	1331	30 ± 2
Incomplete desorption	PETG	53 ± 4	54 ± 3	1350 ± 130	1299	54 ± 3
	PCTG	48 ± 3	220 ± 11	1450 ± 142	1224	57 ± 3
	PET	51 ± 4	200 ± 10	1540 ± 150	1380	56 ± 3
Full desorption	PETG	56 ± 4	14 ± 1	2050 ± 200	1269	61 ± 3
	PCTG	62 ± 5	27 ± 2	2110 ± 210	1253	63 ± 3
	PET	61 ± 5	21 ± 1	2360 ± 205	1356	64 ± 3

The mechanical properties at the maximum sorption level are characterized by a significant increase in tensile elongation and a decrease in tensile strength and tensile modulus, which can be attributed to the plasticization effect. Samples with incomplete desorption of the ester show similar behavior, indicating a moderate decrease in mechanical properties. Compared with the original properties, the level of properties of materials with incomplete desorption is suitable for some functional applications.

Materials after complete desorption of the ether have a low tensile elongation and a higher tensile modulus than the original material. The change in the current properties can be explained by the crystallization caused by the ether. All polyesters used in the current study are semi-crystalline polymers, so sorption of the ester and swelling may cause an increase in molecular mobility, which may facilitate the crystallization process at elevated temperatures.

FTIR spectroscopy and DSC studies were performed for PETG. The FTIR spectra of PETG polyester under stable sorption conditions are shown in Figure S1 (in Supplementary

Materials). Compared with the initial sample, sorption of the ester resulted in a decrease in absorbance at the 3400 cm^{-1} band, which can be attributed to the terminal hydroxyl group of the polyester. One possible explanation for this behavior is hydrogen bonding between the ester and polyester. No other discernible changes in the spectrum were observed.

To investigate the effects of ester mass transfer on the amorphous and crystalline phases of the materials, thermal analyses by DSC were performed on the initial samples and on the samples after the complete desorption of the ester (Figure 7).

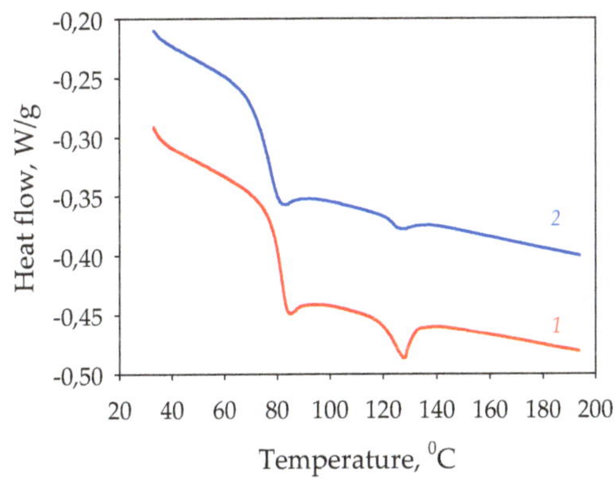

Figure 7. DSC of PETG—ethyl acetate: *1*—sample in equilibrium desorption state, *2*—initial sample.

The DSC heating curves of the original PETG and PETG after the complete desorption of the ester show an increase in the glass transition temperature from 73 to 78 °C. The increase in glass transition temperature and the appearance of a more pronounced melting peak can be explained by the contribution of ester mass transfer through the material to the formation of a crystalline structure in PETG.

3.4. *Melt Processing and Additive Manufacturing*

The melt index was measured on samples of polyesters in the initial state and the state of complete desorption after appropriate drying. The results shown in Table 2 indicate that there is essentially no change in melt viscosity after the desorption of the ester from the polyester.

Table 2. Melt flow index of polyester samples at 260 °C, 2.16 kg.

Polyester Type	Initial Sample	After Full Desorption of Ester
PETG	16.4	16.8
PCTG	17.2	17.6
PET	9.5	9.2

Melt index measurement experiments performed on polyester in the incomplete desorption state show that the material can physically foam during melting and extrusion. By filling such material into the heated chamber, fine foam can be extruded from the capillary nozzle of the melt indexer. This experiment was the starting point for the idea of using ester sorption for physical foaming. Filament samples were introduced into a standard 3D printer and passed through an extruder device. The results obtained in this way confirm the general possibility of the proposed idea.

PETG polyester was selected because of the high degree of equilibrium desorption at room temperature. Ethyl acetate was chosen primarily because its odor is relatively acceptable for indoor use, while the odor of butyl acetate is irritating. Filaments of PETG polyester were extruded with a diameter of 1.62 mm. The PETG filament was wound on a metal spool and dried at 60 °C for 8 h in a forced-air chamber. Sorption of acetic ester was performed in a glass exicator in the vapor phase. The sorption temperature was set at 20 °C to reduce the sorption rate. Sorption at room temperature and higher temperatures is associated with intense cohesive bonding of the material. For filamentary material wound on a spool, this leads to complete adhesion of the filament in many places and difficulties in unwinding. Experimental work provides several ways to eliminate these problems—e.g., by reducing the sorption rate and using a surface treatment of the filament. Surface treatment with potassium soap, silicone oil, and talcum powder is used with some results. The use of a very low winding tension along with a low sorption temperature is the simplest solution to the bonding problem. Filaments with an initial diameter of 1.62 mm absorb 5.5% ethyl acetate after 9 h of sorption at 20 °C and swell to 1.77 mm diameter. The process diagram and parameters for filament fabrication are shown in Figure 8 and Table S1 (in Supplementary Materials).

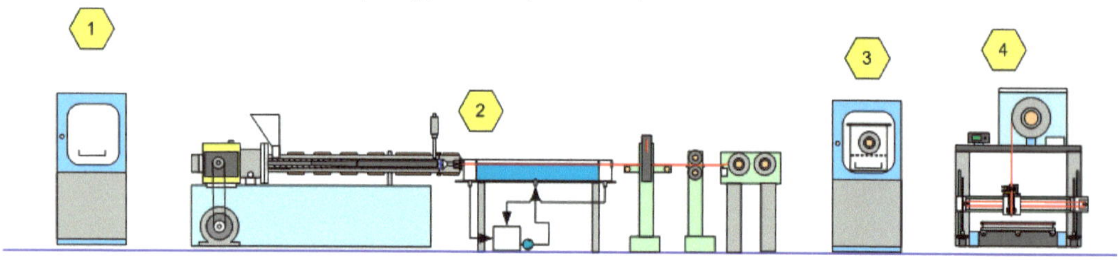

Figure 8. Principal process diagram of AM with PETG foaming: 1—initial material drying, 2—filament extrusion, 3—sorption of acetic acid ester, 4—additive manufacturing.

Additive manufacturing with the filament PETG (ethyl acetate) was performed on a standard Prusa 3D printer, model I3, without any modification. The model of the article with the specific shape was prepared for the AM process in the slicing software Utilmaker Cura 4.1.

The most successful (from the authors' point of view) process and slicing parameters determined during numerous experimental trials are listed in Table 3.

Table 3. AM and slicing process parameters for PETG foaming.

Process Parameter	Value
Nozzle diameter, mm	0.4
Extrusion width, mm	2.5
Layer height, mm	1.5
Infill rate	100
Extrusion rate, %	80
Nozzle temperature, °C	260
Built platform temperature, °C	60
Printing speed, mm/min	4000
First layer speed, %	50
Air cooling	100%
Retraction, mm	1

It was experimentally found that the physical foaming of PETG is accompanied by high swelling of the extrudate, which can reach 500–800% of the original die diameter. The AM process can be carried out at a relatively high printing speed and high temperature. At a low printing speed, most of the ester can escape from the melt zone of the extruder. Compared with the AM process with normal filament, a very high extrusion height and width are used. Nevertheless, width values are able to maintain adequate interlayer cohesion and article consistency. Foam density can be regulated by printing speed (material residual time), die temperature, and extrusion feed coefficient. Extrusion height and width can also significantly affect material density and structure due to the mechanical action of the heated die on the over-expanded foam.

At ambient conditions and the type of material used, the equilibrium desorption value is 6%. This value is sufficient for the foaming process. Experimental tests have shown that ester content of at least 3% is required for successful foaming during the AM process. Regulation of the ester content can be easily achieved by desorption in a vacuum chamber.

It was found that filaments with ethyl acetate can be successfully melted in the extruder die without significant evaporation of ester. Molten polyester with ethyl acetate is physically formed during extrusion through the orifice of the spinneret. The expansion of the foam at the spinneret is significant and suitable for additive manufacturing of foam articles. It is possible to produce foam articles with low density. The polyester foam stream has strong cohesion between layers and streams. The structure of the foam still contains sorbet ester, even after being produced in the molten state. Compared with other foamed materials, PETG foam is not brittle and can be bent to a high degree without cracking.

The foam specimens were tested for tensile strength in two conditions—as printed and after complete desorption of the ester residues in the vacuum chamber. Complete desorption significantly affects the complex properties of the material (Table 4). The specimens for tensile testing were 3D-printed with two material flow orientations (Figure 9), which is a standard procedure for such specimens in many scientific papers. The specimens were 3D-printed with three different density values (Table 4) by adjusting extrusion height and width, extrusion coefficient, and speed with one material and the same ester content. The material properties are listed for general information and cannot be directly compared due to the different printing settings. By changing the pressure parameters, it is possible to achieve a wide range of material density.

Table 4. Typical properties of PETG foam.

Property	Value for PETG Ester Modified Foam Sample					
	1		2		3	
	a	b	a	b	a	b
Material density, kg/m^3	124	118	325	315	650	643
Tensile strength X direction, MPa	4.0 ± 0.3	4.0 ± 0.4	9.0 ± 0.6	8.0 ± 0.4	11.0 ± 0.6	10.0 ± 0.7
Tensile strength Y direction, MPa	5.0 ± 0.3	6.0 ± 0.5	12.0 ± 0.7	13.0 ± 0.7	17.0 ± 1.1	16.0 ± 1.1
Tensile elongation X, %	9.0 ± 0.6	3.0 ± 0.2	12.0 ± 0.7	4.0 ± 0.3	16.0 ± 1.0	4.0 ± 0.3
Tensile elongation Y, %	12.0 ± 0.9	4.0 ± 0.3	16.0 ± 1.2	5.0 ± 0.3	21.0 ± 1.3	4.0 ± 0.3
Ester content, %	1.46	0	1.56	0	1.87	0

a—as a printed sample; b—after full desorption under vacuum.

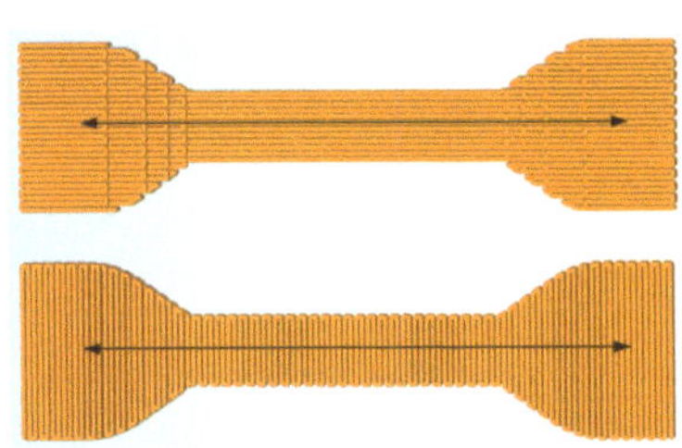

(a)

(b)

Figure 9. Material stream orientation: (**a**)—Y direction, (**b**) X—direction.

The appearance of the surface of foamed PETG in the optical observation camera is shown in Figure S2 (Supplementary Materials). Scanning electron microscope images are shown in Figure 10.

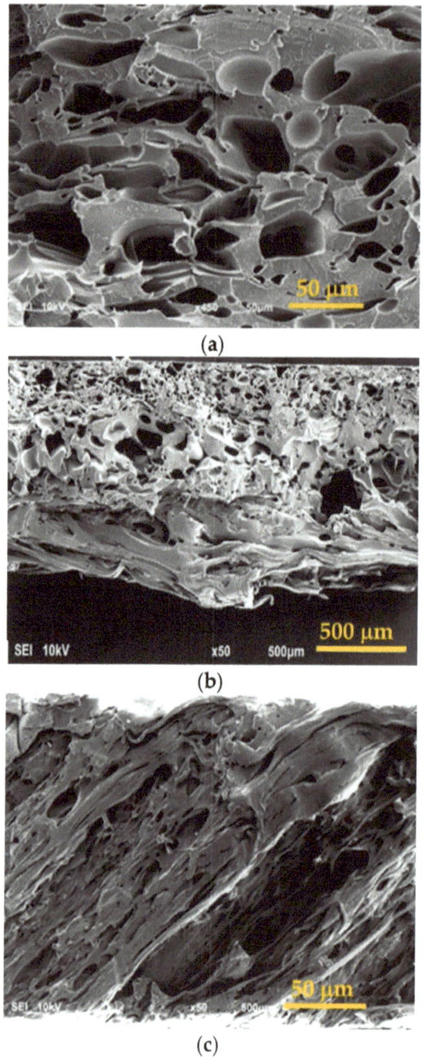

Figure 10. SEM image of PETG foam structure of a sample with different densities: (**a**)—124 kg/m^3, (**b**)—325 kg/m^3, (**c**)—650 kg/m^3.

3.5. Process Limitations and Drawbacks

The most important process limitation is material cohesion or sticking. For filamentary material wound under tension, this results in strong cohesive bonding and limited possibility of demolding. Another special material property is cracking, which occurs after contact with the ester. The filament material shatters at irregular intervals, especially at high winding tension.

A possible solution to this limitation is the goal of further investigation. A possible solution, already found by the authors, is the influence on material orientation during filament production or post-orientation in the solid state.

4. Conclusions

Sorption of esters on polyesters causes plasticizing effect and promotes structural transformation by crystallization. The desorption of ethyl acetate and butyl acetate has a stable equilibrium state, the extent of which depends on temperature, and the rate of desorption is extremely low under these conditions. The sorption of acetic ester affects the complex physical and mechanical properties of materials in a manner typical of plasticization. The sorption equilibrium conditions are stable over a long period of time, which opens the possibility of the practical application of this material property.

PETG polyester with ethyl acetate in the equilibrium state of sorption was used for the extrusion based additive manufacturing process. The remaining ester acts as a physical foaming agent during melt formation and deposition. The conversion of PETG—ethyl acetate in the melt, which is accompanied by foaming, does not result in the complete evacuation of the ester from the polyester. The desorption of the ester from the polyester is relatively slow even in the molten state, allowing the foaming process to occur. The properties of PETG foam are determined by the presence of ester residues. The printed foam samples are soft and not brittle. Foam samples after the complete desorption of the ester show typical brittle foam properties.

Supplementary Materials: The following supporting information can be downloaded at: https://www.mdpi.com/article/10.3390/polym15051138/s1, Figure S1: FTIR spectrum of PETG polyester; 1—initial sample; 2—sample after incomplete desorption; 3—sample after 1 year of incomplete desorption; Table S1: Process parameters for the production of PETG filaments; Figure S2: Appearance of the surface of the 3D-printed PETG foam: a—view of the surface from above; b—view of the surface of the built platform.

Author Contributions: Conceptualization, B.S. and V.B.; methodology, B.S. and N.S.; software, N.S.; validation, B.S., N.S.; formal analysis, B.S., N.S. and I.V.; investigation, B.S. and A.S.; resources, N.S.; data curation, A.S.; writing—original draft preparation, N.S.; writing—review and editing, I.V. and V.B.; visualization, N.S.; supervision, B.S.; project administration, B.S.; and funding acquisition, B.S. All authors have read and agreed to the published version of the manuscript.

Funding: This research received no external funding.

Institutional Review Board Statement: Not applicable.

Data Availability Statement: Not applicable.

Conflicts of Interest: The authors declare no conflict of interest. The funders had no role in the design of the study; in the collection, analyses, or interpretation of data; in the writing of the manuscript; or in the decision to publish the results.

References

1. Rómulo, S.; Domenek, S.; Courgneau, C.; Ducruet, V. Plasticization of poly (lactide) by sorption of volatile organic compounds at low concentration. *Polym. Degrad. Stab.* **2012**, *97*, 1871–1880.
2. Piringer, O.G.; Baner, A.L. Permeation of gases, water vapor and volatile organic compounds. In *Plastic Packaging Materials for Food: Barrier Function, Mass Transport, Quality Assurance and Legislation*; John Wiley & Sons: Hoboken, NJ, USA, 2000; pp. 239–243.
3. Arya, R.K.; Thapliyal, D.; Sharma, J.; Verros, G.D. Glassy Polymers—Diffusion, Sorption, Ageing and Applications. *Coatings* **2021**, *11*, 1049. [CrossRef]
4. Gushterov, N.; Doghieri, F.; Quitmann, D.; Niesing, E.; Katzenberg, F.; Tiller, J.C.; Sadowski, G. VOC sorption in stretched cross-linked natural rubber. *Ind. Eng. Chem. Res.* **2016**, *55*, 7191–7200. [CrossRef]
5. Favre, E.; Schaetzel, P.; Nguygen, Q.T.; Clement, R.; Neel, J. Sorption, diffusion and vapor permeation of various penetrants through dense poly (dimethylsiloxane) membranes: A transport analysis. *J. Membr. Sci.* **1994**, *92*, 169–184. [CrossRef]
6. Chattopadhyay, S. Vapor Permeation—A Thermodynamic Perspective. In *Membrane Processes: Pervaporation, Vapor Permeation and Membrane Distillation for Industrial Scale Separations*; John Wiley & Sons: Hoboken, NJ, USA, 2018; pp. 257–282.
7. Buchhold, R. Swelling behavior of thin anisotropic polymer layers. *Thin Solid Films* **1999**, *350*, 178–185. [CrossRef]
8. Mercado-Colmenero, J.M.; La Rubia, M.D.; Mata-Garcia, E.; Rodriguez-Santiago, M.; Martin-Doñate, C. Experimental and numerical analysis for the mechanical characterization of PETG polymers manufactured with FDM technology under pure uniaxial compression stress states for architectural applications. *Polymers* **2020**, *12*, 2202. [CrossRef]

9. Guang, L.; Jiang, J.; Jin, J.; Yang, S.; Wu, C. Crystallization behavior of modified polyester with varied macromolecular architecture. *J. Macromol. Sci. Part B* **2006**, *45*, 639–652.
10. Ji, L.N. Study on preparation process and properties of polyethylene terephthalate (PET). *Appl. Mech. Mater.* **2013**, *312*, 406–410. [CrossRef]
11. Krista, B.; Regelink, M.; Gaymans, R.J. Crystallization of poly (ethylene terephthalate) modified with codiols. *J. Appl. Polym. Sci.* **2001**, *80*, 2676–2682.
12. El-Farahaty, K.A.; Sadik, A.M.; Hezma, A.M. Study of optical and structure properties of polyester (PET) and copolyester (PETG) fibers by interferometry. *Int. J. Polym. Mater.* **2007**, *56*, 715–728. [CrossRef]
13. Latko-Durałek, P.; Dydek, K.; Boczkowska, A. Thermal, Rheological and Mechanical Properties of PETG/rPETG Blends. *J. Polym. Environ.* **2019**, *27*, 2600–2606. [CrossRef]
14. Turner, S.R. Development of amorphous copolyesters based on 1, 4-cyclohexanedimethanol. *J. Polym. Sci. Part A Polym. Chem.* **2004**, *42*, 5847–5852. [CrossRef]
15. Ramaiah, K.P.; Satyasri, D.; Sridhar, S.; Krishnaiah, A. Removal of hazardous chlorinated VOCs from aqueous solutions using novel ZSM-5 loaded PDMS/PVDF composite membrane consisting of three hydrophobic layers. *J. Hazard. Mater.* **2013**, *261*, 362–371. [CrossRef]
16. Hands, P.J.; Laughlin, P.J.; Bloor, D. Metal–polymer composite sensors for volatile organic compounds: Part 1. Flow-through chemi-resistors. *Sens. Actuators B Chem.* **2012**, *162*, 400–408. [CrossRef]
17. Hsieh, T.-K.S. *Study of Interactions of Selected Organic Solutes with PET and PETG*; Rutgers The State University of New Jersey-New Brunswick: New Brunswick, NJ, USA, 1982.
18. Horst, D.J.; Duvoisin, C.A.; Vieira, R.A. Additive manufacturing at Industry 4.0: A review. *Int. J. Eng. Res.* **2018**, *8*, 3–8.
19. Pastukh, V.; Pavlov, A.; Karpinsky, M.; Karpinska, E.; Sova, N. Experimental Study of the Ultimate Strength of Samples of Material Based on Polylactide and Tricalcium Phosphate, Produced by 3D Printing with Different Porosities, Depending on the Time of Hydration. *Georgian Med. News* **2021**, *316–317*, 173–178.
20. Beloshenko, V.; Beygelzimer, Y.; Chishko, V.; Savchenko, B.; Sova, N.; Verbylo, D.; Voznyak, A.; Vozniak, I. Mechanical Properties of Flexible TPU-Based 3D Printed Lattice Structures: Role of Lattice Cut Direction and Architecture. *Polymers* **2021**, *13*, 2986. [CrossRef]
21. Beloshenko, V.; Chishko, V.; Plavan, V.; Rezanova, N.; Savchenko, B.; Sova, N.; Vozniak, I. Production of Filter Material from Polypropylene/Copolyamide Blend by Material Extrusion-Based Additive Manufacturing: Role of Production Conditions and ZrO_2 Nanoparticles. *3D Print. Addit. Manuf.* **2021**, *1*, 253–262. [CrossRef]
22. Savchenko, B.; Sova, N.; Beloshenko, V.; Debeluy, B.; Slieptsov, A.; Vozniak, I. New Approach for Extrusion Additive Manufacturing of Soft and Elastic Articles from Liquid-PVC-Based Consumable Materials. *Polymers* **2022**, *14*, 4683. [CrossRef]
23. Dave, H.K.; Davim, J.P. *Fused Deposition Modeling Based 3D Printing*; Springer International Publishing: Cham, Switzerland, 2021.
24. Soleyman, E.; Aberoumand, M.; Soltanmohammadi, K.; Rahmatabadi, D.; Ghasemi, I.; Baniassadi, M.; Baghani, M. 4D printing of PET-G via FDM including tailormade excess third shape. *Manuf. Lett.* **2022**, *33*, 1–4. [CrossRef]
25. Aberoumand, M.; Soltanmohammadi, K.; Soleyman, E.; Rahmatabadi, D.; Ghasemi, I.; Baniassadi, M.; Abrinia, K.; Baghani, M. A comprehensive experimental investigation on 4D printing of PET-G under bending. *J. Mater. Res. Technol.* **2022**, *18*, 2552–2569. [CrossRef]
26. Soleyman, E.; Aberoumand, M.; Rahmatabadi, D.; Soltanmohammadi, K.; Ghasemi, I.; Baniassadi, M.; Abrinia, K.; Baghani, M. Assessment of controllable shape transformation, potential applications, and tensile shape memory properties of 3D printed PETG. *J. Mater. Res. Technol.* **2022**, *18*, 4201–4215. [CrossRef]
27. Soleyman, E.; Rahmatabadi, D.; Soltanmohammadi, K.; Aberoumand, M.; Ghasemi, I.; Abrinia, K.; Baniassadi, M.; Wang, K.; Baghani, M. Shape memory performance of PETG 4D printed parts under compression in cold, warm, and hot programming. *Smart Mater. Struct.* **2022**, *31*, 085002. [CrossRef]
28. Niazy, D.; Elsabbagh, A.; Ismail, M.R. Mono–Material 4D Printing of Digital Shape–Memory Components. *Polymers* **2021**, *13*, 3767. [CrossRef] [PubMed]
29. Xia, Y.; He, Y.; Zhang, F.; Liu, Y.; Leng, J. A review of shape memory polymers and composites: Mechanisms, materials and applications. *Adv. Mater.* **2021**, *33*, 2000713. [CrossRef]
30. Shahrubudin, N.; Lee, T.C. Ramlan, R.J.P.M. An overview on 3D printing technology: Technological, materials, and applications. *Procedia Manuf.* **2019**, *35*, 1286–1296. [CrossRef]
31. Ngo, T.D.; Kashani, A.; Imbalzano, G.; Nguyen, K.T.; Hui, D. Additive manufacturing (3D printing): A review of materials, methods, applications and challenges. *Compos. Part B Eng.* **2018**, *143*, 172–196. [CrossRef]
32. Ramya, A.; Vanapalli, S.L. 3D printing technologies in various applications. *Int. J. Mech. Eng.* **2016**, *7*, 396–409.
33. Browne, M.P.; Redondo, E.; Pumera, M. 3D printing for electrochemical energy applications. *Chem. Rev.* **2020**, *120*, 2783–2810. [CrossRef]
34. Walker, M.; Humphries, S. 3D Printing: Applications in evolution and ecology. *Ecol. Evol.* **2019**, *9*, 4289–4301. [CrossRef]
35. Tammaro, D.; Villone, M.M.; Maffettone, P.L. Microfoamed Strands by 3D Foam Printing. *Polymers* **2022**, *14*, 3214. [CrossRef] [PubMed]
36. Kalia, K.; Francoeur, B.; Amirkhizi, A.; Ameli, A. In Situ Foam 3D Printing of Microcellular Structures Using Material Extrusion Additive Manufacturing. *ACS Appl. Mater. Interfaces* **2022**, *14*, 22454–22465. [CrossRef] [PubMed]

37. Yousefi Kanani, A.; Rennie, A.E.W.; Abd Rahim, S.Z.B. Additively manufactured foamed polylactic acid for lightweight structures. *Rapid Prototyp. J.* **2023**, *29*, 50–66. [CrossRef]
38. Mills, N.J.; Gilchrist, A. The effectiveness of foams in bicycle and motorcycle helmets. *Accid. Anal. Prev.* **1991**, *23*, 153–163. [CrossRef] [PubMed]
39. Alperen, B.A.; Neshani, R.; Özerinç, S. *Mechanical Properties of 3D-Printed Elastomers Produced by Fused Deposition Modeling. Fused Deposition Modeling Based 3D Printing*; Springer: Cham, Switzerland, 2021; pp. 107–130.
40. Wypych, G. *Self-Healing Materials: Principles and Technology*, 2nd ed.; Elsevier: Amsterdam, UK, 2022.
41. *ISO 527-2:2012*; Plastics—Determination of Tensile Properties—Part 2: Test Conditions for Moulding and Extrusion Plastics. International Organization for Standardization: Genevan, Switzerland, 2012. Available online: https://www.iso.org/ru/standard/56046.html (accessed on 4 December 2022).
42. *ISO 7619-1:2010*; Rubber, Vulcanized or Thermoplastic—Determination of Indentation Hardness—Part 1: Durometer Method (Shore Hardness). International Organization for Standardization: Genevan, Switzerland, 2010. Available online: https://www.iso.org/ru/standard/50756.html (accessed on 4 December 2022).
43. *ISO 1183-1:2019*; Plastics—Methods for Determining the Density of Non-Cellular Plastics—Part 1: Immersion Method, Liquid Pycnometer Method and Titration Method. International Organization for Standardization: Genevan, Switzerland, 2019. Available online: https://www.iso.org/ru/standard/74990.html (accessed on 4 December 2022).
44. *ISO 1133-1:2011*; Plastics—Determination of the Melt Mass-Flow Rate (MFR) and Melt Volume-Flow Rate (MVR) of Thermoplastics—Part 1: Standard Method. International Organization for Standardization: Genevan, Switzerland, 2011. Available online: https://www.iso.org/standard/44273.html (accessed on 4 December 2022).
45. *ASTM D5477–18*; Standard Practice for Identification of Polymer Layers or Inclusions by Fourier Transform Infrared Microspectroscopy (FT-IR). ASTM International: West Conshohocken, PA, USA, 2018. Available online: https://www.astm.org/Standards/D5477.htm (accessed on 29 January 2023).
46. *ISO 11357-1:2009*; Plastics—Differential Scanning Calorimetry (DSC)—Part 1: General Principles. International Organization for Standardization: Genevan, Switzerland, 2009. Available online: https://www.iso.org/ru/standard/41637.html (accessed on 29 January 2023).
47. Holcomb, G.; Caldona, E.B.; Cheng, X.; Advincula, R.C. On the optimized 3D printing and post-processing of PETG materials. *MRS Commun.* **2022**, *12*, 381–387. [CrossRef]
48. Fakirov, S. *Handbook of Thermoplastic Polyesters*; Weinheim, Wiley-VCH: Weinheim, Germany, 2002; Volume 1.

polymers

Review

Stab-Resistant Polymers—Recent Developments in Materials and Structures

Niklas Panneke and Andrea Ehrmann *

Faculty of Engineering Sciences and Mathematics, Bielefeld University of Applied Sciences, 33619 Bielefeld, Germany
* Correspondence: andrea.ehrmann@fh-bielefeld.de

Abstract: Stab-resistant garments have been used for centuries, utilizing metals, paper, or polymeric structures, often inspired by natural structures such as scales. Nowadays, stab-resistant vests or vest inserts are used by police and security personnel, but also by bus drivers, ambulance officers, and other people who are empirically often attacked on duty. Since stab protection garments are often heavy and thus uncomfortable and not well accepted, whether in the form of chain-mail or metal inserts in protective vests, researchers are striving to find lightweight, drapable alternatives, often based on polymeric materials. These research attempts have recently focused on textile fabrics, mostly with impregnation by shear-thickening fluids (STFs) or ceramic coatings, as well as on lightweight composites. The first studies on 3D printed polymeric objects with tailored shapes, as well as theoretical investigations of the stab-protective effect of different materials, have been published throughout the last years. Here, we discuss different measurement methods, including dynamic and quasistatic methods, and correlations of stab-resistance with other physical properties, before we give an overview of recent developments of stab-resistant polymers, using different materials/material combinations and structures.

Keywords: body armor; additive manufacturing; functional textiles; sensory textiles; shear-thickening fluid; reinforcement; stab protection; VPAM-KDIW; HOSDB

Citation: Panneke, N.; Ehrmann, A. Stab-Resistant Polymers—Recent Developments in Materials and Structures. *Polymers* **2023**, *15*, 983. https://doi.org/10.3390/polym15040983

Academic Editors: Emilia P. Collar and Jesús-María García-Martínez

Received: 30 January 2023
Revised: 11 February 2023
Accepted: 13 February 2023
Published: 16 February 2023

1. Introduction

Stab resistant clothing has been used for a long time, mostly by soldiers and nowadays police, but also by other people who feel potentially endangered of being threatened with knives. The first stab-protective armor was made from wood, leather, or horn, followed by metals [1]. While soft body armor was developed to protect the wearer from low-velocity bullets and protection from fast-velocity bullets was also developed after the First World War and improved since [2–4], stab resistant garments have been less often investigated. Nevertheless, stab protection is often more important since stabbing accounts for more fatal penetrating injuries than gunshot injuries in many studies [5–8].

Stab-resistant vests have been shown to significantly reduce the number of fatal injuries due to stabbing [9]. However, stab-resistant body armor is mostly heavy and uncomfortable [10], besides the problem that it does not necessarily protect all relevant body area from knives and spikes, depending on the wearer's body shape [11]. To motivate police and other people in potentially dangerous situations to wear stab-protection garments, lightweight body armor is necessary, ideally from drapable material with low thermal resistance and high air and water vapor permeability [12–14]. Besides, recent body protectors used by the police and soldiers often do not fulfill the requirements defined in various standards for stab protection [15]. This has inspired researchers to investigate textile and other polymer-based stab-resistant garments. These materials and structures are discussed in this review.

The paper is structured as follows: The next section gives a brief overview of different dynamic and quasistatic measurement methods, followed by recent theoretical discussions of body armor and correlations between stab protection and other physical properties found in the scientific literature. As potential polymeric stab-resistance materials, pure textile fabrics are investigated, followed by coated and impregnated fabrics. Going one step further, rigid composites and reinforced polymers are presented, before 3D printed body armor as well as other new polymers and protective structures are discussed.

2. Stab Resistance Measurement Methods

Generally, quasi-static and dynamic test methods can be found in the literature. Recent test standard differentiate between gloves, e.g., tested according to the EN388, and other garments. One of the dynamic test methods used to evaluate stab resistance of garments is the German VPAM KDIW 2004, in the newest version from 2011 [16].

This test standard was published by the association of test laboratories for bullet resistant materials and constructions (Vereinigung der Prüfstellen für angriffshemmende Materialien und Konstruktionen, VPAM) and describes testing stab resistance against blade ("Klinge"), spike ("Dorn"), injection cannula ("Injektion") as well as impact resistance against a block ("Würfel", combined to the abbreviation KDIW). For stab resistance, the VPAM KDIW defines impact points, strike energies, and angles of incidence which have to be tested to reach different stab resistance classes, such as K1 (knife 1) using a strike energy of 25 J which is reached by letting a blade with a drop mass of 2.5 kg (incl. the test tool) fall from a height of 1.02 m. The test blade and the other test tools are well-defined, as well as the plasticine, the impact locations on the specimen, etc. Figure 1 depicts exemplarily a typical VPAM test stand (Figure 1a) with the standard blade in a drop mass (Figure 1b) and a partly excavated stitch channel (Figure 1c), used to measure the penetration depth. It should be mentioned that stab penetration depths of 20 mm are allowed for blade and spike to pass the test, while injection does not allow any stab penetration.

(a) (b) (c)

Figure 1. Stab resistance against knives: (**a**) test stand; (**b**) drop mass with standard blade; (**c**) stitch channel during excavation.

Another often used standard was published by the British HOSDB (Home Office Scientific Development Branch) [17]. The Home Office Body Armor Standard 2017, based on the HOSDB 2007, is very similar to the aforementioned VPAM KDIW. Instead of the plasticine back, it uses backing foam to test protection against spike and knife. It necessitates two tests with different impact energies for both of the two protection levels, e.g., for level KR1 a maximum penetration depth of 8.0 mm is allowed for an impact energy of 24 J, and a penetration depth of 20 mm for an impact energy of 36 J. Penetration is measured directly by a digital caliper as well as with a synthetic witness paper placed below the stab test sample, allowing the detection the cut length. As opposed to the VPAM KDIW, the spike is not allowed to penetrate the stab resistant material at all. The test sabot in which the knife is embedded contains a damper disk to enhance the realism of the test.

The National Institute of Justice of the USA have published the NIJ standard 0115.00 in 2000 and a draft of the subsequent standard 0115.01 in 2020, defining another test standard for the stab resistance of body armor [18,19]. This standard uses a composite backing material from neoprene sponge, polyethylene foam, and natural rubber, a velocity measuring instrument is located near the test item to measure the velocity of the blade or spike when it reaches the test material. A single-edged blade, similar to the one used in the VPAM, as well as a double-edged blade and a spike are described in the recent draft of the NIJ standard 0115.01. Tests with commercial test threats are performed at energies of 24 J and 36 J, with penetration depths of 7 mm and 20 mm allowed, respectively.

Among the quasi-static test procedures, ASTM F1342 defines a material resistance to puncture [20]. This standard measures the force which is necessary to penetrate a textile fabric, a coated material, or an elastomeric material by a pointed puncture probe. Many reports can be found in the literature about similar tests, using blades, spikes, injection needles, and similar sharp objects.

Especially for gloves, the EN 388 [21] is often used in Europe, in its newest version supplemented by the ISO 13997 which defines mechanical properties of protective clothing [22] and sometimes by an impact resistance measurement according to EN 13594 [23]. The EN388 tests for abrasion, cut, tear and puncture resistance; the performance levels from 1–4 (1–5 in case of cut resistance) are written on the gloves. The ISO 13997 defines a cut test method, sometimes called TDM (tonodynamometer) test, which means that it measures a linear cut with a fixed blade through the examined material, while the coupe test applied in EN 388 for the cut resistance uses a rotating knife. For gloves used in situations where higher impact or vibration hazards may occur, the EN 13594 measures the impact protection. Both of these values can be given with letters behind the aforementioned numbers on a glove, where A-F define the cut resistance level according to ISO 13997, while an additional P informs that the impact protection test according to EN 13594 was passed.

While these tests are standardized, several research groups show experiments according to modified test parameters, such as different blade shapes, different impact energies, or quasi-static tests with different sample holders than specified in the respective standards, leading to the necessity to carefully interpret the results achieved with different test equipment [24]. Generally, quasi-static tests, as also mentioned in the NIJ Standard 0115.00, are performed on universal test machines, with usually the upper clamp holding a standardized blade, while the lower clamp is exchanged by either a sample holder in which the investigated sample is clamped, or by a backing of different foams and sponges as in the dynamic tests. In this way, the load-displacement curve during the quasi-static stabbing process is recorded, before the test is stopped at a defined displacement.

Finally, it should be mentioned that the often used maximum penetration depth of 20 mm is related to the fact that this value is the median distance of vital organs from the skin surface, while the minimum distance was around 10 mm [25], which explains the other often chosen value of 7 mm.

3. Modeling and Correlations with Other Physical Parameters

While many research groups report on experimental investigations of stab resistant garments, few studies can be found in the literature describing the mechanism of penetration of a blade, a spike, or a needle through body armor or potential correlations with other physical parameters, such as thickness, density, elastic modulus, or other properties of protective materials. This is the opposite to body armor against hand-gun bullets, where different parameters play the main role, and which has been investigated intensively [26].

The forces necessary to penetrate the human skin were experimentally found to be in the range of 5–30 N [27], with clothes having different impact on this value, depending on the shape and sharpness of the chosen blade [28]. Horsfall reviewed existing studies regarding the shape and sharpness of the blade and concluded that energy absorption was performed by friction, bending of the cut material, and finally fracture and other mechanisms at the crack tip, with different amounts of friction and bending being reported, depending on the blade geometry [29]. He also mentioned the often large difference between impact energies in dynamic tests and quasi-static energies, leading to identical penetration depths, showing that dynamic and quasi-static test procedures can not always be compared reliably.

An early model of the stab resistance of fibrous systems was suggested by Termonia, taking into account single-ply and also multi-ply fabric systems [30]. He modeled a woven fabric in which the yarns can slip over each other at the crossing positions, held by circular clamps, into which a needle is inserted with constant velocity. Depending on the deformation morphology, as depicted in Figure 2, he calculated different forces for the needle penetrating through a Kevlar plain weave fabric, where the largest force was reached (Figure 2a) directly before the fabric is punctured (Figure 2b), while further displacement of the yarns around the needle let the force increase again (Figure 2d,e), before the cylindrical part of the needle with constant diameter is reached, and force as well as fabric deflection are decreased again (Figure 2f).

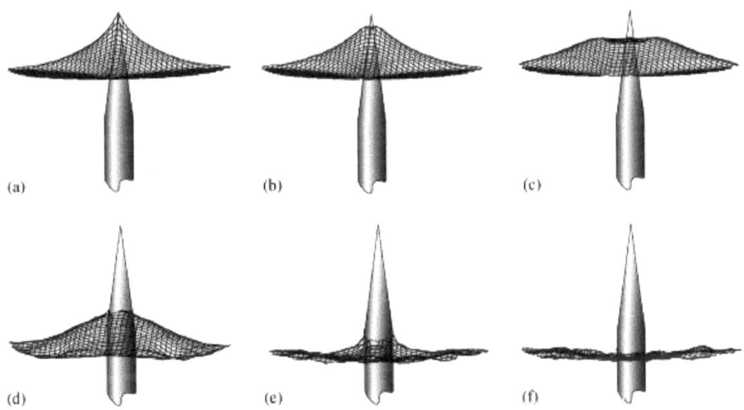

Figure 2. Simulated morphologies of deformation for a needle going through a single ply of plain weave Kevlar fabric. (**a–f**) define increasing displacement. Reprinted from [30], Copyright (2006), with permission from Elsevier.

For weft-knitted aramid fabrics in a flexible Surlyn resin matrix, Liu et al. found that optimization of the hot-pressing temperature, the pressure and holding time, as well as of the resin content, could improve the stab resistance properties of the prepared composites, which they attributed to a combination of shear force, tensile fracture, friction, and deformation of yarns and fabric as potential energy dissipation mechanisms [31].

A numerical model of Barnat et al. for an aramid fabric stack of 35 layers was compared with the optical investigation of the stabbing process, using a high-speed camera with 5000 frames/s [32]. The researchers used the movies to plot the knife velocity and acceleration and showed that the numerically calculated and the measured displacement vs. time were in good agreement. They described subsequent tearing of the roving bundles after first moving them away by the knife, and their calculations showed an increasing contact force between blade and fabrics with increasing penetration into the stack of fabrics.

A high-speed camera was also used by Du et al. who recently used this technique to monitor the damage morphology on the material back during puncture [33]. Opposite to Barnat et al., they worked with quasi-static and dynamic puncture tests on a carbon fiber reinforced polymer instead of pure fabrics. They characterized the local failure modes of a laminate with 12 layers of carbon fiber fabrics, as depicted in Figure 3, as a nearly linearly increasing impact force until the delamination threshold load (DTL) was reached, followed by increasing force with proceeding delamination failure, until the peak force damage is reached shortly before the maximum displacement, characterized by the impact force vanishing. In theoretical calculations and measurements, they showed a disproportionately high increase in the absorbed impact energy with increasing material thickness.

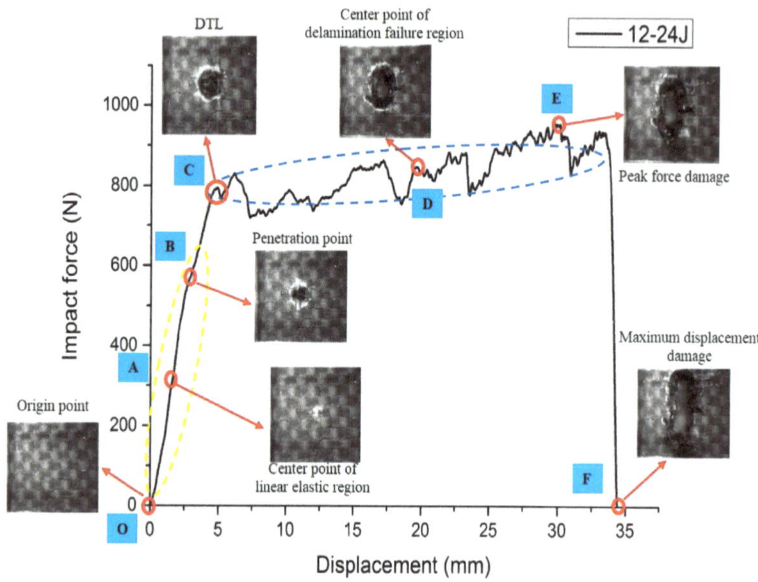

Figure 3. Measured characteristic points and local failure modes of 12-layer composite on the impact force-displacement curve at 24 J impact energy, combined with high-speed camera images of characteristic points. DTL means delamination threshold load. Reprinted from [33], Copyright (2022), with permission from Elsevier.

While the aforementioned thickness dependence of the stab resistance of a fabric or composite seems obvious, Guo et al. studied other parameters of polymers used for stab resistance [34]. Combining experiments and numeral simulations, they found the largest effect caused by the shear strength and the surface hardness of the examined materials, making polycarbonate (PC) ideal for stab resistance and polyethylene (PE) the worst.

As these examples show, the general process of stabbing into a material is generally mostly understood, with many authors subdividing it into the three phases of cavitation, penetration, and perforation [34], partly using other nomenclature. Nevertheless, in a real textile fabric, polymer plate, composite, or similar polymeric object, many interactions occur, hampering the simple calculation of the stab resistance by measuring some other parameters. The following sections thus give an overview of recent approaches to improve the stab resistance of polymeric objects, including new polymers, new shapes, new techniques such as additive manufacturing (3D printing), and more and more improved textile fabrics with different coatings and impregnations.

4. New Shapes and Polymers

One possible approach to improving body armor for stab resistance is based on developing new shapes or new polymers, e.g., inspired by nature. Such bio-inspired structures can be found in many areas, such as energy absorption [35], impact resistance [36], or biological armor design [37]. While 3D printing (described in Section 5) offers the largest degree of freedom and is thus often used to prepare such bio-inspired stab-resistant structures, there are nevertheless a few other approaches to preparing bio-inspired armor without additive manufacturing.

Liu et al. prepared a scale-like structure composite from a knitted fabric reinforced with resin, inspired by pangolin scales [38]. Figure 4 depicts the pangolin with its scales (Figure 4a), the idea of reinforcing the double-layer scale-like knitted fabric with Surlyn and polyethylene (PE) resins only at the scales, while the lower fabric parts stay untreated, the measured hardness values compared to real pangolin scales, and an untreated scale-like knitted fabric from front and back (Figure 4d–g). By restricting the resin to the movable scale structure, the authors showed that the composite stays relatively flexible, while the stab-resistance could significantly be increased by infusing the upper fabric part with resin.

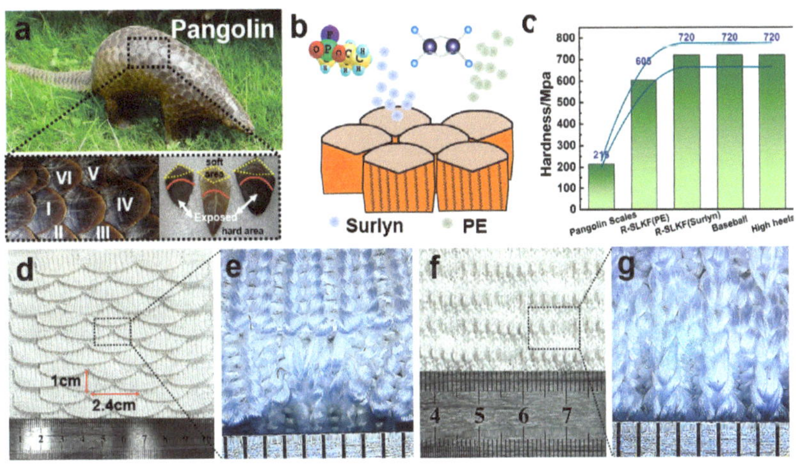

Figure 4. (**a**) Pangolin and its scales. (**b**) Preparation diagram of reinforced scale-like knitted fabric (R-SLKF). (**c**) Hardness of R-SLKF and comparison with other materials. (**d**) Optical image on the front of scale-like knitted fabric (SLKF) and (**e**) enlarged view. (**f**) Optical image on the back of SLKF and (**g**) enlarged view. Reprinted from [38], Copyright (2022), with permission from Elsevier.

A warp-knitted scale structure was prepared in the same lab, in this case using epoxy resin with SiC particles to impregnate the scales [39]. They reported an increase in the stab resistance due to the addition of SiC particles, but also mention that the knitted material (here polyester and Spandex) should be changed to a technical yarn, and washing of the composite structure should be tested.

Besides these ideas to improve stab resistance by sophisticated structures, other research groups concentrated on developing new polymers or blends to enable the production of better body armor, e.g., by making it flexible. Yong prepared flexible composites from rubber wood fiber mats in a rubber matrix including silica filler to improve the fiber-matrix adhesion [40]. He reported good elastomeric behavior at a high fiber loading of about 33 wt% and medium stab-resistance level, tested according to NIJ 0115.00, passing the tests for knives with one and two cutting edges at threat levels 1 (impact energy 24 J) and 2 (33 J), while level 3 (43 J) was not reached. The test with the spike was only passed under an angle of incidence of 0°, not of 45°, at level 1, a finding which is often reported since pointed weapons are generally harder to stop with textile-based objects.

A non-textile approach to building flexible, often even stretchable stab-resistant objects is based on hydrogels. Nakahata et al. prepared a hydrogel containing β-cyclodextrin-acrylamide and adamantane-acrylamide which was not broken by strong compression after crosslinking, could be stretched over a pencil or a cutter-blade, and showed self-healing properties [41]. Similarly, Tan et al. used free-radical copolymerization of acrylamide and adamantane-2-isocyanatoethyl acrylate-β-cyclodextrin to prepare a supramolecular hydrogel with high notch resistance, stab resistance tested by sharp scissors, and self-healing properties [42]. Using a poly(acrylamide)/poly(ethylene oxide)/LiCl hydrogel, Li et al. prepared strain sensors which also showed a high stab-resistance and self-healing properties [43].

It must be mentioned that these pure hydrogels are not tested according to standards for body armor, but are used in other stab-resistance applications, where much smaller impact energies or quasi-static forces are expected. This is often different for the 3D printed structures discussed in the next section.

5. 3D Printed Polymer-Based Body Armor

3D printing, or additive manufacturing, summarizes several different technologies, typically based on material extrusion, such as the most often fused deposition modeling (FDM) technique, on photopolymerization of a resin, such as stereolithography (SLA), or on melting defined positions inside a powder bed, such as selective laser sintering (SLS), while a few techniques have other principles, e.g., combining powder with a glue [44].

Using FDM, Cicek et al. printed square specimens of lateral dimensions 40 mm–80 mm with thickness from 6–10 mm from acrylonitrile butadiene styrene (ABS), poly(lactic acid) (PLA), and other materials with an infill of 100% [45]. They tested the samples according to the HOSDB level KR1-E1, meaning an impact energy of 24 J. While they found suitable mean penetration depths around 8 mm for ABS, PLA was found to fracture in some tests, as shown in Figure 5. The fracture lines are oriented along the 45° direction, i.e., along one of the printing orientations (in FDM, the molten polymer is typically placed under ±45° in subsequent layers). Besides, the authors report that polycarbonate (PC) and tough PLA (TPLA) were found most suitable for stab resistance, necessitating structures of only 5 mm thickness. Similar results were reported by Maidin et al. who used FDM printing of ABS and PC-ABS samples which were tested according to HOSDB KR1-E1 with an impact energy of 24 J, and fractured samples were found in many cases, where PC-ABS showed higher stab resistance than ABS, and an optimum thickness of 8.0 mm was defined [46].

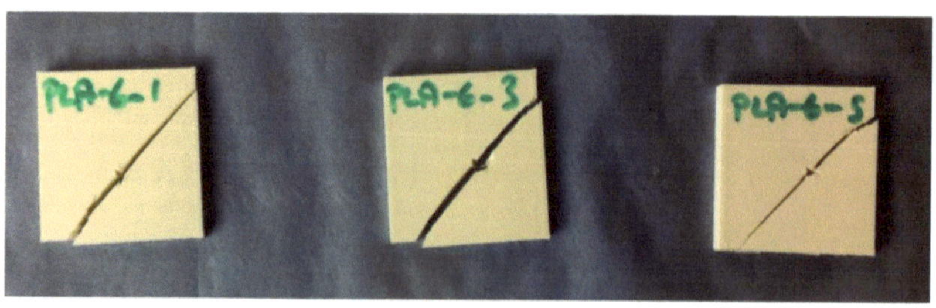

Figure 5. Fractured PLA specimens after stab test. Reprinted from [45], originally published under a CC-BY-NC license.

Sample fracture was also reported by Jiang et al. who used SLS to produce plates with a pyramid structure with different tilt angles from polyamide (PA) [47]. They found plate thicknesses around 7.5 mm to be sufficient to fully block a knife which impinged with an impact energy of 24 J for tilt angles from 20–30°, fulfilling the Chinese GA 68-2008 National Standard, while for a tilt angle of 35°, even a thickness of 8 mm was not sufficient. The same structures were investigated by Gong et al. who tested different PA materials, partly including glass fiber, with very similar results [48]. Including PA/carbon fiber, the same structures were tested again, where the PA/carbon fiber specimens needed only a thickness of 6.5 mm to show sufficient stab resistance [49].

Another shape was investigated by He et al. who used laser sintering of PA to produce egg-shell-like scale structures [50]. In their experiments, according to GA 68-2008 National Standard with an impact energy of 24 J, fractures were sometimes found along the weaker areas between the egg-shell parts, as visible in Figure 6. A larger body armor part was built by partly overlapping sections with such egg-shell structures in different ways, leading to sufficient stab protection of the optimized design which had an areal density of 7.3 kg/m^2, which is more than one-third less than common stab resistance body armor vests.

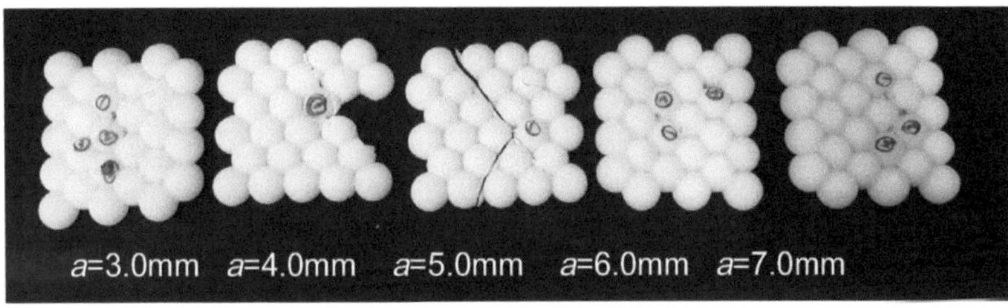

Figure 6. Stab resistant test results of samples with increasingly steeper egg-shell cores (from left to right). Reprinted from [50], Copyright (2018), with permission from Elsevier.

A special FDM apparatus, the Markforged Mark Two, was used by Sitotaw et al. to produce nylon/aramid specimens with well-defined aramid filament orientations in each layer [51]. Stab tests were performed with an impact energy of 25 J according to class K1 of the VPAM-KDIW. While pure nylon samples of a thickness of 3 mm always showed penetration larger than 70 mm and thus clearly failed, samples of identical thickness with unidirectional fiber orientation interestingly also failed, independent from the orientation of the impacting knife with respect to the fiber orientation. Combining fiber orientations of 0°/30°/60°/90°/120°/150° as well as 0°/45°/90°/135°, however, resulted in mean penetration depths of 17 mm and 15 mm, respectively, showing the severe influence of fiber oriented in such a composite.

A similar printer, the MarkOne by Markforged, was used to prepare different scale structures as well as fiber-reinforced polymer samples [52]. Ahrendt et al. found that 4 mm thick 3D printed fiber-reinforced polymers specimens with different fiber orientations in subsequent layers fulfilled the stab protection level K2 of the VPAM-KDIW, meaning that the penetration depth for an impact energy of 25 J was smaller than 5 mm on average and 10 mm maximum.

Generally, 3D printing with its diverse technologies offers many possibilities to prepare interesting shapes for stab-resistant body armor; however, the best results are reached with composites that contain fibers or filaments in defined orientations. This is why the next section gives an overview of reinforced polymers and composites which are prepared with traditional techniques.

6. Reinforced Polymers and Composites for Stab Resistance

Among the textile fibers, yarns, woven fabrics and nonwovens used in fiber-reinforced polymers as well as in composites, aramid belongs to the most often reported materials. Aramid, or more exactly *p*-aramid (poly(*p*-phenylene terephthalamide), PPTA), is an aromatic polyamide showing a very high crystallinity, which together with strong intermolecular hydrogen bonds makes the material very strong and thermally stable [53]. Kim and Nam investigated composites from *p*-aramid fabrics in thermoplastic low-density poly(ethylene) (LDPE) and epoxy resin with thicknesses of 11–16 mm [54]. Applying quasi-static and drop-tower tests according to NIJ standard 0115.00, epoxy resin of 43 wt%

add-on to the woven aramid fabric was proven to optimally stop the impinging knife in the drop tower tests.

Stojanovic et al. combined multiaxial and woven *p*-aramid fabrics, impregnated with polyurethane (PU), and laminated them with thermoplastic PU from one side [55]. Afterwards, they were impregnated with silica nanoparticles in a poly(vinyl butyral) (PVB) matrix. Stab resistance was tested with a quasi-static procedure according to NIJ standard 0115.00. The authors reported improved mechanical and stab-resistance properties upon addition of amino-modified SiO_2.

A matrix of PVB with small fractions of inorganic fullerene-like WS_2 nanoparticles and WS_2 nanotubes was chosen to produce composites with woven and cross-plied Kevlar [56]. Simic et al. found in quasi-static tests according to NIJ 0115.00 a significant increase in the absorbed energy of a knife stab and of the deformation depth for the samples with these nano-reinforements.

Stab and puncture resistance of a woven Kevlar (brand name of a *p*-aramid) fabric were investigated by Zhao et al. who concentrated on the impact of a sizing agent on the composite's mechanical properties [57]. The authors prepared a sizing agent from water-borne epoxy resin, poly(acrylamide) and fatty alcohol polyoxyethylene phosphate potassium salt in which they immersed the fabrics. The resulting composites were investigated by quasi-static tests according to ASTM F1342-05, applying a spike, a knife blade and a bursting impact head. Thy found an optimum sizing rate of 10 wt% for puncture and stab resistance as well as burst strength, with an increase by a factor of approx. 3–7 as compared to the pure fabrics which they attributed to the immobilized fiber bundles in the impregnated fabrics.

Polyethylene (PE), Surlyn® (poly(ethylene-*co*-methacrylic acid) and Surlyn/PE bilayer films were used to prepare composites with Kevlar in different thicknesses between 2.67 mm and 7.95 mm [58]. Mayo Jr. et al. performed quasi-static and dynamic tests according to NIJ standard 0115.00 to investigate stab and puncture resistance of these specimens. For the quasi-static and dynamic knife tests, all samples showed damage zones with fiber fracture, as depicted in Figure 7, while the overall stab-resistance was best for Surlyn impregnation and significantly lowest for pure aramid.

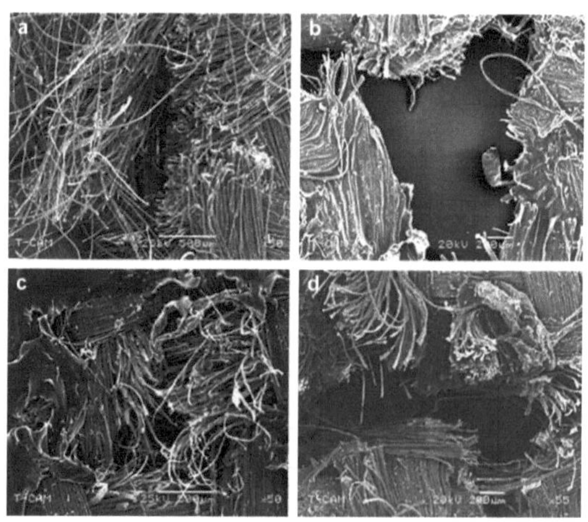

Figure 7. Scanning electron micrographs of (**a**) pure Kevlar, (**b**) PE composite, (**c**) Surlyn/PE composite, and (**d**) Surlyn composite after dynamic stab testing. Reprinted from [58], Copyright (2009), with permission from Elsevier.

Besides aramid, other technical fibers or filaments have been used in stab-resistant composites. Li et al. used ultrahigh molecular weight polyethylene (UHMWPE) impregnated with different thermoplastic films and found especially polyethylene terephthalate

(PET) and polypropylene (PP) films better suited than polyethylene (PE) [59]. For stab energies of 24 J, the penetration depth of samples with an areal weight of 8 kg/m^2 with PET and PP films was below 20 mm. Firouzi et al. suggested nylon 6,6 and nylon 6,12 coatings on UHMWPE fabrics to improve their stab resistance in dynamic and quasi-static tests [60].

Carbon, E-glass, and *p*-aramid fibers were compared by Cheon et al. who prepared fiber-reinforced polymer composites with an epoxy matrix and 6–24 fabric layers and tested their stab-resistance according to the NIJ standard with a drop tower as well as in a quasi-static test [61]. Besides these samples, hybrid composites from carbon/aramid and carbon/glass were investigated with 8 layers per material in different orders. Depending on the material, each layer had a thickness of 0.22 mm to 0.26 mm and an areal weight of 313 g/m^2 to 426 g/m^2. The authors found a clear dependence of the penetration depth on the thickness and areal weight, i.e., on the number of layers for all reinforcement materials, as expected. From the quasi-static tests, they described the blade penetration and stab resistance mechanism of the samples, as depicted in Figure 8. Besides, they found the carbon fiber reinforced polymer composite to show optimum stab resistance, with a thickness of 2.6 mm being sufficient to reach level 1 of the NIJ standard (<7 mm penetration depth), while the glass composites needed a thickness of 3.2 mm and the aramid composite needed a thickness of 3.9 mm, making carbon an interesting material for this application.

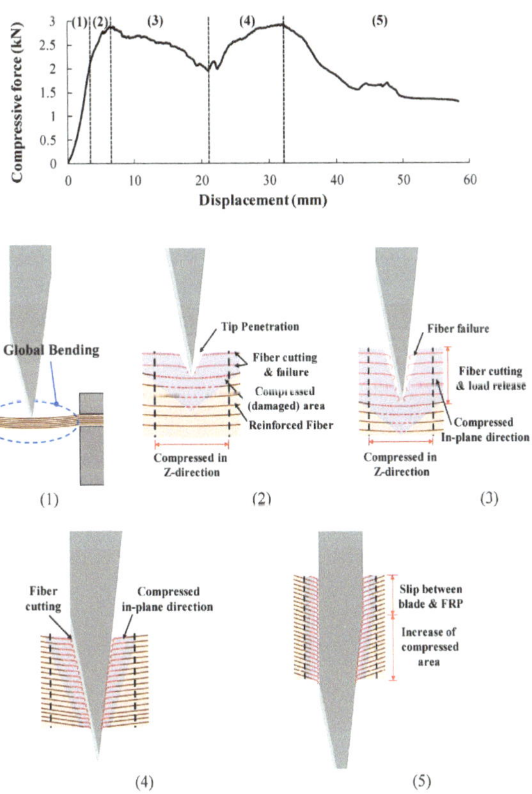

Figure 8. Blade penetration and stab resistance mechanism of the fiber-reinforced polymer composites. Numbers (1)–(5) correspond to the positions marked in the upper diagram. Reprinted from [61], Copyright (2020), with permission from Elsevier.

An interesting material combination was suggested by Chuang et al. who prepared PET/PET fiber/matrix composites with woven carbon, aramid, and basalt fabrics [62].

They combined recycled high-strength polyester with low-melting point polyester, needle-bonded them to form multilayer fabrics on both sides of a woven fabric, and used hot-pressing to form a composite bonded by the low-melting point polyester. Quasi-static puncture tests were performed according to ASTM F1342, showing highest puncture resistance for the highest amounts of recycled high strength PET fibers, while no significant differences were visible comparing basalt, carbon, and Kevlar woven fabrics. It should be mentioned that these experiments, although declared as stab tests, are actually puncture tests and thus only partly comparable with real stab tests.

Khuyen et al. avoided technical fibers and investigated the stab-resistance of composites with linen and silk plain weave fabrics in water-based PU, urea formaldehyde (UF) and poly(vinyl alcohol) (PVA) matrices [63]. The highest tensile strength was found for linen/UF; manual (and thus highly subjective) stabbing tests were performed with an undefined blade, where 20 layers of hard silk/UF, resulting in a thickness of 7.9 mm, showed sufficient results. Unfortunately, the different blade and the manual stabbing make a comparison with other experiments highly unreliable.

Another way to optimize stab-resistant body armor is to concentrate on the structures in which common materials are used, as described in the previous sections. For this purpose, composites can contain auxetic structures, i.e., structures with negative Poisson's ratio which give them special mechanical properties [64]. Xu et al. used auxetic warp-knitted spacer fabrics with two-component silicone rubbers with different filling rates and showed that a larger auxetic effect resulted in less damage which they attributed to denser face layer structures and more deformable units of fabrics with larger negative Poisson's ratio [65]. Novak et al. combined auxetic and non-auxetic layers, laminating non-auxetic cotton/elastane weft-knitted fabrics with a structured ethylene-vinyl acetate (EVA) foam with two different auxetic cellular structures (Figure 9), bonded with rubber-based adhesive [66]. While tensile force-displacement tests showed partly auxetic behavior of the laminates, stab resistance was here only suggested as a potential application of such structures and not measured. For a re-entrant auxetic weft-knitted aramid fabric without lamination, however, quasi-static tests showed a significantly higher energy absorption by the auxetic fabrics, as compared to plain weft-knitted fabrics [67]. In spite of these interesting results, composites with auxetic properties are only scarcely investigated with respect to their stab-resistant properties.

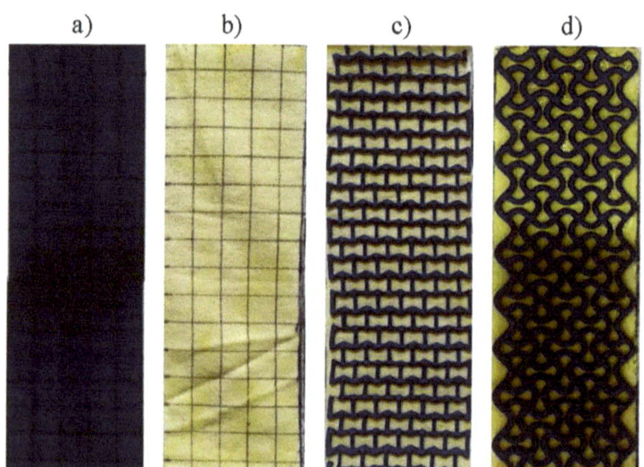

Figure 9. Analyzed specimens: (**a**) EVA foam, (**b**) fabric (with an adhesive), (**c**) re-entrant auxetic laminate, and (**d**) chiral auxetic laminate. Reprinted from [66], Copyright (2020), with permission from Elsevier.

Generally, different polymers can be used to glue the fibers or filaments of a textile fabric together and thus increase the fiber/fiber friction, leading to improved stab resistance [68–70]. On the other hand, many textile fabrics with a coating of silica or other ceramic materials are reported as lightweight stab resistance materials. The next section gives an overview of such coated textile fabrics.

7. Textile Fabrics with Ceramic Coatings

Ceramic coatings can be used to increase the fiber/fiber friction as well as the hardness and wear resistance, in this way damaging the penetrating blade and thus reducing its ability to cut subsequent fibers [71,72]. Manaee et al. suggested an Al_2O_3/TiO_2 plasma-sprayed ceramic coating to increase the stab resistance of aramid fabrics [73]. They applied copper and aluminum metal powder as bond coat material and Al_2O_3–13%TiO_2 as the top coat, as visible in Figure 10. While the tensile strength remained unchanged by these coatings, quasi-static stab tests showed a severely improved penetration work by Al_2O_3–13%TiO_2 with either metallic bond coat, in both cases approx. five times the value of the uncoated fabric, which the authors attributed to fixing the aramid fabric and increasing the fabric hardness, or in other words improving abrasion and friction by the coatings.

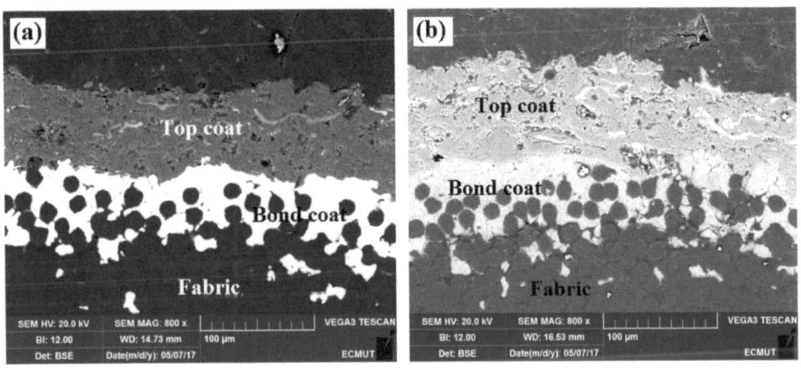

Figure 10. SEM image of the cross-section of aramid fabric coated with (**a**) Cu/ Al_2O_3–13%TiO_2 and (**b**) Al/ Al_2O_3–13%TiO_2. Reprinted from [73], Copyright (2020), with permission from Elsevier.

Boron carbide (B_4C) coatings on aramid/ballistic nylon were investigated by Nayak et al. who showed in quasi-static tests according to NIJ 0115.00 a significant increase in the stab resistance, as compared to the pure textile fabric [14]. Similarly, a boron carbide coating in epoxy resin on aramid fabrics reached an approximately five times puncture load in quasi-static puncture tests, as compared to the uncoated textiles [74]. Applied on UHMWPE fabrics, B_4C could also improve the stab resistance approx. by a factor of five [75].

Besides boron carbide, silicon carbide was investigated as a coating on stab-resistant plain-weave thermoset-aramid composites [76]. Wei et al. sprayed SiC powder in water with dispersing agents and PVA as a binder on the fabrics and finally impregnated them with vinyl ester resin. Dynamic stab tests according to NIJ standard 0115.00 showed a clear correlation of decreasing penetration depth with increasing SiC concentration up to a concentration of 20 wt%.

Applying SiO_2 coatings on woven aramid fabrics, Javaid et al. showed more than doubled knife penetration resistance as compared to the pure fabrics, which they attributed to an increase in the yarn-yarn friction [77]. A PU/*p*-aramid multiaxial fabric was coated by PVB with SiO_2 nanoparticles and carbon nanotubes, resulting in improved wear resistance and 35% higher absorbed energy as compared to the PVB coated samples [78]. Similarly, Kanesalingam et al. observed an increase in quasi-static stab and puncture resistance for silica coatings on Kevlar/wool and Kevlar/wool-nylon fabrics, but at the same time an increase in fabric stiffness, which reduces ergonomics and wearability [79].

It should be mentioned that ceramic aerogels have been suggested as temperature sensor materials for firefighters, so that a combination of improved stab resistance and sensory properties may also been taken into account, depending on the specific application of a stab-resistant garment [80–82].

Besides such ceramic coatings, an interesting approach which is heavily investigated in recent years is based on coatings with shear thickening fluids (STFs), which will be discussed in the next section.

8. Textile Fabrics with Shear-Thickening Fluid

Shear-thickening fluids are non-Newtonian fluids which show an increase in viscosity with increasing stress or shear rate [83,84]. Such STFs are prepared by dispersing concentrated colloidal suspensions of solid particles in a liquid, where the transition from low to very high viscosity is based on forming transient aggregates upon shear [85,86]. They are thus expected to not significantly impact the wearing comfort of an STF-coated stab-resistant garment, while stiffening in case of an impact and thus offering higher stab protection than uncoated textile fabrics [87]. This promising material class has been investigated deeply in recent years.

Decker et al., e.g., prepared STFs by dispersing colloidal silica particles with average diameter 450 nm in poly(ethylene glycol) (PEG) with molecular weight 200 Dalton and coated plain-weave Kevlar and nylon fabrics with them by diluting the STFs in ethanol and soaking the fabrics in this fluid [88]. The resulting coated fabrics are depicted in Figure 11, where PEG is mostly evaporated in the vacuum of the scanning electron microscope (SEM) chamber. Drop tower tests according to NIJ standard 0115.00 with knife and spike showed that a 12-layer STF/Kevlar specimen provided better protection than a 15-layer pure Kevlar sample, while both had similar areal weights, but the STF coating increased neither thickness nor bending rigidity, so that an additional STF coating could lead to preparing thinner, more flexible body armor. In quasi-static tests, STFs significantly increased the cut resistance of the Kevlar sample. Similar results were found for STF-coated nylon fabrics, which in general showed less stab resistance than Kevlar samples.

Kang et al. prepared SFT by dispersing fumed silica particles in methanol and blending this dispersion with medium fluid ethylene glycol [89]. Kevlar plain weave fabrics were immersed in this STF dispersion, squeezed, and dried. While stress-strain curves of untreated and coated Kevlar fabrics, bending angle and thickness were similar, quasi-static stab tests showed significantly higher loads and less damage for the coated fabric. Similar experiments, coating aramid or UHMWPE with varying silica/PEG STFs with a typical amount of 35–55% silica, with comparable results were also reported by other research groups [90–95]. The STF preparation process was slightly varied in some studies, e.g., by adding a silane coupling agent to improve silica/PEG bonding and thus form siloxane (Si-O-Si) bonds [96], comparing different silica particle shapes [97] and particle sizes [98], or varying the impregnation pressure to increase the STF loading of the fabrics [99]. Asija et al. used poly(propylene glycol) (PPG) instead of PEG as a base material for STF preparation [100]. Different ionic liquids instead of PEG were tested by Qin et al. who prepared STFs from silica microspheres in these ionic liquids by ultrasonication [101]. They found an optimum stab resistance for an STF loading of 35 wt% on the tested Kevlar fabrics and measured that the yarn-yarn friction was increased by more than one order of magnitude by the addition of the STF, as compared to the pure aramid fabric.

To further improve STF coatings, several studies combined them with additional nanoparticles, carbon nanotubes etc. Li et al. suggested STFs combined with multi-walled carbon nanotubes (CNTs) for improved stab resistance of Kevlar plain-weave fabrics, as depicted in Figure 12 [102]. In quasi-static stab tests, a significant increase in the peak force for STF-coated Kevlar fabrics was found, while CNT/STF coatings further increased the peak force values. The authors attributed this result to increased rigidity and decreased yarn mobility upon the coating.

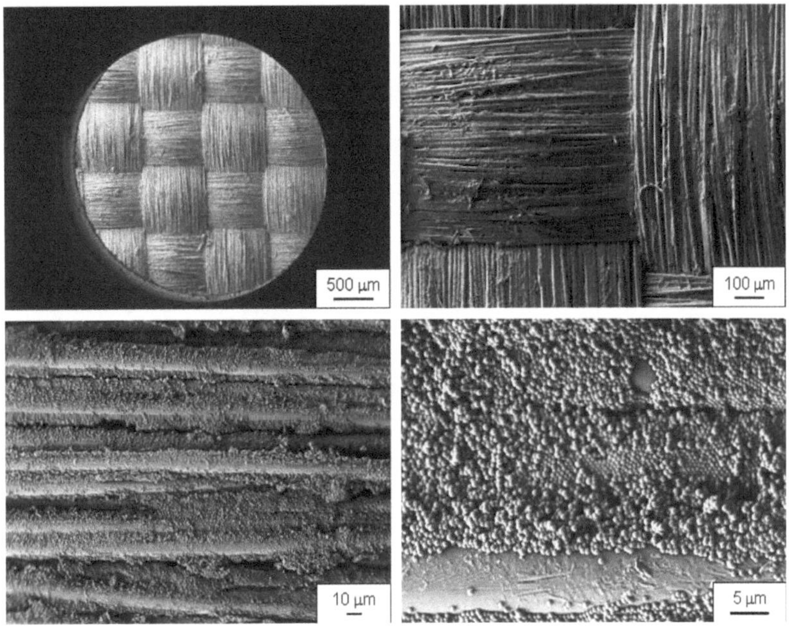

Figure 11. SEM images of undamaged Kevlar coated with STFs in different magnifications. Reprinted from [88], Copyright (2007), with permission from Elsevier.

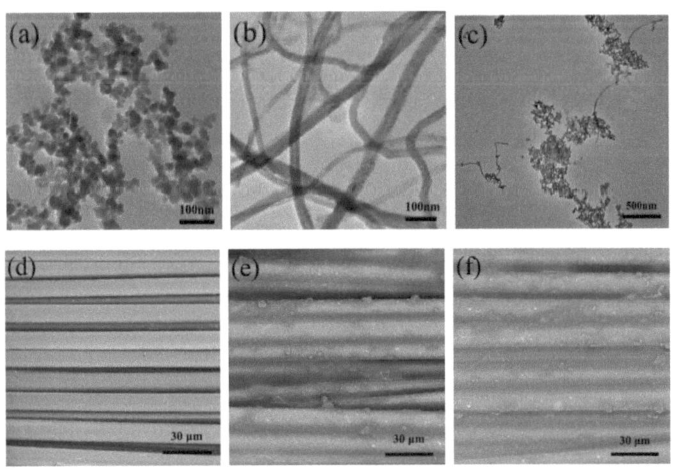

Figure 12. TEM images of (**a**) fumed silica nanoparticles, (**b**) O_2-plasma treatment MWNTs and (**c**) fumed silica nanoparticles and M-MWNTs, SEM images of (**d**) neat Kevlar fabric, (**e**) STF/Kevlar fabric, and (**f**) M-MWNT/STF/Kevlar fabric. Reprinted from [102], Copyright (2018) by the authors, originally published under a CC-BY license.

Carbon nanotubes were not only shown to increase the stab resistance properties of STF/Kevlar specimens [103], but were also sometimes discussed in terms of their conductivity, enabling the use of such fabrics as body movement sensors, etc. [104,105].

Different nanoparticles were also added to STF-coatings on stab-resistant fabrics. Gürgen and Kushan as well as Gürgen and Yildiz added SiC particles of different sizes, leading to an increased viscosity profile of the STFs and thus higher energy dissipation, as well as to a reduced penetration depth in drop-tower tests with spike and knife [106–108].

Combinations of STF and polymer coatings were suggested, e.g., by Zhang et al. who combined nonwovens from recycled Kevlar and nylon fibers on both sides of an aramid fabric by needle-punching, impregnated them by STF, and finally coated them with thermoplastic polyurethane (TPU) [109]. In dynamic and quasi-static tests according to ASTM F1342-05 with spike and knife, they found an increase in the stab resistance due to STF coating as well as due to the TPU coating, with the highest stab resistance reached by combining both treatments. Combining STF with a TPU coating including fumed silica was suggested by the same group and led to a significantly improved quasi-static stab and puncture performance, with the optimum stab resistance reached for a coating with 3% fumed silica [110].

As these examples show, a large part of recent research in stab-resistance reached with polymeric materials is based on finding new, optimized coatings for aramid and other high-tenacity fabrics. However, investigations of the fabrics themselves are also important to further improve the textile components in composites and coated fabrics. The next section gives an overview of recent studies of pure stab-resistant textiles.

9. Pure Textile Fabrics for Stab Resistance

Pure textile fabrics receive their stab-resistant properties partly from the fiber, filament or yarn properties, i.e., from the fiber/filament material and titer as well as the production of the yarn. Mayo Jr. and Wetzel compared different technical fibers, such as different *p*-aramids and UHMWPEs, carbon, and S-glass, and pressed an industrial cutting blade laterally on the single fibers to investigate the angle-dependent failure [111]. They found similar cut resistance levels for all fiber types, with generally higher average cut resistance of the inorganic fibers due to their hardness.

Tian et al. suggested using Kevlar- and UHMWPE-covered yarns and reported Kevlar fiber wrapping around a core fiber leading to enhanced cut resistance, as compared to pure Kevlar or pure UHMWPE yarns [112]. On the other hand, the cut resistance decreased with increasing twists of the covered yarns. The authors thus suggested testing such covered yarns for stab-resistant garments.

Diverse technologies exist to prepare fabrics from these and other yarns or fibers. Needle-punching is an often used method to create nonwovens. However, needle-punched composites are usually applied in the form of composites with epoxy resin, thermal bonding of low-melting point fibers, or similar matrices, when they are to be used in stab resistance applications [113,114], and investigations of pure nonwovens for this purpose are hard to find.

Knitted fabrics are also used only scarcely for stab-resistance applications, which can be explained by their elasticity. Sun et al. nevertheless performed quasi-static stab resistance tests on auxetic weft-knitted Kevlar fabrics and found a higher stab resistance than on plain weft-knitted Kevlar fabrics, where the break-points in the auxetic fabric could be significantly reduced by alternating face and reverse loops [66].

Liu et al. combined UHMWPE with a polyamide/elastane core-spun yarn to shrink the produced double-layer weft-knitted fabric [115]. They found puncture failure of this double-layer fabric to be accompanied by fiber cutting and stretching. However, impregnation with epoxy resin increased the maximum load during quasi-static stab tests by more than one order of magnitude, showing that the pure knitted fabric is not really suitable for stab resistance.

As an effort to reduce the stretchability of knitted fabrics, Zhang et al. prepared co-woven knitted fabrics, as depicted in Figure 13, with E-glass yarn as warp and weft and a polyester knitting yarn [116]. In this way, they found maximum penetration forces around 200 N, which is much higher than values found for pure knitted fabrics [115].

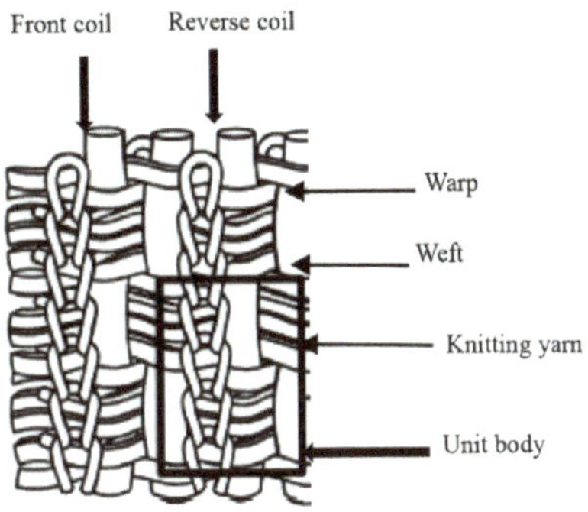

(a)

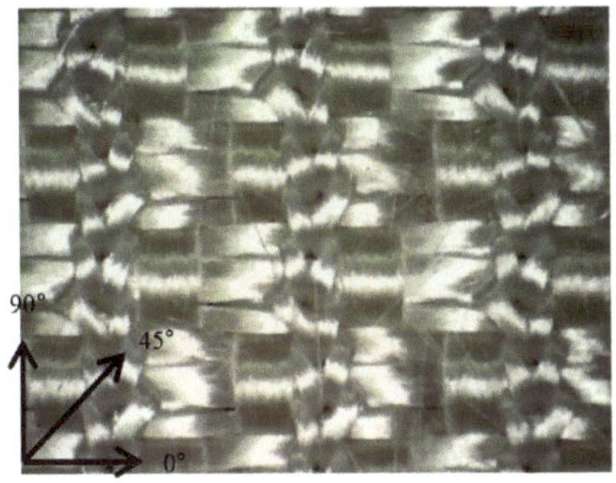

(b)

Figure 13. Co-woven-knitted fabric: (**a**) structure diagram, (**b**) fabric map. Reprinted from [116], Copyright (2022) by the authors, originally published under a CC-BY license.

Most often, woven fabrics are used for stab protection, often using aramid or aramid hybrid yarns. Tien et al. prepared aramid-core spun yarns from aramid filaments in the core and cotton staples wrapping them, using a ring-spinning machine [117]. Two strands were plied to avoid snarls in the yarn and corresponding problems during weaving. Plain-woven fabrics of these yarns were investigated in dynamic stab tests according to NIJ standard 0115.00, testing for level 1 with an impact energy of 24 J. The penetration depths were lowest for the densest weaves. The allowed penetration depth of max. 7 mm was reached by 17 layers of the core-spun yarns or more than 60 layers for pure aramid yarns, clearly showing the advantage of the core-spun yarns. The same authors later investigated more hybrid yarns and found basalt/cotton woven fabrics to have the optimum stab-resistant properties [118].

Basalt was also investigated by Li et al., here in the form of a needle-punched, laminated composite with low-melting point PET [119]. The authors found a significant impact of the punching density and the needle punch depth as well as the areal weight of the fabrics.

Fibrilized aramid yarn was the base for plain-weave fabrics investigated by Nasser et al. [120]. The authors found significantly increased yarn-yarn friction in the fibrilized aramid, as compared to the native material, and peak loads six times higher than in the original aramid material, which they attributed to the mechanical interlocking between the fibrilized fibers.

Besides the yarn, the fabric construction method strongly influences the stab resistance of a textile fabric. A typical fabric type is 3D warp interlock fabrics, which combine woven layers with through-the-thickness interlock structures and thus form a 3D structure of a certain thickness. Li et al. compared different 3D warp interlock structures from UHMWPE yarns and found a strong dependence on the structure as well as on the number of layers and their respective orientation [121]. Such fabrics were also investigated in pre-deformed shapes, especially aiming at modeling molded armor panels, as they are used in women's body armor, where the authors found a significant impact of the stab localization on the fabric [122]. Comparing different orientations of the impinging blade with respect to warp and weft threads, the authors also investigated the depth of trauma in addition to the usual depth of penetration and showed again the strong influence not only of the number of layers, but also of the chosen fabric construction on the depth of penetration as well as the depth of trauma [123].

Another interesting possibility for preparing stab-resistant fabrics is given by triaxial woven fabrics, i.e., fabrics woven from three sets of yarns, oriented approx. 60° with respect to each other. While they are usually not as dense as conventional woven fabrics, the interyarn friction can be increased, thus supporting energy dissipation in such fabrics. Stab resistance of triaxial woven fabrics was described by El Messiry and Eltahan who tested different materials and compared these fabrics with conventional plain weave as well as weft-knitted single jersey fabrics, applying the drop tower test according to NIJ standard 0115.00 [124]. They found the best stab resistance for Vectran triaxial fabrics and only slightly lower values for Kevlar triaxial samples, while the knitted fabrics and even the plain-weave fabrics had significantly lower stab protection. Recently, El Messiry and El-Tarfawy combined triaxial fabrics with weft-knitted fabrics and suggested knitted/triaxial Kevlar multilayer fabrics as optimum regarding the cutting force, suggesting such fabrics also for stab resistance tests [125].

10. Conclusions and Outlook

As this review shows, many materials and structures can be used to develop polymer-based stab resistant body armor further. One approach is based on developing new, often bio-inspired structures which can, for example, be produced by different 3D printing techniques. Besides such special structures, most polymer-based stab-resistant garments are based on textile fabrics, either solely or, in most cases, with an additional lamination or coating. Alternatively, composites with embedded fibers or textile fabrics can be prepared, which are stiffer than pure or coated fabrics, but more lightweight than metallic body armor.

While recent research approaches are very often based on impregnating textile fabrics with shear thickening fluids, many more possibilities exist to improve the stab resistant properties of body armor on the yarn or fabric level, by developing materials, structures, and production processes further. In the authors' opinion, especially combinations of new materials and new shapes, as they can be produced by 3D printing in combination with textile fibers or fabrics [126], offer further improvements for lightweight, yet efficient body armor. With this paper, we hope to inspire more researchers working in these fields to contribute new ideas and experiments to this interesting field of research.

Author Contributions: Conceptualization, N.P. and A.E.; methodology, A.E.; investigation, N.P. and A.E.; writing—original draft preparation, N.P. and A.E. All authors have read and agreed to the published version of the manuscript.

Funding: The study was partly funded by the German Federal Ministry for Economic Affairs and Climate Action via the AiF, based on a resolution of the German Bundestag, grant number KK5129708TA1. The APC was funded by the Deutsche Forschungsgemeinschaft (DFG, German Research Foundation)—490988677—and Bielefeld University of Applied Sciences.

Institutional Review Board Statement: Not applicable.

Data Availability Statement: No new data were created for this review paper.

Conflicts of Interest: The authors declare no conflict of interest.

References

1. Reiners, P. Investigation about the Stab Resistance of Textile Structures, Methods for Their Testing and Improvements. Dissertation Thesis, Université de Haute Alsace, Alsace, France, 2016.
2. Yadav, R.; Naebe, M.; Wang, X.G.; Kandasubramanian, B. Body armour materials: From steel to contemporary biomimetic systems. *RSC Adv.* **2016**, *6*, 115145–115174. [CrossRef]
3. Mawkhlieng, U.; Majumdar, A. Soft body armour. *Text. Prog.* **2019**, *51*, 139–224. [CrossRef]
4. Mawkhlieng, U.; Majumdar, A.; Laha, A. A review of fibrous materials for soft body armour applications. *RSC Adv.* **2020**, *10*, 1066–1086. [CrossRef]
5. Scott, K.W.M. Homicide patterns in the West Midlands. *Med. Sci. Law* **1990**, *30*, 234–238. [CrossRef] [PubMed]

6. Behera, C.; Sikary, A.K.; Gupta, S.K. Homicide patterns for the last 20 years in South and South East Delhi, India. *Med. Sci. Law* **2019**, *59*, 83–94. [CrossRef]
7. Rikken, Q.G.H.; Chadid, A.; Peters, J.; Geeraedts, L.M.G.; Giannakooulos, G.F.; Tan, E.C.T.H. Epidemiology of penetrating injury in an urban versus rural level 1 trauma center in the Netherlands. *Hong Kong J. Emerg. Med.* **2022**, *29*, 38–45. [CrossRef]
8. Hakkenbrak, N.A.G.; Bakkum, E.R.; Zuidema, W.P.; Halm, J.A.; Dorn, T.; Reijnders, U.J.L.; Giannakopoulos, G.F. Characteristics of fatal penetrating injury; data from a retrospective cohort study in three urban regions in the Netherlands. *Injury* **2023**, *54*, 256–260. [CrossRef]
9. LaTourrette, T. The life-saving effectiveness of body armor for police officers. *J. Occup. Environ. Hyg.* **2010**, *7*, 557–562. [CrossRef]
10. Sitotaw, D.B.; Ahrendt, D.; Kyosev, Y.; Kabish, A.K. A review on the performance and comfort of stab protection armor. *AUTEX Res. J.* **2022**, *22*, 96–107. [CrossRef]
11. Muenks, D.; Pilgrim, J.; Kyosev, Y. Possibilities for qualitative evaluation of the protection area of protective clothing. *Commun. Dev. Assem. Text. Prod.* **2022**, *3*, 156–162. [CrossRef]
12. Ricciardi, R.; Deuster, P.A.; Talbot, L.A. Metabolic Demands of Body Armor on Physical Performance in Simulated Conditions. *Mil. Med.* **2008**, *173*, 817. [CrossRef] [PubMed]
13. Matusiak, M. Thermal Comfort Index as a Method of Assessing the Thermal Comfort of Textile Materials. *Fibres Text. East. Eur.* **2010**, *18*, 45–50.
14. Nayak, R.; Kanesalingam, S.; Wang, L.; Padhye, R. Stab resistance and thermophysiological comfort properties of boron carbide coated aramid and ballistic nylon fabrics. *J. Text. Inst.* **2019**, *110*, 1159–1168. [CrossRef]
15. Parmar, M.S.; Kapil, N.; Sisodia, N. Development of a Unique Stab and Impact Resistant Material for Anti-riot Body Protector. In *Functional Textiles and Clothing*; Majumdar, A., Gupta, D., Gupta, S., Eds.; Springer: Singapore, 2020; pp. 55–66.
16. KDIW 2004. Available online: https://www.vpam.eu/pruefrichtlinien/aktuell/kdiw-2004/ (accessed on 14 January 2023).
17. Home Office Body Armor Standard 2017 (Knife and Spike). Available online: https://protectiongroupdenmark.com/articles/14-home-office-body-armor-standard-2017-knife-and-spike/ (accessed on 14 January 2023).
18. *NIJ Standard–0115.00*; Stab Resistance of Personal Body Armor. National Institute of Justice: Washington, DC, USA, 2000.
19. *NIJ Standard–0115.01*; Draft: Stab Resistance of Personal Body Armor. National Institute of Standards and Technology: Gaithersburg, MD, USA, 2020.
20. *ASTMF1342/F1342M-05(2022)*; Standard Test Method for Protective Clothing Material Resistance to Puncture. ASTM International: West Conshohocken, PA, USA, 2022. Available online: https://www.astm.org/f1342_f1342m-05r22.html (accessed on 14 January 2023).
21. *DIN EN 388:2016+A1:2018*; Protective Gloves against Mechanical Risks; with an Amendment from 2018.
22. *ISO 13997:1999*; Protective Clothing—Mechanical properties—Determination of Resistance to Cutting by Sharp Objects.
23. *EN 13594:2015*; Protective Gloves for Motorcyclists.
24. Nayak, R.; Crouch, I.; Kanesalingam, S.; Wang, L.J.; Ding, J.; Tan, P.; Lee, B.; Miao, M.H.; Ganga, D.; Padhye, R. Body armor for stab and spike protection, Part 2: A review of test methods. *Text. Res. J.* **2018**, *89*, 3411–3430. [CrossRef]
25. Bleetman, A. *Determining the Protective Requirements of Stab-Resistant Body Armour: The Vulnerability of the Internal Organs to Penetrating Edged Weapons. Proc. Sharp Weapons Armour Technology Symposium*; Cranfield University: Bedford, UK, 1999.
26. Nayak, R.; Crouch, I.; Kanesalingam, S.; Ding, J.; Tan, P.; Lee, B.; Miao, M.H.; Ganga, D.; Wang, L.J. Body armor for stab and spike protection, Part 1: Scientific literature review. *Text. Res. J.* **2017**, *88*, 812–832. [CrossRef]
27. Knight, B. The dynamics of stab wounds. *Forensic Sci.* **1975**, *6*, 249–255. [CrossRef] [PubMed]
28. Jones, S.; Noakes, L.; Leadbetter, S. The mechanics of stab wounding. *Forensic Sci. Int.* **1994**, *67*, 59–63. [CrossRef]
29. Horsfall, I. Stab Resistant Body Armour. Ph.D. Thesis, Cranfield University, Bedford, UK, 2000.
30. Termonia, Y. Puncture resistance of fibrous structures. *Int. J. Impact Eng.* **2006**, *32*, 1512–1520. [CrossRef]
31. Liu, Q.; Sun, Y.X.; Zhao, J.Z.; Ma, P.B. Failure mechanism of weft-knitted insertion fabric/Surlyn resin flexible composite for stab resistance. *Text. Res. J.* 2022, *online first*. [CrossRef]
32. Bartat, W.; Sokolowski, D.; Gieleta, R. Numerical and experimental research on stab resistance of a body armour package. *Fibers Text. East. Eur.* **2014**, *22*, 90–96.
33. Du, Z.; Chen, C.J.; Wang, X.H. The mechanism of stab resistance of carbon fiber reinforced polymer. *Eng. Fail. Anal.* **2022**, *142*, 106817. [CrossRef]
34. Guo, Y.X.; Yuan, M.Q.; Qian, X.M.; Wei, Y.C.; Liu, Y. Rapid prediction of polymer stab resistance performance. *Mater. Des.* **2020**, *192*, 108721. [CrossRef]
35. Ramakrishna, D.; Murali, G.B. Bio-inspired 3D-printed lattice structures for energy absorption applications: A review. *Proc. Inst. Mech. Eng. L J. Mater. Des. Appl.* 2022, *online first*. [CrossRef]
36. Yang, J.K.; Gu, D.D.; Lin, K.J.; Ma, C.L.; Wang, R.; Zhang, H.M.; Guo, M. Laser 3D printed bio-inspired impact resistant structure: Failure mechanism under compressive loading. *Virtual Phys. Prototyp.* **2020**, *15*, 75–86. [CrossRef]
37. Islam, M.K.; Hazell, P.J.; Escobedo, J.P.; Wang, H.X. Biomimetic armour design strategies for additive manufacturing: A review. *Mater. Des.* **2021**, *205*, 109730. [CrossRef]
38. Liu, Q.; Mao, H.W.; Niu, L.; Chen, F.X.; Ma, P.B. Excellent flexibility and stab-resistance on pangolin-inspired scale-like structure composite for versatile protection. *Compos. Commun.* **2022**, *35*, 101266. [CrossRef]

39. Liu, Q.; Lang, L.L.; Luo, M.; Wu, Q.; Kang, Y.; Ma, P.B. Stab resistance of flexible composite reinforced with warp-knitted fabric like scale structure at quasi-static loading. *J. Ind. Text.* **2022**, *51*, 7983S–7998S. [CrossRef]
40. Yong, K.C. Rubber Wood Fibre Based Flexible Composites: Their Preparation, Physical Strength Reinforcing and Stab Resistance Behaviour. *Polym. Comp.* **2014**, *22*, 375–380. [CrossRef]
41. Nakahata, M.; Takashima, Y.; Harada, A. Highly Flexible, Tough, and Self-Healing Supramolecular Polymeric Materials Using Host–Guest Interaction. *Macromol. Rap. Comm.* **2016**, *37*, 86–92. [CrossRef]
42. Tan, M.; Cui, Y.L.; Zhu, A.D.; Han, H.; Guo, M.Y.; Jiang, M. Ultraductile, notch and stab resistant supramolecular hydrogels via host-guest interactions. *Polym. Chem.* **2015**, *6*, 7543. [CrossRef]
43. Li, R.; Zhang, K.L.; Cai, L.; Chen, G.X.; He, M.H. Highly stretchable ionic conducting hydrogels for strain/tactile sensors. *Polymer* **2019**, *167*, 154–158. [CrossRef]
44. Blachowicz, T.; Ehrmann, G.; Ehrmann, A. Optical elements from 3D printed polymers. *e-Polymers* **2021**, *21*, 549–565. [CrossRef]
45. Cicek, U.I.; Southee, D.J.; Johnson, A.A. Assessing the stab resistive performance of material extruded body armour specimens. *Int. J. Prot. Struct.* 2022, *online first*. [CrossRef]
46. Maidin, S.; Chong, S.Y.; Heing, T.K.; Abdullah, Z.; Alkahari, R. Stab resistant analysis of body armour design features manufactured via fused deposition modelling process. In *Textile Manufacturing Processes*; Uddin, F., Ed.; IntechOpen: London, UK, 2019; pp. 69–83.
47. Jiang, J.H.; Yuan, M.Q.; Ji, T.C. Investigations on laser sintered textiles for stab-resistant application. Solid Freeform Fabrication. In Proceedings of the 26th Annual International Solid Freeform Fabrication Symposium—An Additive Manufacturing Conference, Austin, TX, USA, 10–12 August 2015; pp. 2155–2164.
48. Gong, Z.; Qian, X.M.; Yuan, M.Q. Structural design of a 3-D printed stab resistant body armor. *Rapid Prototyp. J.* **2019**, *25*, 143–151. [CrossRef]
49. Yuan, M.Q.; Liu, Y.; Gong, Z.; Qian, X.M. The application of PA/CF in stab resistance body armor. *IOP Conf. Ser. Mater. Sci. Eng.* **2017**, *213*, 012027. [CrossRef]
50. He, J.J.; Yuan, M.Q.; Gong, Z.; Qian, X.M. Egg-shell structure design for stab resistant body armor. *Mater. Today Comm.* **2018**, *16*, 26–36. [CrossRef]
51. Sitotaw, D.B.; Ahrendt, D.; Kyosev, Y.; Kabish, A.K. Investigation of Stab Protection Properties of Aramid Fibre-Reinforced 3D Printed Elements. *Fibres Text. East. Eur.* **2021**, *29*, 67–73. [CrossRef]
52. Ahrendt, D.; Krzywinski, S.; Justo i Massot, E.; Krzywinski, J. Hybrid material designs by the example of additive manufacturing for novel customized stab protective clothing. In Proceedings of the Light-Weight Armour for Defense & Security (LWAG), Roubaix, France, 8–9 October 2019; pp. 286–294.
53. Miao, L.; Jiang, T.T.; Lin, S.D.; Jin, T.; Hu, J.W.; Zhang, M.; Tu, Y.Y.; Liu, G.J. Asymmetric forward osmosis membranes from *p*-aramid nanofibers. *Mater. Des.* **2020**, *191*, 108591. [CrossRef]
54. Kim, H.D.; Nam, I.W. Stab Resisting Behavior of Polymeric Resin Reinforced *p*-Aramid Fabrics. *J. Appl. Polym. Sci.* **2012**, *123*, 2733–2742. [CrossRef]
55. Stojanovic, D.B.; Zrilic, M.; Jancic-Heinemann, R.; Zivkovic, I.; Kojovic, A.; Uskokovic, P.S.; Aleksic, R. Mechanical and anti-stabbing properties of modified thermoplastic polymers impregnated multiaxial p-aramid fabrics. *Polym. Adv. Technol.* **2013**, *24*, 772–776. [CrossRef]
56. Simic, D.M.; Stojanovic, D.B.; Dimic, M.; Miskovic, K.; Marjanovic, M.; Burzic, Z.; Uskokovic, P.S.; Zak, A.; Tenne, R. Impact resistant hybrid composites reinforced with inorganic nanoparticles and nanotubes of WS_2. *Comp. B Eng.* **2019**, *176*, 107222. [CrossRef]
57. Zhao, H.-Y.; Qiang, Y.-Q.; Peng, H.-K.; Xing, M.-F.; Zhang, X.-Y.; Lou, C.-W. Enhancement of a Novel Sizing Agent in Mechanical Properties and Stab/Puncture Resistance of Kevlar Fabrics. *Fibers Polym.* **2021**, *22*, 3309–3316. [CrossRef]
58. Mayo, J.B., Jr.; Wetzel, E.D.; Hosur, M.V.; Jeelani, S. Stab and puncture characterization of thermoplastic-impregnated aramid fabrics. *Int. J. Impact Eng.* **2009**, *36*, 1095–1105. [CrossRef]
59. Li, C.-S.; Huang, X.-C.; Yang, N.C.; Shen, Z.H.; Fan, X.-H. Stab resistance of UHMWPE fiber composites impregnated with thermoplastics. *Polym. Adv. Technol.* **2014**, *25*, 1014–1019. [CrossRef]
60. Firouzi, D.; Foucher, D.A.; Bougherara, H. Nylon-Coated Ultra High Molecular Weight Polyethylene Fabric for Enhanced Penetration Resistance. *J. Appl. Polym. Sci.* **2014**, *131*, 40350. [CrossRef]
61. Cheon, J.S.; Lee, M.W.; Kim, M.K. Study on the stab resistance mechanism and performance of the carbon, glass and aramid fiber reinforced polymer and hybrid composites. *Compos. Struct.* **2020**, *234*, 111690. [CrossRef]
62. Chuang, Y.-C.; Bao, L.M.; Lin, M.-C.; Lou, C.-W.; Lin, T.A. Mechanical and Static Stab Resistant Properties of Hybrid-Fabric Fibrous Planks: Manufacturing Process of Nonwoven Fabrics Made of Recycled Fibers. *Polymers* **2019**, *11*, 1140. [CrossRef]
63. Khuyen, N.Q.; Han, P.V.D.; Nguyen, N.T.; Le, Q.B.; Harjo, M.; Anbarjafari, G.; Kiefer, R.; Tamm, T. The Use of Laminates of Commercially Available Fabrics for Anti-Stab Body-Armor. *Polymers* **2021**, *13*, 1077. [CrossRef]
64. Grimmelsmann, N.; Meissner, H.; Ehrmann, A. 3D printed auxetic forms on knitted fabrics for adjustable permeability and mechanical properties. *IOP Conf. Ser. Mater. Sci. Eng.* **2016**, *137*, 012011. [CrossRef]
65. Xu, W.L.; Sun, Y.X.; Lin, H.T.; Wie, C.Y.; Ma, P.B.; Xia, F.L. Preparation of soft composite reinforced with auxetic warp-knitted spacer fabric for stab resistance. *Text. Res. J.* **2019**, *90*, 323–332. [CrossRef]

66. Novak, N.; Dubrovski, P.D.; Borovinsek, M.; Vesenjak, M.; Ren, Z. Deformation behaviour of advanced textile composites with auxetic structure. *Compos. Struct.* **2020**, *252*, 112761. [CrossRef]
67. Sun, Y.X.; Xu, W.L.; Wei, W.F.; Ma, P.B.; Xia, F.L. Stab-resistance of auxetic weft-knitted fabric with Kevlar fibers at quasi-static loading. *J. Ind. Text.* **2019**, *5*, 1384–1396. [CrossRef]
68. Rodríguez-Millán, M.; Díaz-Álvarez, A.; Aranda-Ruiz, J.; Díaz-Álvarez, J.; Loya, J.A. Experimental analysis for stabbing resistance of different aramid composite architectures. *Compos. Struct.* **2019**, *208*, 525–534. [CrossRef]
69. Zhang, X.Y.; Li, T.-T.; Wang, Y.X.; Shiu, B.-C.; Peng, H.-K.; Lou, C.-W.; Lin, J.-H. Hydrogel with high toughness and strength for fabricating high performance stab-resistant aramid composite fabric. *J. Mater. Res. Technol.* **2021**, *15*, 1630–1641. [CrossRef]
70. Lu, Z.Q.; Li, D.Y.; Yuan, Z.S. Polypyrrole coating on aramid fabrics for improved stab resistance and multifunction. *J. Eng. Fibers Fabr.* **2022**, *17*, 15589250221081856. [CrossRef]
71. Gadow, R.; von Niessen, K. Ceramic coatings on fiber woven fabrics. *Ceramic Eng. Sci. Proc.* **2002**, *23*, 277–285.
72. Gadow, R.; von Niessen, K. Lightweight Ballistic with Additional Stab Protection Made of Thermally Sprayed Ceramic and Cermet Coatings on Aramide Fabrics. *Int. J. Appl. Ceram. Technol.* **2006**, *3*, 284–292. [CrossRef]
73. Manaee, P.; Valefi, Z.; Goodarz, M. The effect of bond coat type on the stab resistance of Al_2O_3–13 wt% TiO_2 plasma sprayed ceramic coating on aramid fabrics. *Surf. Interf.* **2020**, *18*, 100432. [CrossRef]
74. Xia, M.M.; Quan, Z.Z.; Wang, X.L.; Yu, J.Y. Preparation and characterization of B_4C particle coated composites for stab-resistance. *Compos. Struct.* **2019**, *228*, 111370. [CrossRef]
75. Yang, W.-Q.; Liu, X.-Y.; Yu, Y.-P.; Yu, W.-D. Evaluation of Stab Resistance of Coated UHMWPE Fabric. *Fibres Text. East. Eur.* **2020**, *28*, 76–79.
76. Wei, R.; Zhai, W.; Li, F.; Li, Y.J.; Zhang, W.T.; Liang, Y.F.; Dong, B.; Wang, X.W. Enhancing stab resistance of thermoset–aramid composite fabrics by coating with SiC particles. *J. Ind. Text.* **2018**, *48*, 1228–1241.
77. Javaid, M.U.; Jabbar, A.; Irfan, M.; Javed, Z. Investigation of the stab resistance mechanism and performance of uncoated and SiO_2 coated high-performance aramid fabrics. *J. Text. Inst.* **2022**, *113*, 2143–2158. [CrossRef]
78. Obradovic, V.; Stojanovic, D.B.; Jokic, B.; Zrilic, M.; Radojevic, V.; Uskokovic, P.S.; Aleksic, R. Nanomechanical and anti-stabbing properties of Kolon fabric composites reinforced with hybrid nanoparticles. *Comp. B Eng.* **2017**, *108*, 143–152. [CrossRef]
79. Kanesalingam, S.; Nayak, R.; Wang, L.J.; Padhye, R.; Arnold, L. Stab and puncture resistance of silica-coated Kevlar–wool and Kevlar–wool–nylon fabrics in quasistatic conditions. *Text. Res. J.* **2019**, *89*, 2219–2235. [CrossRef]
80. Wang, Y.S.; Liu, J.R.; Zhao, Y.H.; Qin, Y.; Zhu, Z.Y.; Yu, Z.C.; He, H.L. Temperature-triggered fire warning PEG@wood powder/carbon nanotube/calcium alginate composite aerogel and the application for firefighting clothing. *Comp. B Eng.* **2022**, *247*, 110348. [CrossRef]
81. He, H.L.; Liu, J.R.; Wang, Y.S.; Zhao, Y.H.; Qin, Y.; Zhu, Z.Y.; Yu, Z.C.; Wang, J.F. An Ultralight Self-Powered Fire Alarm e-Textile Based on Conductive Aerogel Fiber with Repeatable Temperature Monitoring Performance Used in Firefighting Clothing. *ACS Nano* **2022**, *16*, 2953–2967. [CrossRef] [PubMed]
82. He, H.L.; Qin, Y.; Liu, J.R.; Wang, Y.S.; Wang, J.F.; Zhao, Y.H.; Zhu, Z.Y.; Jiang, Q.; Wang, Y.H.; Qu, X.R.; et al. A wearable self-powered fire warning e-textile enabled by aramid nanofibers/MXene/silver nanowires aerogel fiber for fire protection used in firefighting clothing. *Chem. Eng. J.* **2023**, *460*, 141661. [CrossRef]
83. Barnes, H.A. Shear-thickening ("dilatancy") in suspensions of nonaggregating solid particles dispersed in Newtonian liquids. *J. Rheol.* **1989**, *33*, 329–366. [CrossRef]
84. Lee, Y.S.; Wagner, N.J. Dynamic properties of shear thickening colloidal suspensions. *Rheol. Acta* **2003**, *42*, 199–208. [CrossRef]
85. Catherall, A.A.; Melrose, J.R.; Ball, R.C. Shear thickening and order–disorder effects in concentrated colloids at high shear rates. *J. Rheol.* **2000**, *44*, 1–25. [CrossRef]
86. Maranzano, B.J.; Wagner, N.J. Flow-small angle neutron scattering measurements of colloidal dispersion microstructure evolution through the shear thickening transition. *J. Chem. Phys.* **2002**, *117*, 10291–10302. [CrossRef]
87. Lee, Y.S.; Wetzel, E.D.; Wagner, N.J. The ballistic impact characteristics of Kevlar woven fabrics impregnated with a colloidal shear thickening fluid. *J. Mater. Sci.* **2003**, *38*, 2825–2833. [CrossRef]
88. Decker, M.J.; Halbach, C.J.; Nam, C.H.; Wagner, N.J.; Wetzel, E.D. Stab resistance of shear thickening fluid (STF)-treated fabrics. *Compos. Sci. Technol.* **2007**, *67*, 565–578. [CrossRef]
89. Kang, T.J.; Hong, K.H.; Yoo, M.R. Preparation and Properties of Fumed Silica/Kevlar Composite Fabrics for Application of Stab Resistant Material. *Fibers Polym.* **2010**, *11*, 719–724. [CrossRef]
90. Kang, T.J.; Kim, C.Y.; Hong, K.W. Rheological Behavior of Concentrated Silica Suspension and Its Application to Soft Armor. *J. Appl. Polym. Sci.* **2012**, *124*, 1534–1541. [CrossRef]
91. Li, T.-T.; Dai, W.N.; Wu, L.W.; Peng, H.-K.; Zhang, X.Y.; Shiu, B.-C.; Lin, J.-H.; Lou, C.-W. Effects of STF and Fiber Characteristics on Quasi-Static Stab Resistant Properties of Shear Thickening Fluid (STF)-Impregnated UHMWPE/Kevlar Composite Fabrics. *Fibers Polym.* **2019**, *20*, 328–336. [CrossRef]
92. Li, W.; Xiong, D.S.; Zhao, X.D.; Sun, L.L.; Liu, J. Dynamic stab resistance of ultra-high molecular weight polyethylene fabric impregnated with shear thickening fluid. *Mater. Des.* **2016**, *102*, 162–167. [CrossRef]
93. Sun, L.-L.; Xiong, D.-S.; Xu, C.-Y. Application of Shear Thickening Fluid in Ultra High Molecular Weight Polyethylene Fabric. *J. Appl. Polym. Sci.* **2013**, *129*, 1922–1928. [CrossRef]

94. Khodadadi, A.; Liaghat, G.h.; Vahid, S.; Sabet, A.R.; Hadavinia, H. Ballistic performance of Kevlar fabric impregnated with nanosilica/PEG shear thickening fluid. *Compos. B Eng.* **2019**, *162*, 643–652. [CrossRef]
95. Hassan, T.A.; Rangari, V.K.; Jeelani, S. Synthesis, processing and characterization of shear thickening fluid (STF) impregnated fabric composites. *Mater. Sci. Eng. A* **2010**, *527*, 2892–2899. [CrossRef]
96. Mahfuz, H.; Clements, F.; Rangari, V.; Dhanak, V.; Beamson, G. Enhanced stab resistance of armor composites with functionalized silica nanoparticles. *J. Appl. Phys.* **2009**, *105*, 064307. [CrossRef]
97. Wei, R.B.; Dong, B.; Wang, F.L.; Yang, J.J.; Jiang, Y.Y.; Zhai, W.; Li, H. Effects of silica morphology on the shear-thickening behavior of shear thickening fluids and stabbing resistance of fabric composites. *J. Appl. Polym. Sci.* **2020**, *137*, 48809. [CrossRef]
98. Zhang, X.Y.; Li, T.-T.; Peng, H.-K.; Wang, Z.K.; Huo, H.L.; Lou, C.-W.; Lin, J.-H. Effects of bi-particlesized shear thickening fluid on rheological behaviors and stab resistance of Kevlar fabrics. *J. Ind. Text.* **2022**, *51*, 3014S–3029S. [CrossRef]
99. Majumdar, A.; Butola, B.S.; Srivastava, A.; Bhattacharjee, D.; Biswas, I.; Laha, A.; Arora, S.; Ghosh, A. Improving the impact resistance of p-aramid fabrics by sequential impregnation with shear thickening fluid. *Fibers Polym.* **2016**, *17*, 199–204. [CrossRef]
100. Asija, N.; Chouhan, H.; Gebremeskel, S.A.; Bhatnagar, N. Impact Response of Shear Thickening Fluid (STF) Treated High Strength Polymer Composites—Effect of STF Intercalation Method. *Proc. Eng.* **2017**, *173*, 655–662. [CrossRef]
101. Qin, J.B.; Guo, B.R.; Zhang, L.; Wang, T.W.; Zhang, G.C.; Shi, X.T. Soft armor materials constructed with Kevlar fabric and a novel shear thickening fluid. *Comp. B Eng.* **2020**, *183*, 107686. [CrossRef]
102. Li, D.Y.; Wang, R.; Liu, X.; Fang, S.; Sun, Y.L. Shear-Thickening Fluid Using Oxygen-Plasma-Modified Multi-Walled Carbon Nanotubes to Improve the Quasi-Static Stab Resistance of Kevlar Fabrics. *Polymers* **2018**, *10*, 1356. [CrossRef]
103. Li, D.Y.; Wang, R.; Guan, F.W.; Zhu, Y.J.; You, F.F. Enhancement of the quasi-static stab resistance of Kevlar fabrics impregnated with shear thickening fluid. *J. Mater. Res. Technol.* **2022**, *18*, 3673–3683. [CrossRef]
104. Liu, M.; Zhang, S.S.; Liu, S.; Cao, S.S.; Wang, S.; Bai, L.F.; Sang, M.; Xuan, S.H.; Jiang, W.Q.; Gong, X.L. CNT/STF/Kevlar-based wearable electronic textile with excellent anti-impact and sensing performance. *Compos. A Appl. Sci. Manuf.* **2019**, *126*, 105612. [CrossRef]
105. Wang, S.; Xuan, S.H.; Liu, M.; Bai, L.F.; Zhang, S.S.; Sang, M.; Jiang, W.Q.; Gong, X.L. Smart wearable Kevlar-based safeguarding electronic textile with excellent sensing performance. *Soft Matter* **2017**, *13*, 2483–2491. [CrossRef]
106. Gürgen, S.; Kushan, M.C. The stab resistance of fabrics impregnated with shear thickening fluids including various particle size of additives. *Comp. A Appl. Sci. Manuf.* **2017**, *94*, 50–60. [CrossRef]
107. Gürgen, S.; Kushan, M.C. The ballistic performance of aramid based fabrics impregnated with multi-phase shear thickening fluids. *Polym. Test.* **2017**, *64*, 296–306. [CrossRef]
108. Gürgen, S.; Yildi, T. Stab resistance of smart polymer coated textiles reinforced with particle additives. *Comp. Struct.* **2020**, *235*, 111812. [CrossRef]
109. Zhang, X.Y.; Li, T.-T.; Peng, H.-K.; Lou, C.-W.; Lin, J.-H. Enhanced sandwich structure composite with shear thickening fluid and thermoplastic polyurethanes for High-performance stab resistance. *Comp. Struct.* **2022**, *280*, 114930. [CrossRef]
110. Zhang, X.Y.; Li, T.-T.; Wang, Z.K.; Peng, H.-K.; Lou, C.-W.; Lin, J.-H. Facile fabrication and mass production of TPU/Silica/STF coated aramid fabric with excellent flexibility and quasi-static stab resistance for versatile protection. *Progr. Org. Coat.* **2021**, *151*, 106088. [CrossRef]
111. Mayo, J.B., Jr.; Wetzel, E.D. Cut resistance and failure of high-performance single fibers. *Text. Res. J.* **2014**, *84*, 1233–1246. [CrossRef]
112. Tian, L.X.; Shi, J.J.; Chen, H.X.; Huang, X.M.; Cao, H.J. Cut-resistant performance of Kevlar and UHMWPE covered yarn fabrics with different structures. *J. Text. Inst.* **2022**, *113*, 1457–1463. [CrossRef]
113. Li, T.-T.; Wang, Z.K.; Zhang, X.Y.; Wu, L.W.; Lou, C.-W.; Lin, J.-H. Dynamic cushion, quasi-static stab resistance, and acoustic absorption analyses of flexible multifunctional inter-/intra-bonded sandwich-structured composites. *J. Text. Inst.* **2021**, *112*, 47–55. [CrossRef]
114. Zhang, X.Y.; Li, T.-T.; Sun, F.; Peng, H.-K.; Wang, Z.K.; Lin, J.-H.; Lou, C.-W. Stab/Puncture Resistance Performance of Needle Punched Nonwoven Fabrics: Effects of Filament Reinforcement and Thermal Bonding. *Fibers Polym.* **2022**, *23*, 2330–2339. [CrossRef]
115. Liu, Q.; Wang, L.L.; Luo, Q.; Sun, Z.L.; He, H.J.; Chen, C.Y.; Ma, P.B. Stab-resistance improvement of short carbon fiber reinforced UHMWPE knitted composites with plasma/oxidation treatment. *J. Ind. Text.* 2022, *online first*. [CrossRef]
116. Zhang, W.; Liu, S.; Ma, P.B. Experimental investigation on stab-resistant properties of co-woven-knitted fabric. *J. Eng. Fibers Fabr.* **2022**, *17*, 15589250221090502. [CrossRef]
117. Tien, D.T.; Kim, J.S.; Huh, Y. Stab-resistant Property of the Fabrics Woven with the Aramid/Cotton Core-spun Yarns. *Fibers Polym.* **2010**, *11*, 500–506. [CrossRef]
118. Tien, D.T.; Kim, J.S.; Huh, Y. Evaluation of anti-stabbing performance of fabric layers woven with various hybrid yarns under different fabric conditions. *Fibers Polym.* **2011**, *12*, 808. [CrossRef]
119. Li, T.-T.; Zhang, X.Y.; Wu, L.W.; Peng, H.K.; Shiu, B.-C.; Lou, C.-W.; Lin, J.-H. Polyethylene terephthalate/basalt stab-resistant sandwich composites based on the Box–Behnken design: Parameter optimization and empirical regression model. *J. Sandw. Struct. Mater.* **2020**, *22*, 2391–2407. [CrossRef]
120. Nasser, J.; Steinke, K.; Groo, L.A.; Sodano, H.A. Improved Interyarn Friction, Impact Response, and Stab Resistance of Surface Fibrilized Aramid Fabric. *Adv. Mater. Interfaces* **2019**, *6*, 1900881. [CrossRef]

121. Li, M.R.; Wang, P.; Boussu, F.; Soulat, D. Dynamic stab resistance of multi-ply three-dimensional warp interlock fabrics with high-performance high-molecular-weight polyethylene yarns for protective applications. *J. Ind. Text.* **2022**, *51*, 2391S–2411S. [CrossRef]
122. Li, M.R.; Boussu, F.; Soulat, D.; Luo, J.; Wang, P. Impact resistance of pre-deformed stab of multi-ply three-dimensional interlock polymeric fabrics. *J. Ind. Text.* **2022**, *51*, 4818S–4841S. [CrossRef]
123. Li, M.R.; Wang, P.; Boussu, F.; Soulat, D. Investigation of impact performance of 3-dimensional interlock polymer fabrics in double and multi-angle pass stabbing. *Mater. Des.* **2021**, *206*, 109775. [CrossRef]
124. El Messiry, M.; Eltahan, E. Stab resistance of triaxial woven fabrics for soft body armor. *J. Ind. Text.* **2016**, *45*, 1062–1082. [CrossRef]
125. El Messiry, M.; El-Tarfawy, S.Y. Cutting resistance of flexible armour using multiple layers of triaxial Kevlar fabric. *J. Ind. Text.* **2022**, *51*, 1564S–1591S. [CrossRef]
126. Kozior, T.; Blachowicz, T.; Ehrmann, A. Adhesion of 3D printing on textiles fabrics—Inspiration from and for other research areas. *J. Eng. Fibers Fabr.* **2020**, *15*, 1558925020910875.

polymers

MDPI

Article

Mechanical Properties Degradation of Fiberglass Tubes during Biaxial Proportional Cyclic Loading

Valeriy Wildemann *, Oleg Staroverov, Elena Strungar, Artur Mugatarov and Artur Kuchukov

Center of Experimental Mechanics, Perm National Research Polytechnic University, 614990 Perm, Russia; cem_staroverov@mail.ru (O.S.); cem.spaskova@mail.ru (E.S.); cem_mugatarov@mail.ru (A.M.); artur.kuchukov.59@mail.ru (A.K.)

* Correspondence: wildemann@pstu.ru; Tel.: +7-342-239-10-01

Abstract: Composite structures during an operation are subjected to various types of external loading (impact, vibration, cyclic, etc.), which may lead to a decrease in mechanical properties. Previously, many experimental investigations of the mechanical behavior of composites under uniaxial cyclic loading were carried out. Acquisition of new data on the reduction of composite materials' mechanical characteristics under conditions of multiaxial cyclic loading, as well as verification of existing models for calculation of the residual properties, are relevant. Therefore, this work is devoted to the experimental investigation of the mechanical behavior of fiberglass tubes under proportional cyclic loading. Static and fatigue tests were carried out under tension with torsion conditions. Inhomogeneous strain fields were obtained using a non-contact optical video system VIC-3D. The structural damage accumulation processes were analyzed by an AMSY-6 acoustic emission signals recording system. Surface defects were determined using a DinoLite microscope. Residual dynamic elastic modules were calculated during fatigue tests, and fatigue sensitivity curves were built. Data was approximated using various models, and their high descriptive capability was revealed. Damage accumulation stages were determined. The dependence of the models' parameters on a stress state were observed. It was concluded that multiaxial cyclic loading leads to a significant decrease in mechanical properties, which should be taken into account in composite structure design.

Keywords: composite; tubular sample; multiaxial fatigue; damage accumulation; residual mechanical properties

Citation: Wildemann, V.; Staroverov, O.; Strungar, E.; Mugatarov, A.; Kuchukov, A. Mechanical Properties Degradation of Fiberglass Tubes during Biaxial Proportional Cyclic Loading. *Polymers* **2023**, *15*, 2017. https://doi.org/10.3390/polym15092017

Academic Editors: Emilia P. Collar and Jesús-María García-Martínez

Received: 31 March 2023
Revised: 20 April 2023
Accepted: 21 April 2023
Published: 24 April 2023

1. Introduction

Composite materials allow the problem of maintaining a high level of physical and mechanical characteristics while reducing the structure weight to be solved in industries such as space, aviation, automotive, construction, etc. Trends demonstrate that the development of technology for the production of the composite materials based on a polymer matrix and the growth of their usage in designs [1–6].

Every unforeseen impact, vibration, or other load, even of low intensity and duration, can lead to a significant reduction in the exploitation time [7–12]. In this regard, it is relevant to investigate patterns of the degradation of composites' mechanical properties under complex and combined loads. The study of the cyclic exposure influence on the residual mechanical characteristics of composites, called fatigue sensitivity, is one of the most significant.

A large number of works on this topic are devoted to the experimental investigation of the degradation of the polymer composites' mechanical properties (elasticity modulus and tensile strength) during uniaxial cyclic loading [13–19]. On the other hand, the overwhelming majority of composite structures are in a complex stress–strain state, so it is necessary to study fatigue sensitivity of composites under multiaxial cyclic loading. In this case, many loading parameters should be taken into account, including the ratio of the amplitudes of the stress tensor components [20–22], the phase shift angle between the loading modes [23], the frequencies ratio [24,25], the mean stress, and the stress amplitude in the cycle [26–29].

Recent testing systems allow the implementation of a complex stress–strain state by loading cruciform specimens in two orthogonal axes [30–32], the tension of tubular specimens under the internal pressure [33,34], the torsion of tubular specimens with bending [35–38], and tension–compression with torsion [39–41]. Experimental studies of the degradation of the composites' mechanical properties under multiaxial cyclic loading have been presented previously [42–44]. The residual characteristics assessment is carried out by calculation of the elastic modulus during the fatigue test.

The relationship between the mechanical characteristics and preliminary cyclic exposure can be represented as the fatigue sensitivity curve. As demonstrated previously [45], these curves have three characteristic stages: initiation, where intense damage accumulation processes occur, leading to rapid degradation of properties; stabilization, where damage accumulation is slow, and the stage is almost linear; and aggravation, where intense damage accumulation processes lead to a complete loss of bearing capacity. Previously [45,46], it was proposed to determine the boundaries of these stages by the points where the damage value function derivative reaches a certain value (characteristic to the material).

The experimentally obtained fatigue sensitivity curves are necessary to build and verify residual mechanical properties prediction models. Some of them were proposed previously [47–51]. However, these models are often only suitable for a particular class of composite and require the definition of many parameters. Mao and Mahadevan [52] proposed the convenient approximation of the fatigue sensitivity curve as a sum of two power functions. The authors of this study previously proposed two models based on the usage of cumulative probability distribution functions [46]. The advantages of these models are simplicity and few parameters, which means a reduction in the required experimental tests.

Nevertheless, an understanding of the damage accumulation processes requires usage of additional diagnostic systems. At first, the non-contact optical video system and digital image correlation method (DIC) allow the observation of inhomogeneous strain fields on the sample surface. DIC technology showed an advantage in the analysis of the stress–strain state of composite objects [53,54], the formation and propagation of cracks [55], the detection of defects, and the quantitative assessment of displacements and deformations [56]. Compared to traditional strain gauges [57] and finite element simulations [58], DIC can obtain more accurate and complete information about displacement and strain fields across the entire surface, not limited by sample geometry [59]. Moreover, since the destruction of polymer composites is a complex process that occurs at different scale levels, the identification of damages is necessary to predict the failure mechanism. The acoustic emission method allows the estimation of the intensity of damage accumulation and establishment of a connection with the composites' structural destruction mechanisms by analyzing the frequency spectrum of the signals [60–65]. For verification of the acoustic emission signal analysis, optical microscopy is also required.

In this study, we investigate the degradation of mechanical properties during biaxial static and cyclic loading. The main goals of this work were to obtain new experimental data on stiffness degradation of fiberglass thin-walled tubes (produced by oblique transverse longitudinal continuous winding) and to verify previously proposed fatigue sensitivity curve approximation models. The investigation includes usage of the non-contact video system, acoustic emission system, and optical microscope.

The paper is organized as follows: the material, equipment, methods, and approximation models are described in Section 2; Section 3.1 presents the quasistatic tests results; fatigue experiments results and approximations of fatigue sensitivity curves are discussed in Section 3.2; and Section 4 closes the paper with the main conclusions of the work.

2. Material and Methods

2.1. Material

Tubes based on EC glass fiber (1200 tex, ≈580–600 N breaking load value) and cold curing epoxy resin KER 828 were made by oblique transverse longitudinal continuous winding at an angle of 85° to the axis (this angle is often used in the production of pipes, which work under internal pressure). The winding tapes consisted of fibers lying along the tape and had few fibers, which were oriented across it. For testing, thin-walled fiberglass tubular samples were prepared. The sample length was 140 mm with the working part length $L \approx 60$ mm, inner diameter $d = 25.4$ mm, and outer diameter $D \approx 30$ mm.

2.2. Equipment

Experimental studies were carried out using the large-scale research facilities "Complex of testing and diagnostic equipment for studying properties of structural and functional materials under complex thermomechanical loading" at the Center of Experimental Mechanics of the Perm National Research Polytechnic University (PNRPU).

Quasistatic and cyclic tests were carried out using the Instron 8802 (±100 kN) universal servohydraulic testing system. The loading was recorded by a load cell up to 100 kN and 1000 N·m. The loading measurement accuracy is 0.5%. The testing system included a FastTrack controller. The WaveMatrix V. 1.4 software allowed the implementation of biaxial loading modes.

Collet grips of the test system Instron 8802 were used to set samples (Figure 1, left side). Cylindrical steel plugs were inserted for the entire length of the gripping parts to prevent specimen crushing. When the sample was set, the pressure in the servo-hydraulic grip circuit gradually increased from 50 to 200 bar. To prevent the occurrence of overloads during fixation, the values of the axial load and torsional were kept at zero ("Specimen Protect" was on).

Displacements and strains of the specimen surface were recorded using the VIC-3D contactless optical video system (Correlated Solutions, Irmo, SC, USA) and the digital image correlation (DIC) method. A video recording was carried out using the camera Prosilica GE4900 50 mm with 16 MP resolution (Allied Vision, Stadtroda, Germany). The recording frequency was 1 frame per 3 s. The normalized sum of the squared difference (NSSD) was used as the correlation criterion for the mathematical assessment of the digital image correspondence. The strains were calculated using the Lagrange finite strain tensor. Transition from the Cartesian to cylindrical coordinate system, associated with specimen axis, was carried out using VIC-3D software (Vic-Snap Image Acquisition V. 9).

The acoustic emission signals were recorded using the AMSY-6 system (Vallen Systeme GmbH, Wolfratshausen, Germany). We employed a wideband sensor AE144A (Fujicera, Fujinomiya, Japan) with a frequency range of 100–500 kHz, a M31 (Fujicera, Fujinomiya, Japan) sensor with the frequency range of 300–800 kHz, and a preamplifier with a gain of 34 dB. The sensors were attached to the sample using a rubber fixture. The data sampling frequency was 10 MHz, and the threshold value for recording AE signals was 40 dB. The energy parameter and the frequency of the spectral maximum (characteristic of the fast Fourier transform) were considered as informative. The energy parameter of AE signals was calculated using a special software option in energy units (eu), 1 eu = 10^{-14} $V^2 \cdot s$. The AMSY-6 and the video system were synchronized with the test system controller using a 16-bit high-speed NI USB-6251 ADC unit. A photo of the diagnostic systems is shown in Figure 1 (right side).

After testing, we used a Dino-Lite microscope (AnMo Electronics Corporation, New Taipei City, Taiwan) with DinoCapture 2.0 software in order to determine the sample surface defects.

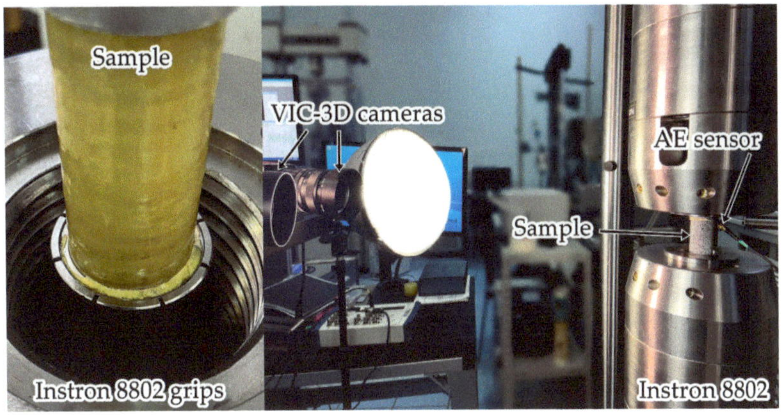

Figure 1. Sample in the testing machine grips and diagnostic systems.

2.3. Methods

Eighteen fiberglass tubular samples were divided into two groups for quasi-static and cyclic tensile, torsion, and proportional tension–torsion tests with various ratios of normal and shear stress tensor components. To determine the nominal values of the fracture load and torque, 6 samples were tested. The speed of the movable grip was 2 mm/min in the tensile (strain rate ≈ 0.033 min^{-1}), and 20 deg/min in the torsion test (strain rate ≈ 0.087 min^{-1}). In the proportional loading tests, 2 samples were fractured for each loading mode. The Proportional 1 mode corresponds to the speed of movement grip 1 mm/min (strain rate ≈ 0.017 min^{-1}) and 20 deg/min; for the Proportional 2 mode, it is 2 mm/min and 20 deg/min. On some samples, video recording of the displacement and strain fields was carried out using the VIC-3D system, as well as the recording of acoustic emission signals using the AMSY-6 system (Table 1).

As a result of quasi-static tests, the dependences of load, P, on grip displacement, u, and torque, M, on torsion angle, φ, were determined. Normal stress, σ, longitudinal strain, ε, shear stress, τ, and shear strain, γ (taking into account the length change during tension), were calculated as:

$$\sigma = \frac{4P}{\pi(D^2 - d^2)};\ \varepsilon = \frac{u}{L};\ \tau = \frac{16DM}{\pi(D^4 - d^4)};\ \gamma = \frac{\varphi D}{2(L+u)} = \frac{\varphi D}{2L(1+\varepsilon)} \quad (1)$$

For each of the cyclic loading modes, 2 samples were tested to fatigue failure ($R = 0.1$, frequency $v = 1$ Hz). Acoustic emission signals were recorded on one sample in each loading mode (Table 1). The maximum stress value during the cycle was equal to a half of the average maximum normal and shear stress values from the quasi-static tests.

To assess the mechanical properties reduction by changing the dynamic elasticity modulus, the peak maximum and minimum values of the load, torque, displacement, and torsion angle were recorded for every 1, 10, and 100 cycles. The values of the secant dynamic Young's modulus, E', and the shear modulus, G', were determined by the formulas:

$$E' = \frac{P_{max} - P_{min}}{u_{max} - u_{min}} \frac{4L}{\pi(D^2 - d^2)};\ G' = \frac{M_{max} - M_{min}}{\varphi_{max} - \varphi_{min}} \frac{32L}{\pi(D^4 - d^4)} \quad (2)$$

Table 1. Experimental program.

Sample Number	Experiment Type	Loading	VIC-3D	AMSY-6
1	Quasi-static	Tension	+	+
2			−	−
3			−	−
4		Torsion	+	+
5			−	−
6			−	−
7		Proportional 1	+	+
8			−	−
9		Proportional 2	+	+
10			−	−
11	Cyclic	Tension	−	+
12			−	−
13		Torsion	−	+
14			−	−
15		Proportional 1	−	+
16			−	−
17		Proportional 2	−	+
18			−	−

2.4. Approximation Models

According to the calculated dynamic elasticity moduli for the first loading cycle, which can be designated as E'_0 and G'_0, and durability for current loading mode, N_0, fatigue sensitivity coefficients, K_E and K_G, as well as the relative value of preliminary cyclic exposure, n, can be calculated for cycle number, N:

$$K_E = \frac{E'}{E'_0};\ K_G = \frac{G'}{G'_0};\ n = \frac{N}{N_0} \tag{3}$$

The approximation of experimentally obtained dependences, $K_E(n)$ and $K_G(n)$, was carried out by three models based on cumulative distribution function usage (Weibull law and beta distribution) [46] and that proposed by Mao and Mahadevan [52]. The approximating formulas, respectively, had the form:

$$K_E = 1 - \lambda(-\ln(1-n))^{\frac{1}{\kappa}};\ K_E = 1 - \frac{B_n(\alpha,\beta)}{B(\alpha,\beta)};\ K_E = 1 - qn^{m_1} - (1-q)n^{m_2} \tag{4}$$

The damage accumulation rate can be defined as the damage value function derivative ω'_E (or ω'_G):

$$\omega'_E = \frac{\lambda}{\kappa}(-\ln(1-n))^{\frac{1}{\kappa}-1}\frac{1}{1-n};\ \omega'_E = \frac{n^{\alpha-1}(1-n)^{\beta-1}}{B(\alpha,\beta)};\ \omega'_E = qm_1n^{m_1-1} + (1-q)m_2n^{m_2-1} \tag{5}$$

The model parameters were obtained numerically.

3. Results and Discussion

3.1. Static Loading

3.1.1. Loading Curves

Loading curves (Figure 2) were built using the data from the test machine. It was noted that under tension, the diagram was almost linear to the maximum load value, and after the peak, a sharp drop was observed. This behavior corresponds to an elastic–brittle fracture. During torsion, the diagram was non-linear; the presence of a stage similar to plastic was noted. In addition, after reaching the peak, in some cases, a gradual torque decrease was observed, which means a postcritical stage realization [66]. Similar non-linear behavior of polymer composites under off-axis loading was considered previously [67–69]. For proportional loading modes, both the maximum load and the maximum torque value decrease. A two-fold slower grip movement in the Proportional 1 loading mode (in comparison with the Proportional 2 loading mode) led to a change in the ratio between normal and shear stresses; therefore, a lower load value and higher torque value were observed. Material sensitivity to the complex stress–strain state can be concluded.

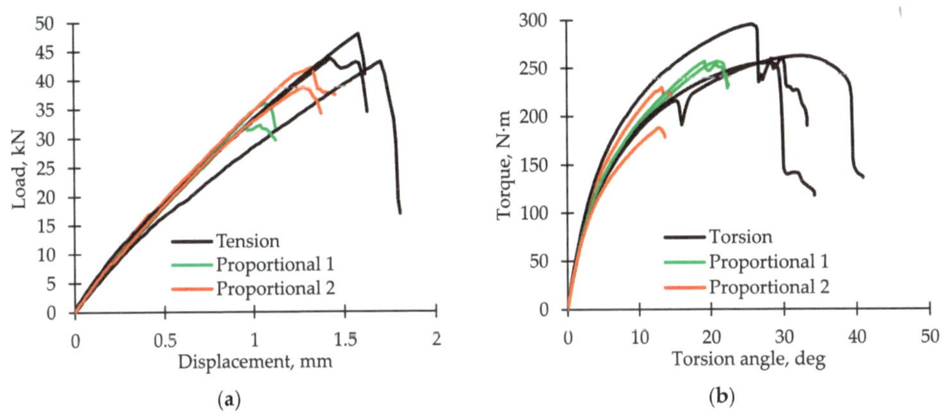

Figure 2. Loading curves. (**a**) Dependence of load on displacement. (**b**) Dependence of torque on torsion angle.

3.1.2. Failure Criterion

Maximum values of normal stress, σ_{max}, and shear stress, τ_{max}, were calculated in the coordinate system, which was associated with sample axis, from the maximum values of the load and torque using Formula (1) (Figure 3a). However, since pipes were obtained by continuous winding, and their characteristic types of destruction is matrix cracking across the tape reinforcement and violations of adhesion between the fibers and tapes, it is rational to recalculate the stresses into a coordinate system associated with the direction of winding [70]. The formulas (for winding angle $\chi = 85°$) are:

$$\sigma_{n\ max} = \sigma_{max} \sin^2 \chi + \tau_{max} \sin 2\chi; \ \tau_{n\ max} = \left| \frac{1}{2} \sigma_{max} \sin 2\chi + \tau_{max} \cos 2\chi \right| \quad (6)$$

If $\tau_{n\ max} = 0$, $\sigma_{n\ max}$ at the failure moment will be equal to the strength value denoted as $\sigma_{0n\ max}$. Likewise, if $\sigma_{n\ max} = 0$, $\tau_{n\ max}$ will be equal to $\tau_{0n\ max}$ strength value. The failure criterion (similar to the 2-dimensional Hashin failure criterion in matrix tension mode [71]) was proposed as:

$$\left(\frac{\sigma_{n\ max}}{\sigma_{0n\ max}} \right)^2 + \left(\frac{\tau_{n\ max}}{\tau_{0n\ max}} \right)^2 = 1 \quad (7)$$

The criterion parameters were obtained numerically by experimental data approximation and were: $\sigma_{0n\,max}$ = 243.4 MPa and $\tau_{0n\,max}$ = 107.7 MPa. A determination coefficient, R^2, was 0.808, which means a good descriptive capability of this model. The experimental points and line that corresponded to criteria (7) are shown in Figure 3b. The proposed criteria can be used to define the failure of composite pipes obtained by winding.

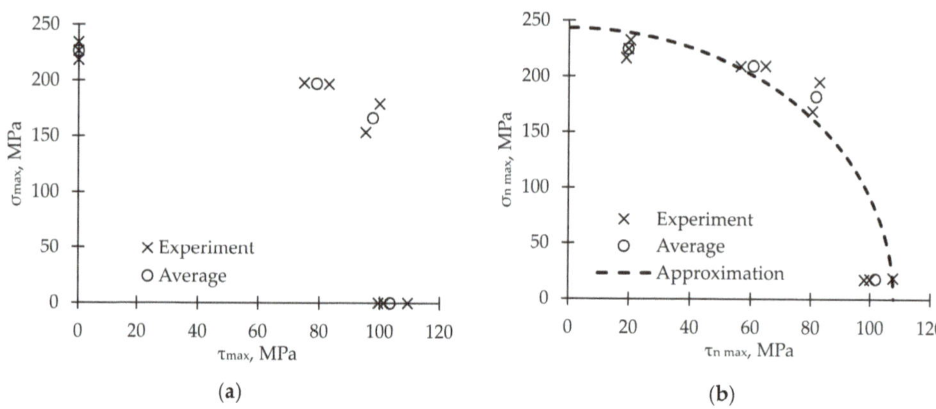

Figure 3. Maximum stress values. (**a**) In the global cylindrical coordinate system. (**b**) In a coordinate system associated with the winding angle.

3.1.3. Inhomogeneous Displacement and Strain Field Analysis

For a composite sample tested under tension, an analysis of the longitudinal strain, ε_{zz}, field evolution on the surface was carried out. As an example, shown in Figure 4a, fields were obtained for the points marked on the loading diagram in Figure 4b. A convenient analysis tool is a strain diagram along the line, *Lz*, as shown in Figure 4c. The evolution of these diagrams as the load increased was considered (Figure 4d). The plots corresponding to the states p1–p5 are highlighted in color, the other are shown for clarity. It was found out that the primary localization was formed at a load of approximately 30% of the maximum value, P_{max}, as evidenced by single bursts in longitudinal strain (peaks 3 and 5 in Figure 4d). As the loading progressed, bursts along the entire sample length were observed; peak 4 appeared at a load of 35% P_{max}, then peak 6 at a load of 36% P_{max}, and peaks 1 and 2 at a load of 38% P_{max}. The diagrams were similar and gradually increased in the peaks, while the minimum values were changing slightly. Fracture occurred in the region with the greatest strain localization. Six strain peaks were recorded, corresponding to the stripes depicted in Figure 4a. Since these stripes were directed along the winding direction, it can be assumed that the matrix between the outer reinforcement tapes was most deformed.

(**a**)

Figure 4. *Cont.*

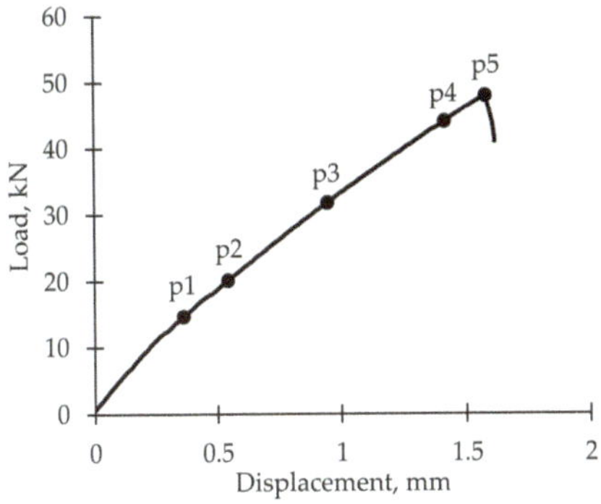

(b)

(c)

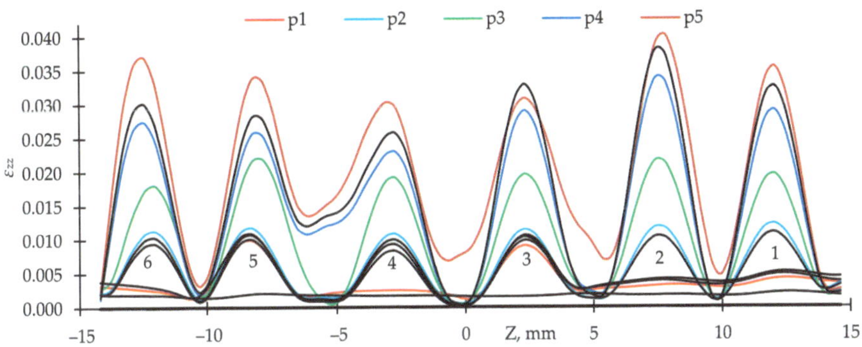

(d)

Figure 4. Inhomogeneous fields of longitudinal strain, ε_{zz}, analysis. (**a**) Fields for load values p1–p5. (**b**) Loading diagram. (**c**) The line to build the strain diagram. (**d**) Evolution of the strain diagram.

For other loading modes, using a non-contact optical video system, the fields of axial displacement, u_z, longitudinal strain, ε_{zz}, circumferential displacement, u_θ, and shear deformations, $\varepsilon_{\theta z}$, at the maximum torque value were obtained (Figure 5). The presence of significant localization was noted, such as for the tension loading mode; however, there is no obvious formation of stripes directed along the reinforcement. The symmetry of the displacement field was maintained during loading. The necessity of the video system usage in the pipes destruction analysis was conclude since it allows us to evaluate the dangerous zones of strain localization.

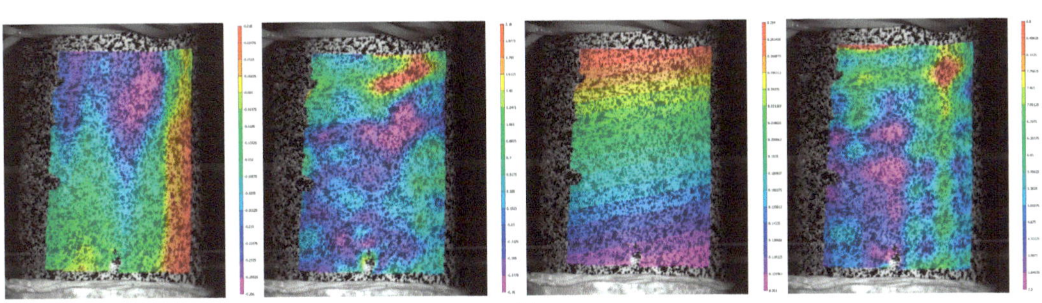

(a)

Figure 5. *Cont.*

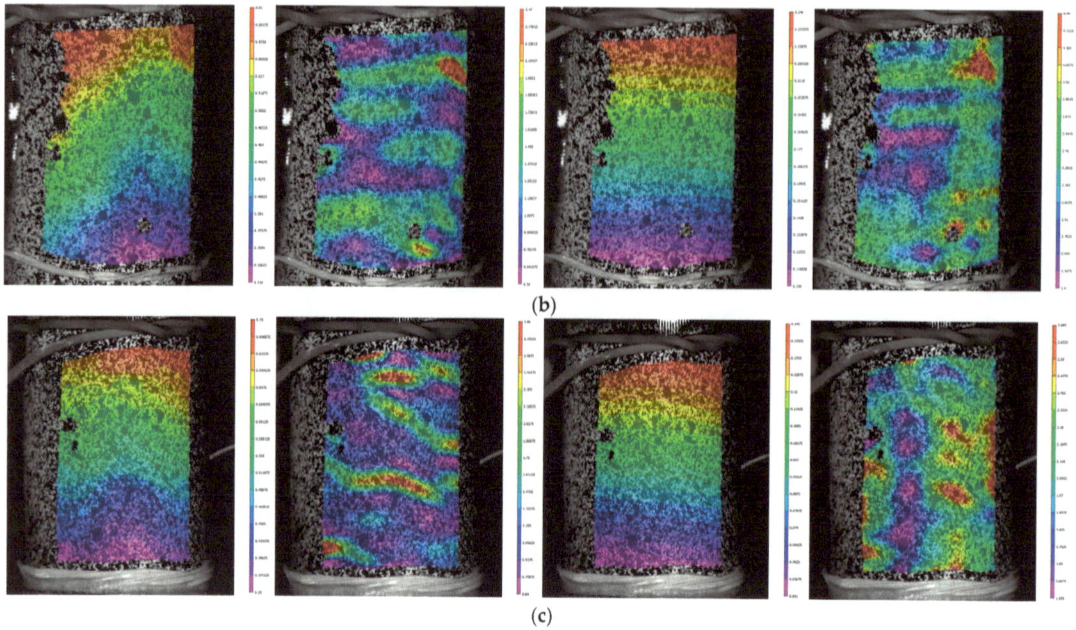

Figure 5. From left to right: fields of axial displacement, u_z, longitudinal strain, ε_{zz}, circumferential displacement, u_θ, shear strain, $\varepsilon_{\theta z}$, and loading modes. (**a**) Torsion. (**b**) Proportional 1. (**c**) Proportional 2.

Stress–strain curves were built according to the data from the video system (using the built-in module "virtual extensometer") and from the testing machine (Figure 6). The usage of data from the machine led to an underestimation of the Young's modulus by 1.5–1.65 times and the shear modulus by 2.1–2.25 times. However, this ratio is quite stable; therefore, it is possible to compare various loading modes of cyclic action on the change in dynamic elastic moduli.

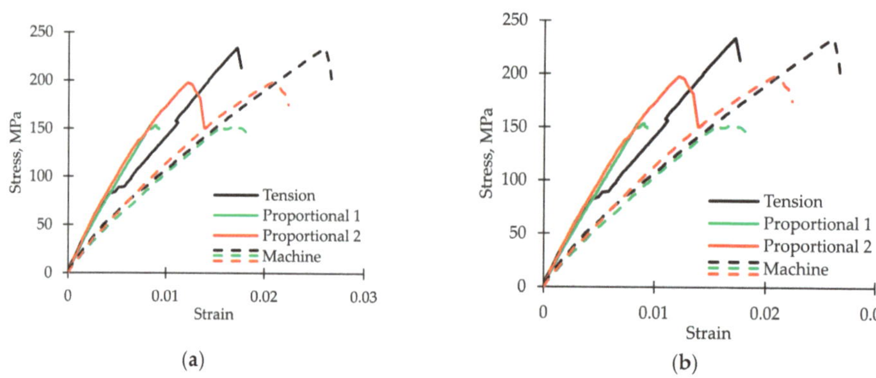

Figure 6. Comparison of stress–strain curves built using data from the video system and from the testing machine. (**a**) In tension. (**b**) In torsion.

It should be noted that in torsion tests, the material demonstrated non-linear behavior, so usage of Equation (2) in secant elastic moduli calculation seems to not be applicable [72]. However, we assumed that plastic strains were slightly influenced by the dynamic elasticity moduli, and Equation (2) still could be used in the stiffness degradation regularities. The influence of composites' non-linear behavior on stiffness and fatigue behavior will be considered in further studies.

3.1.4. Analysis of Acoustic Emission Signals

Figure 7 shows graphs of the AE signals spectral maximum frequency distribution for the fiberglass samples tested under static tension, torsion, and proportional tension with torsion. For all specimens, three frequency ranges were obtained throughout the test by sensors AE144A (No. 1, green points) and M31 (No. 2, blue points). Figure 8 represents AE signals frequency distributions. A significant part of the signals was located in the ranges up to 100 kHz (matrix cracking) and 250–300 kHz (violation of adhesion between matrix and fiber). A small number of signals were recorded in the range of 600–800 kHz (fiber breakage) by the M31 sensor. Similar frequency ranges were observed in other GFRP tests [73,74].

To assess the damage accumulation intensity, graphs of the complete cumulative energy (from sensor No. 1) and cumulative energy for frequency ranges of 1–100 kHz, 250–300 kHz, and 600–800 kHz dependences on displacement (or torsion angle) were constructed, combined with a loading diagram for all specimens (Figure 9). Various nature and rate of damage accumulations were noted for each loading mode. The complete cumulative energy value corresponded with the maximum normal stress value; an increase from $\approx 5 \times 10^8$ eu for pure torsion up to $\approx 150 \times 10^8$ eu for pure tension was observed. The cumulative energy curves for various frequency ranges had similar characteristics for the proportional loading modes. The greatest contribution to the AE signals complete cumulative energy in the torsion loading mode was made by the violations of adhesion (250–300 kHz range). In other loading modes, the greatest contribution was made by matrix cracking (up to 100 kHz range). In comparison, fiber breakage (600–800 kHz range) was almost absent in all of the loading modes. It can be concluded that the failure mechanism depends on the loading mode.

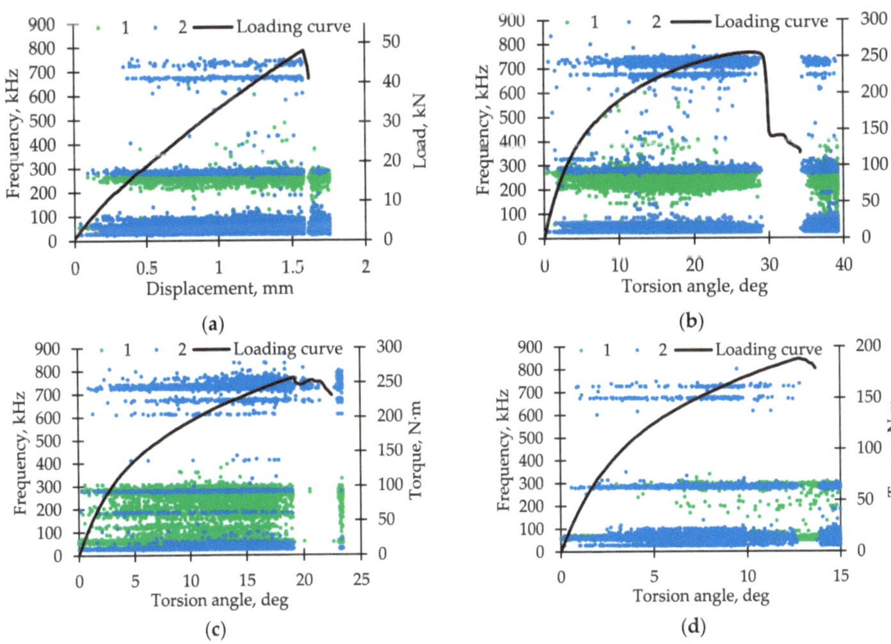

Figure 7. Distribution diagram of AE signal frequency parameters for loading modes. (**a**) Tension. (**b**) Torsion. (**c**) Proportional 1. (**d**) Proportional 2.

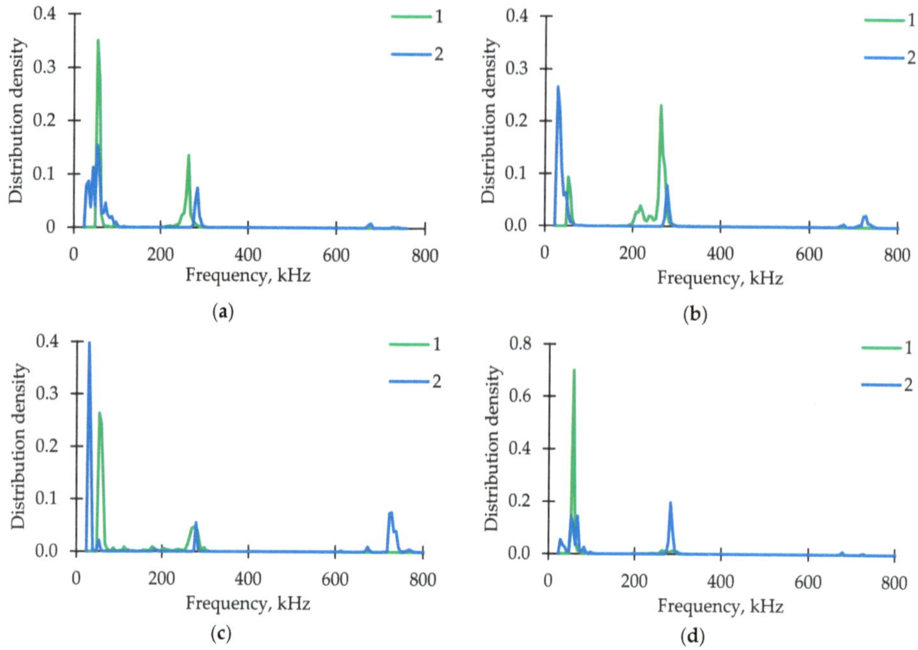

Figure 8. AE signal frequency distribution for loading modes. (**a**) Tension. (**b**) Torsion. (**c**) Proportional 1. (**d**) Proportional 2.

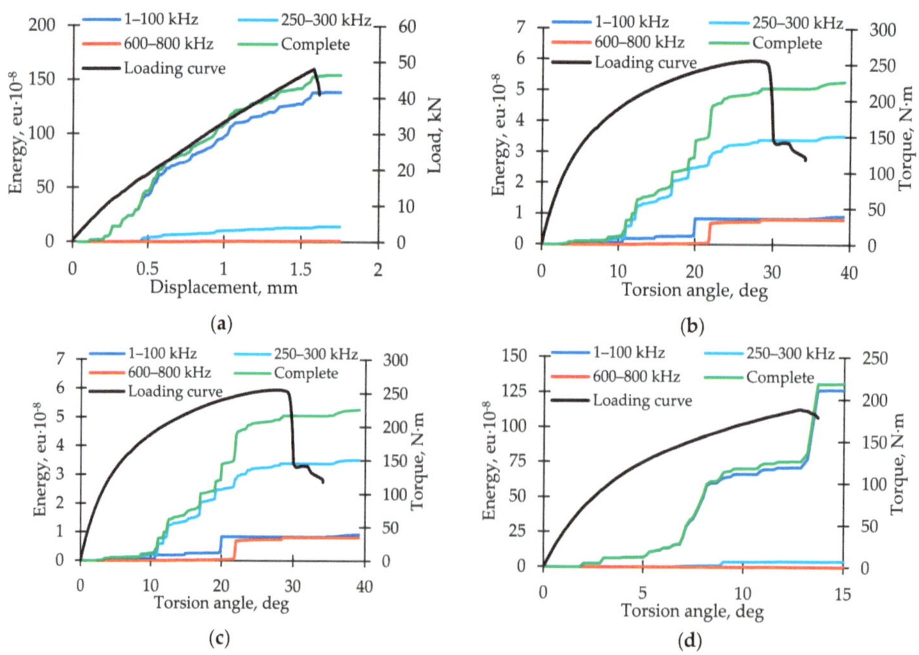

Figure 9. Cumulative energy for AE signals with various frequency ranges for loading modes. (**a**) Tension. (**b**) Torsion. (**c**) Proportional 1. (**d**) Proportional 2.

3.1.5. Surface Defects after Static Failure

The analysis of defects occurring on the sample surface after static failure was carried out (Figure 10). The formation of many type I cracks due to the destruction of the matrix inside the winding tape during the tension test was revealed, which corresponds with the data obtained by the acoustic emission system. During torsion, a lot of damage was obtained associated with a violation of adhesion between the matrix and the fibers, which led to the formation of type III shear cracks on the surface. A mixed-type crack was formed near the sample gripping part in the Proportional 1 loading mode due to matrix destruction and adhesion violation. Next, the destruction of the tape fibers (placed along the sample axis) occurred, which explains the peak in the 700–750 kHz frequency range of acoustic emission signals. In the Proportional 2 loading mode, mixed-type cracks and few destroyed fibers (along the winding angle and along the tube axis) were also observed. A good correspondence between acoustic emission signals analysis and optical microscopy was concluded.

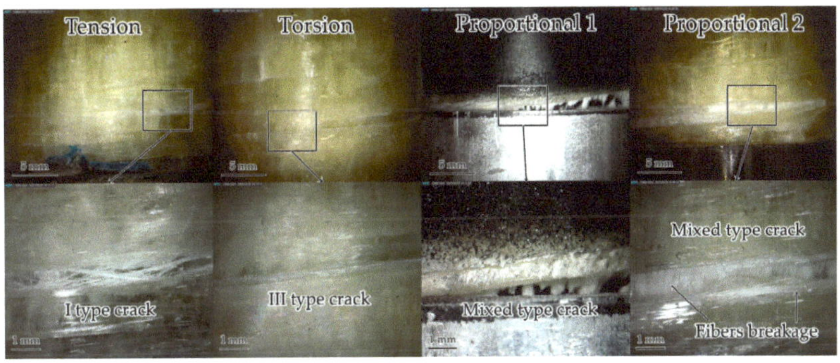

Figure 10. Samples after static failure.

3.2. Cyclic Loading

3.2.1. Fatigue Test Results

The results of the fatigue tests are shown in Table 2. The average durability values were as follows: for uniaxial tension N_0 = 40,255, for torsion N_0 = 29,972, for Proportional 1 mode N_0 = 4066, and for Proportional 2 loading mode N_0 = 16,024. In case of equal ratios of the stress tensor components maximum values in the cycle to their values during static tests, proportional loading led to a significant decrease in the composite material durability. Besides, the Proportional 1 mode was more dangerous than Proportional 2, where the normal stress was greater. A number of cycles less than 50,000 is explained by many structural defects; maximum stress values must be reduced to increase durability. In addition, significant variations in durability were obtained; therefore, additional tests are required. These data will be used in further investigations of the reduction in the strength properties of fiberglass tubes.

Table 2. Fatigue test results.

Sample Number	Loading	N_0, Cycles
11	Tension	49,681
12		30,828
13	Torsion	29,306
14		30,638

Table 2. *Cont.*

Sample Number	Loading	N_0, Cycles
15	Proportional 1	4274
16		3858
17	Proportional 2	11,523
18		20,525

3.2.2. Approximation and Analysis of Fatigue Sensitivity Curves

Figure 11 represents the experimentally obtained curves using Formula (3) dependences $K_E(n)$ and $K_G(n)$ and their approximations by models based on the Weibull distribution law (WL), beta distribution (BD), and that proposed by Mao and Mahadevan (MM) (given in formula (4)). In addition, to determine the boundaries of the damage accumulation stages, diagrams of the damage value growth rate, $\omega'_E(n)$ and $\omega'_G(n)$, determined by Formula (5), were plotted. For all approximation variants, high values of the determination coefficient were noted: for the WL model, R^2 was not lower than 0.957; for the BD model, it was not lower than 0.920; and for the MM model, it was not lower than 0.885. The high descriptive capability of approximations was concluded.

The dynamic shear modulus is sensitive to the preliminary cyclic exposure; it decreased down to 60–65% of the initial value before the onset of failure. However, this decrease was not sensitive to the stress–strain state. In addition, the sharpest drop at the initiation stage was also observed for the shear modulus in all tests (up to 90% of the initial value). On the contrary, the dynamic Young's modulus decrease during the cyclic loading was not so sharp: down to 80% from the initial value before the failure. This dissimilarity in elastic moduli reduction can be explained, firstly, by the presence of fibers oriented along the sample axis, and secondly, by weaker adhesion between the fiber and the matrix inside the tape.

The stabilization stage predicted by the models was quite long; its length ranged from 0.6–0.7 for the shear modulus up to 0.9 for the Young's modulus of the material. The WL and BD models gave close experimental data approximations (WL model slightly shifted the start of the stabilization stage to the right). The MM model described the experimental data well; however, it had a significant disadvantage as it strongly shifted the start of the stabilization stage to the right, but at the same time, predicted a sharp increase in the damage accumulation rate at the aggravation stage. This can lead to an over-prediction of the material elastic properties at the end of the stabilization stage. In this regard, usage of models based on the cumulative probability distribution functions is preferred.

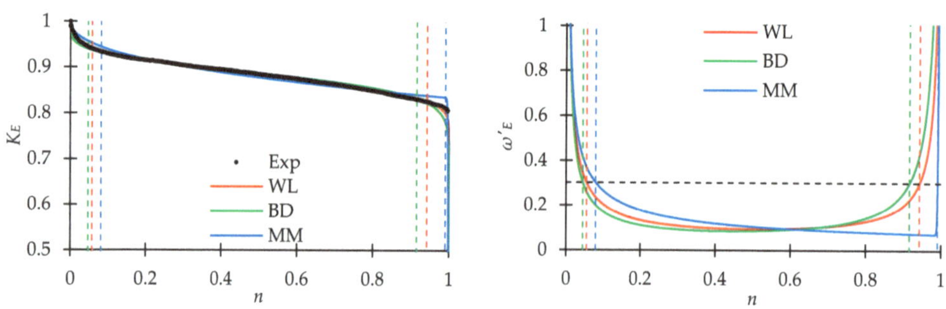

(a)

Figure 11. *Cont.*

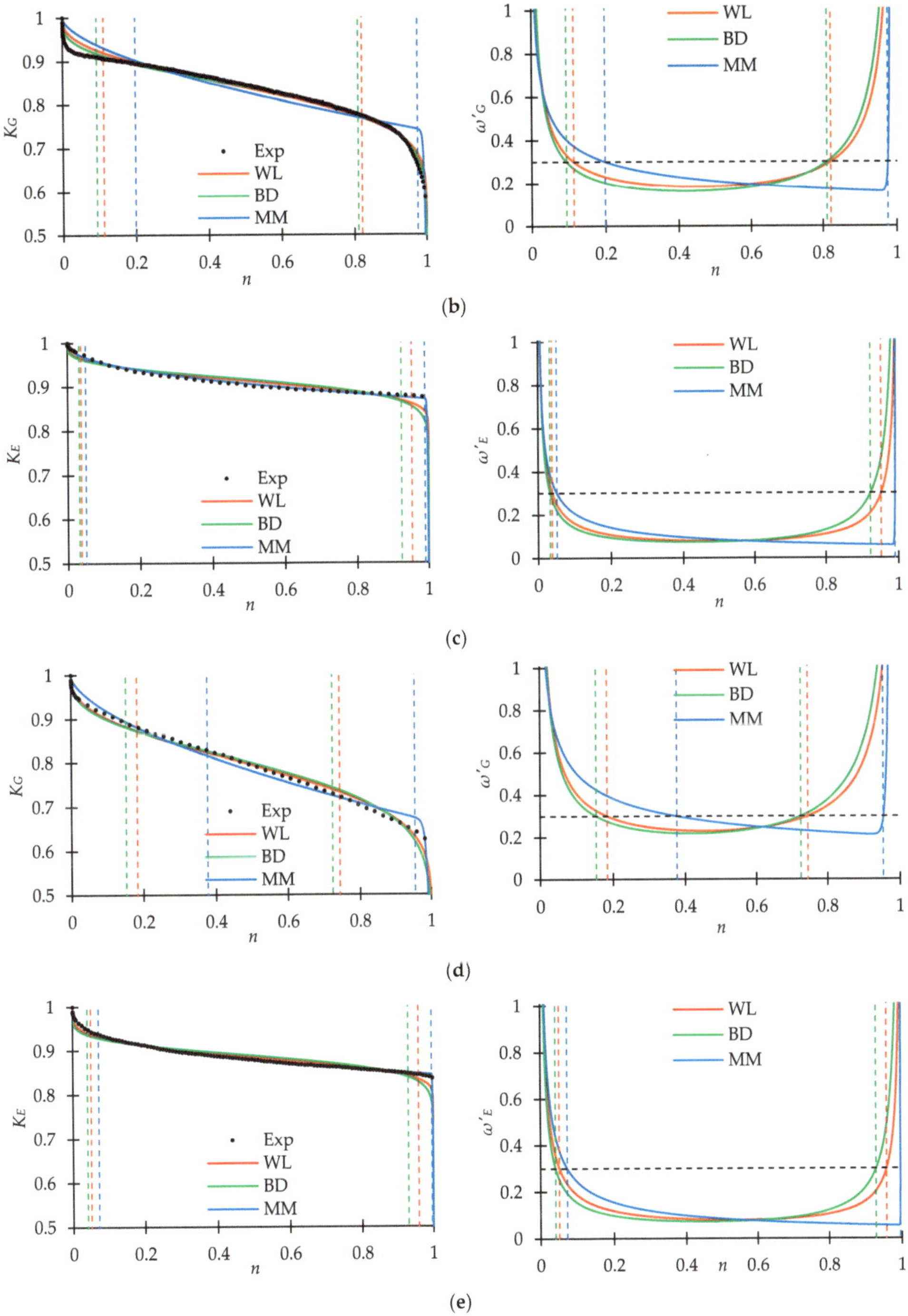

Figure 11. *Cont.*

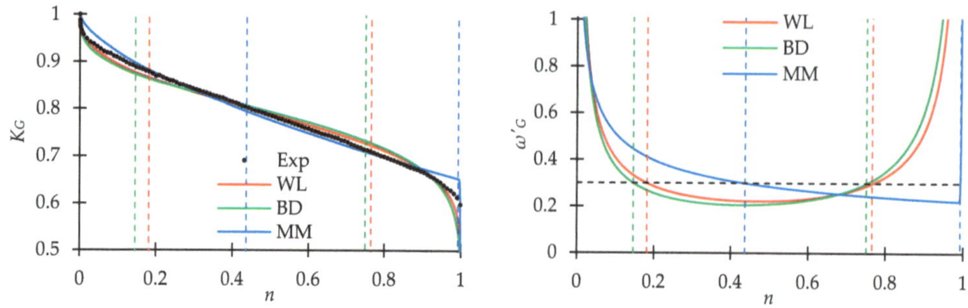

(f)

Figure 11. Fatigue sensitivity curves (left side) and damage value function derivatives (right side) for the dynamic elasticity modulus and loading mode. (**a**) Young's modulus, Tension. (**b**) Shear modulus, Torsion. (**c**) Young's modulus, Proportional 1. (**d**) Shear modulus, Proportional 1. (**e**) Young's modulus, Proportional 2. (**f**) Shear modulus, Proportional 2.

The ratio between the maximum values of normal and shear stresses in a cycle was expressed through the parameter η, which lay in the range from 0° to 90° and was determined by the formula:

$$\eta = \arctan\left(\frac{\sigma_{\max}}{\tau_{\max}}\right) \tag{8}$$

The models' parameter dependences on the η value are shown in Figure 12 (the parameter m_2 of the MM model in all cases was a large number, and its change by 10 or more times slightly affects the shape of the fatigue sensitivity curve, so it was not shown in the figure). For the same loading mode on different samples, very close parameters values were obtained. Hence, identified patterns were reliable. The analysis shows that in some cases, the models' parameters slightly depended on the stress state (κ_G, α_G, β_E, m_{1G}). On the other hand, several dependences were not only nonlinear, but also nonmonotonic (λ_G, β_G). Therefore, it is necessary to carry out additional research of the parameters dependences on the ratio between the components of the stress tensor ($\kappa(\eta)$, etc.).

Figure 12 also represents the dependences of the average rates of dynamic elasticity moduli reduction at the stabilization stage, ψ [46], on the parameter η. All models gave approximately the same value of this parameter: $\psi \approx 0.1$ for the Young's modulus and $\psi \approx 0.2$ for the shear modulus. It seems convenient to use the average rate of dynamic elasticity moduli reduction in composite structure design to estimate the mechanical properties during fatigue damage accumulation.

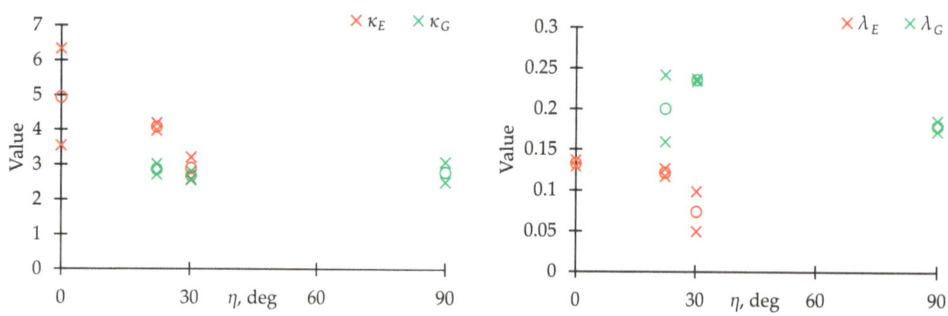

(a) (b)

Figure 12. *Cont.*

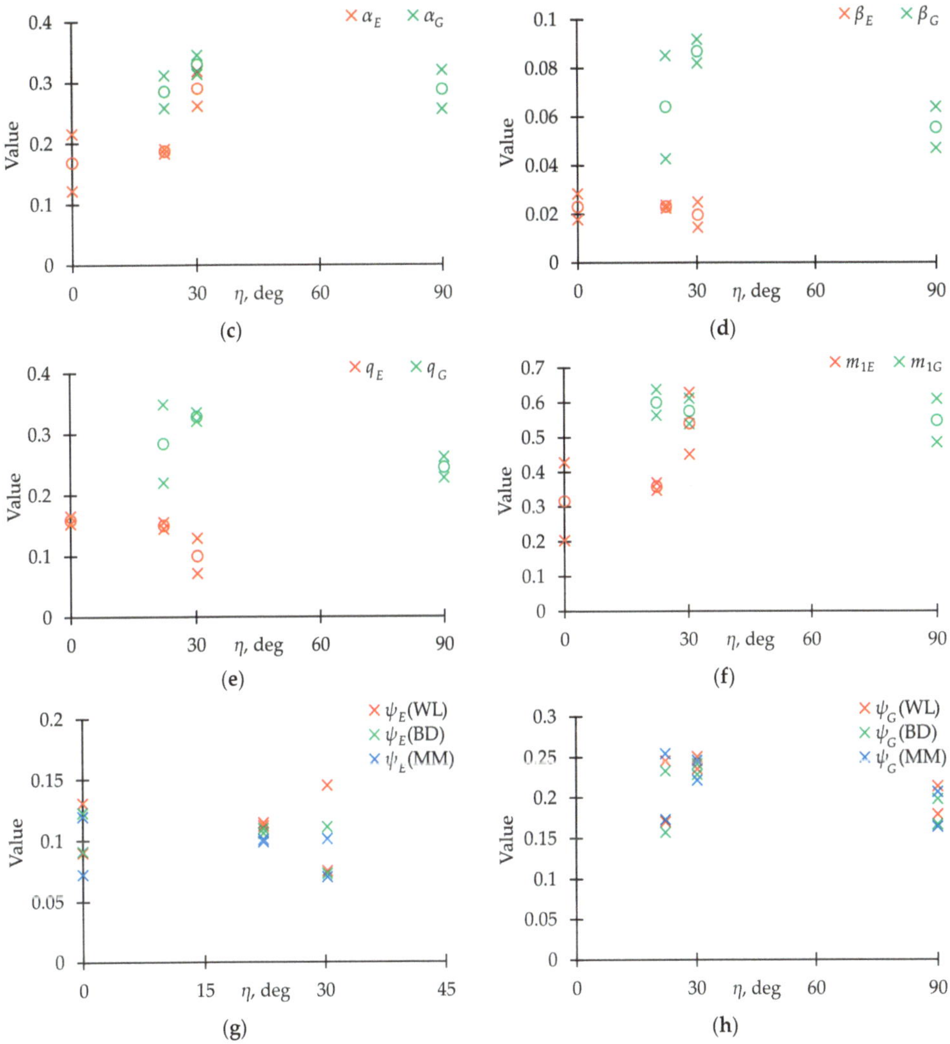

Figure 12. Dependence of approximation models' parameters on η. (**a–f**) Corresponds to κ, λ, α, β, q, m_1. (**g**) Dependence of the average rate of dynamic Young's modulus reduction at the stabilization stage, ψ, on parameter η. (**h**) The same dependence for dynamic shear modulus. ×—experimental values, ○—average values.

3.2.3. Analysis of Acoustic Emission Signals

Figure 13 represents the signal frequency spectral maximum distribution for fiberglass specimens tested under cyclic tension, torsion, and proportional tension with torsion. Only the AE144A sensor (100–500 kHz) was used. Two typical frequency ranges were observed: up to 220 kHz (corresponds to matrix cracking) and 220–300 kHz (violations of adhesion). In comparison with static loading, frequency distributions were smooth. The largest number of signals was in the range of 220–300 kHz for each loading mode. Figure 14 shows dependences of the total cumulative energy (complete and for two described frequency ranges) on preliminary cyclic exposure. The maximum complete cumulative energy was observed for specimens tested under cyclic tension. In other cases, the maximum complete cumulative energy was similar. In addition, under proportional loading, contributions of matrix cracking and violations of adhesion to the AE signals complete cumulative energy were practically equal. The stage-by-stage nature of damage accumulation was noted. During pure tension and torsion cyclic loading, the cumulative energy growth rate was gradually increasing. On the other hand, during proportional loading, the cumulative energy growth rate was decreasing up until failure. However, transition from one stage to another was observed in all cases after preliminary cyclic exposure ≈ 0.25 N_0. It seems appropriate to conduct further studies to establish the relationship between the damage accumulation stages and acoustic emission signals.

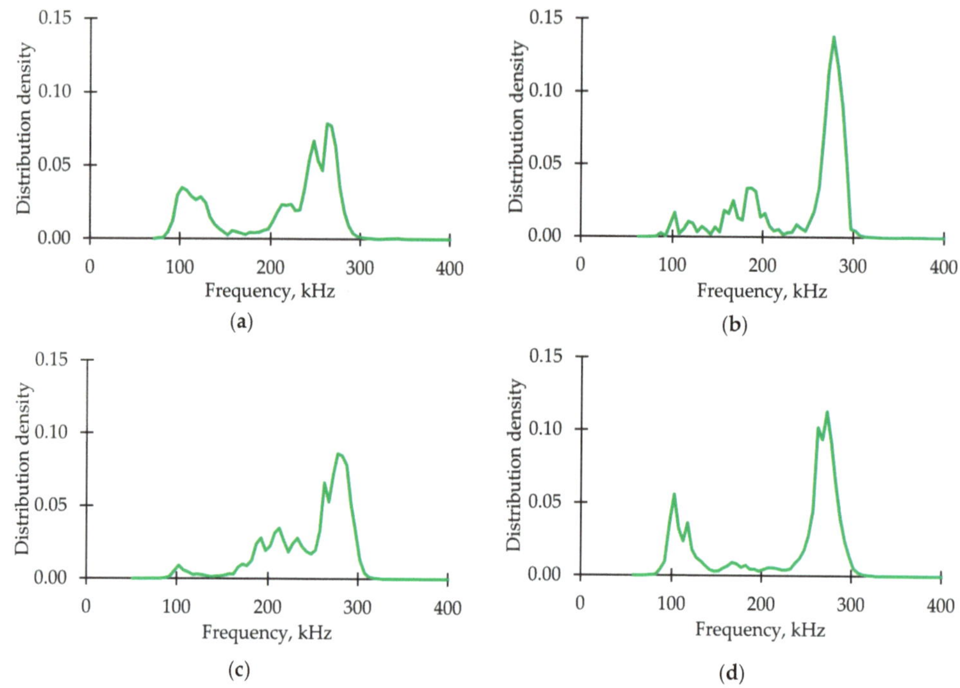

Figure 13. AE signals frequencies distribution for loading modes. (**a**) Tension. (**b**) Torsion. (**c**) Proportional 1. (**d**) Proportional 2.

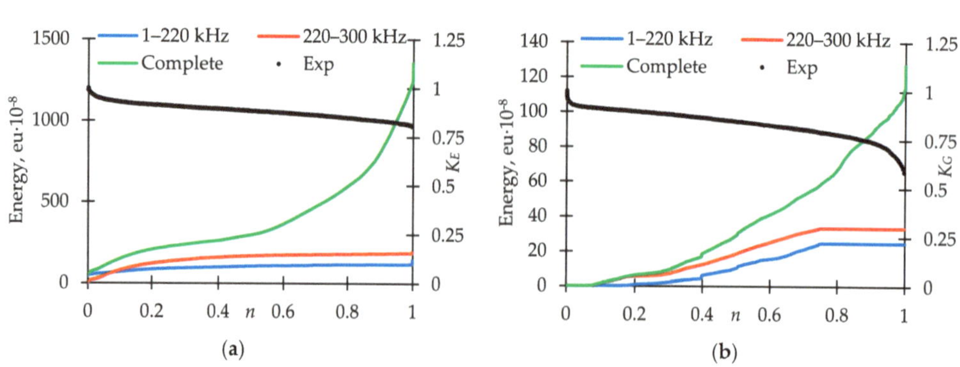

Figure 14. *Cont.*

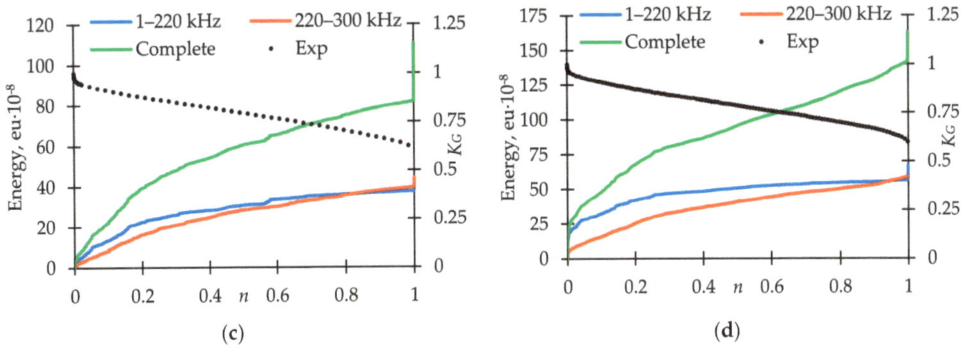

Figure 14. Cumulative energy for AE signals with various frequency ranges for loading modes. (**a**) Tension. (**b**) Torsion. (**c**) Proportional 1. (**d**) Proportional 2.

3.2.4. Surface Defects after Fatigue Failure

The analysis of surface defects was carried out (Figure 15). In all loading modes, the destruction occurred in the working part. The formation of multiple type I cracks during cyclic tension was revealed; it corresponded to the static failure. During cyclic torsion, a lot of adhesion violations occurred between the matrix and the reinforcing fiber (type III shear cracks). In addition, winding tapes detached one from another and broke. In the Proportional 1 loading mode, mixed type cracks were observed due to the matrix destruction and the violation of adhesion; the winding tapes also detached and broke. During the Proportional 2 cyclic loading, mixed type cracks occurred. Besides, fibers, which were placed along the sample axis, were destructed. The data on surface damage was in good agreement with the acoustic emission signals. Therefore, the acoustic emission system and optical microscopy should be used for further investigations of composites' mechanical behavior under preliminary cyclic exposure.

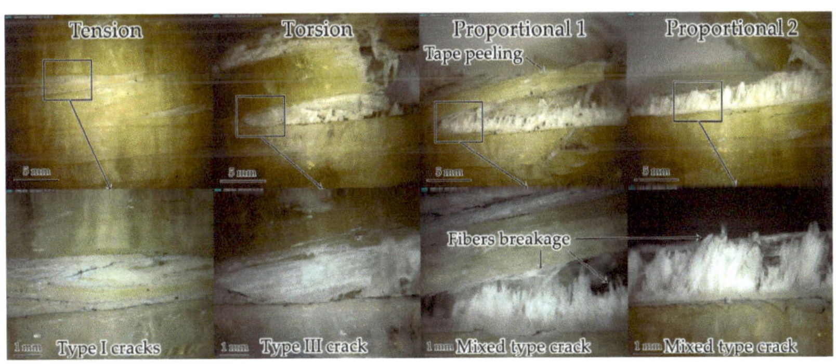

Figure 15. Samples after fatigue failure.

4. Conclusions

This paper presents the results of the investigation of the mechanical behavior of thin-walled fiberglass tubes under proportional multiaxial cyclic loading. The sensitivity to the complex stress–strain state and non-linear behavior were observed. The inhomogeneous displacement and strain fields were analyzed using a non-contact optical video system. The necessity of video system usage in defining the mechanical properties of the composites was concluded. Fatigue tests were carried out under loading modes similar to those in static tests. Dependences of the fatigue sensitivity curves approximation models' parameters on the stress–strain state were revealed. We recommend using previously proposed models [46], which are based on cumulative distribution functions, to calculate composites' residual mechanical properties. Sharing the optical microscope and acoustic emission system established a connection between the occurring defects and acoustic response signals. The stage-by-stage nature of damage accumulation was noted. Further investigations will be aimed at defining the connection between damage accumulation stages and occurring defects.

Based on the above, we conclude the necessity and rationality of the fatigue sensitivity investigation of composites under multiaxial cyclic loading; it must be taken into account in composite structure design for accurate life prediction.

Author Contributions: Conceptualization, V.W., O.S. and A.M.; methodology, O.S. and E.S.; software, E.S. and O.S.; validation, V.W.; formal analysis, A.M. and O.S.; investigation, A.M., E.S., O.S. and A.K.; resources, O.S. and A.K.; data curation, A.M.; writing—original draft preparation, O.S., A.M., E.S. and A.K.; writing—review and editing, V.W., O.S. and A.M.; visualization, A.M.; supervision, V.W.; project administration, O.S.; funding acquisition, O.S. All authors have read and agreed to the published version of the manuscript.

Funding: This research was carried out with the support of the Russian Science Foundation (Project No 22-79-00136, https://rscf.ru/en/project/22-79-00136/).

Institutional Review Board Statement: Not applicable.

Data Availability Statement: Not applicable.

Acknowledgments: The authors are grateful to Ekaterina M. Lunegova (Perm National Research Polytechnic University) for the scientific support.

Conflicts of Interest: The authors declare no conflict of interest.

References

1. Geier, N.; Patra, K.; Anand, R.; Ashworth, S.; Balázs, B.Z.; Lukács, T.; Magyar, G.; Tamás-Bényei, P.; Xu, J.; Davim, J.P. A critical review on mechanical micro-drilling of glass and carbon fibre reinforced polymer (GFRP and CFRP) composites. *Compos. Part B Eng.* **2023**, *254*, 110589. [CrossRef]
2. Gao, H.; Sun, Y.; Jian, J.; Dong, Y.; Liu, H. Study on mechanical properties and application in communication pole line engineering of glass fiber reinforced polyurethane composites (GFRP). *Case Stud. Constr. Mater.* **2023**, *18*, e01942. [CrossRef]
3. Liu, Y.; Zhang, H.; Tafsirojjaman, T.; Dogar, A.R.; AlAjarmeh, O.; Yue, Q.; Manalo, A. A novel technique to improve the compressive strength and ductility of glass fiber reinforced polymer (GFRP) composite bars. *Constr. Build. Mater.* **2022**, *326*, 126782. [CrossRef]
4. Sarfraz, M.S.; Hong, H.; Kim, S.S. Recent developments in the manufacturing technologies of composite components and their cost-effectiveness in the automotive industry: A review study. *Compos. Struct.* **2021**, *266*, 113864. [CrossRef]
5. Collinson, M.G.; Bower, M.P.; Swait, T.J.; Atkins, C.P.; Hayes, S.A.; Nuhiji, B. Novel composite curing methods for sustainable manufacture: A review. *Compos. Part C Open Access* **2022**, *9*, 100293. [CrossRef]
6. Hunt, C.J.; Morabito, F.; Grace, C.; Zhao, Y.; Woods, B.K. A review of composite lattice structures. *Compos. Struct.* **2022**, *284*, 115120. [CrossRef]
7. Cheng, X.; Gong, Y.; Liu, Y.; Wu, Z.; Hu, X. Prediction of residual mechanical properties in flexure-after-impact of woven composite beams through electrical resistance measurement. *Compos. Struct.* **2020**, *240*, 112066. [CrossRef]
8. Shim, V.P.W.; Yang, L.M. Characterization of the residual mechanical properties of woven fabric reinforced composites after low-velocity impact. *Int. J. Mech. Sci.* **2005**, *47*, 647–665. [CrossRef]
9. Whitworth, H.A. Evaluation of the residual strength degradation in composite laminates under fatigue loading. *Compos. Struct.* **2000**, *48*, 261–264. [CrossRef]
10. Staroverov, O.; Lobanov, D. Estimation of residual fatigue life of polymer composites after preliminary low-velocity impact. *Proc. Struct. Integr.* **2022**, *37*, 804–810. [CrossRef]
11. Mirbagheri, M.; Rahmani, O.; Mirbagheri, Y. Estimation of residual tensile strength of composite laminate after low-velocity impact using visually inspection. *Eng. Fail. Anal.* **2022**, *131*, 105898. [CrossRef]
12. Staroverov, O.; Lobanov, D.; Strungar, E.; Lunegova, E. Evaluation of the influence of preliminary low-velocity impacts on the residual fatigue life of CFRP composites. *Int. J. Struct. Integr.* **2023**, *14*, 44–56. [CrossRef]
13. Vassilopoulos, A.P. *Fatigue Life Prediction of Composites and Composite Structures*, 2nd ed.; Woodhead Publishing: Sawston, UK, 2019.
14. Kaminski, M.; Laurin, F.; Maire, F.J.; Rakotoarisoa, C.; Hemon, E. Fatigue damage modeling of composite structures: The ONERA viewpoint. *AerospaceLab* **2015**, *9*, 1–12. [CrossRef]

15. Philippidis, T.P.; Passipoularidis, V.A. Residual strength after fatigue in composites: Theory vs. experiment. *Int. J. Fatigue* **2007**, *29*, 2104–2116. [CrossRef]
16. Shiri, S.; Yazdani, M.; Pourgol-Mohammad, M. A fatigue damage accumulation model based on stiffness degradation of composite materials. *Mater. Des.* **2015**, *88*, 1290–1295. [CrossRef]
17. Khoshmanesh, S.; Watson, S.J.; Zarouchas, D. The effect of the fatigue damage accumulation process on the damping and stiffness properties of adhesively bonded composite structures. *Compos. Struct.* **2022**, *287*, 115328. [CrossRef]
18. Gao, J.; Zhu, P.; Yuan, Y.; Wu, Z.; Xu, R. Strength and stiffness degradation modeling and fatigue life prediction of composite materials based on a unified fatigue damage model. *Eng. Fail. Anal.* **2022**, *137*, 106290. [CrossRef]
19. Vassilopoulos, A.P. The history of fiber-reinforced polymer composite laminate fatigue. *Int. J. Fatigue* **2020**, *134*, 105512. [CrossRef]
20. Liu, T.; Shi, X.; Zhang, J.; Fei, B. Crack initiation and propagation of 30CrMnSiA steel under uniaxial and multiaxial cyclic loading. *Int. J. Fatigue* **2019**, *122*, 240–255. [CrossRef]
21. Wang, Q.; Xin, C.; Sun, Q.; Xiao, L.; Sun, J. Biaxial fatigue behavior of gradient structural purity titanium under in-phase and out-of-phase loading. *Int. J. Fatigue* **2018**, *116*, 602–609. [CrossRef]
22. Zhang, J.; Shi, X.; Fei, B. High cycle fatigue and fracture mode analysis of 2A12-T4 aluminum alloy under out-of-phase axial-torsion constant amplitude loading. *Int. J. Fatigue* **2012**, *38*, 144–154. [CrossRef]
23. Wang, Y.; Yao, W. A multiaxial fatigue criterion for various metallic materials under proportional and nonproportional loading. *Int. J. Fatigue* **2006**, *28*, 401–408. [CrossRef]
24. Pejkowski, L.; Skibicki, D.; Seyda, J. Stress-strain response and fatigue life of a material subjected to asynchronous loadings. *AIP Conf. Proc.* **2018**, *2028*, 020016. [CrossRef]
25. Skibicki, D.; Pejkowski, L. Low-cycle multiaxial fatigue behaviour and fatigue life prediction for CuZn37 brass using the stress-strain models. *Int. J. Fatigue* **2017**, *102*, 18–36. [CrossRef]
26. Gates, N.R.; Fatemi, A. On the consideration of normal and shear stress interaction in multiaxial fatigue damage analysis. *Int. J. Fatigue* **2017**, *100*, 322–336. [CrossRef]
27. Gu, A.; Luo, Y.; Xu, B. Continuous condition monitoring of reinforced concrete using an active diagnosis method. *Struct. Health Monit.* **2016**, *15*, 104–111. [CrossRef]
28. Papuga, J.; Halama, R. Mean stress effect in multiaxial fatigue limit criteria. *Arch. Appl. Mech.* **2018**, *89*, 823–834. [CrossRef]
29. Sines, G. *Failure of Materials under Combined Repeated Stresses with Superimposed Static Stress*; National Advisory Committee for Aeronautics: Washington, DC, USA, 1955.
30. Mitukiewicz, G.; Głogowski, M.; Stelmach, J.; Leyko, J.; Dimitrova, Z.; Batory, D. Strengthening of cruciform sample arms for large strains during biaxial stretching. *Mater. Today Commun.* **2019**, *21*, 100692. [CrossRef]
31. Serna Moreno, M.C.; Muñoz, S.H. Elastic stability in biaxial testing with cruciform specimens subjected to compressive loading. *Compos. Struct.* **2020**, *234*, 111697. [CrossRef]
32. Mateen, M.A.; Ravi Shankar, D.V.; Hussain, M.M. A Review on Cruciform Shaped FRP Composites Laminates under Biaxial Testing. *Mater. Today Proc.* **2018**, *5*, 20837–20841. [CrossRef]
33. Fang, P.; Xu, Y.; Gao, Y.; Ali, L.; Bai, Y. Mechanical responses of a fiberglass flexible pipe subject to tension & internal pressure. *Thin-Walled Struct.* **2022**, *181*, 110107. [CrossRef]
34. Lee, P.A.; Kim, S.; Stakenborghs, B.; Suh, Y.; Choi, S. Development of hydro-axial tension method for whole pipe butt-fusion joint tensile test. *Polym. Test.* **2022**, *109*, 107553. [CrossRef]
35. Du, F.; Wang, J.J.; Tan, T. Using Spiral Notch Torsional Test to Investigate Fracture of Unidirectional Fiber Reinforced Composites. In *Reference Module in Materials Science and Materials Engineering*; Elsevier: Amsterdam, The Netherlands, 2023. [CrossRef]
36. Capela, C.; Ferreira, J.A.M.; Febra, T.; Costa, J.D. Fatigue strength of tubular carbon fibre composites under bending/torsion loading. *Int. J. Fatigue* **2015**, *70*, 216–222. [CrossRef]
37. Wang, Y.; Wang, Y.; Hou, C.; Zhou, X.; Deng, R.; Lan, Y.; Luo, W.; Kong, W. Combined compression-bending-torsion behaviour of CFST columns confined by CFRP for marine structures. *Compos. Struct.* **2020**, *242*, 112181. [CrossRef]
38. Demiral, M.; Abbassi, F.; Muhammad, R.; Akpinar, S. Service Life Modelling of Single Lap Joint Subjected to Cyclic Bending Load. *Aerospace* **2023**, *10*, 8. [CrossRef]
39. Armanfard, A.; Melenka, G.W. Experimental evaluation of carbon fibre, fibreglass and aramid tubular braided composites under combined tension–torsion loading. *Compos. Struct.* **2021**, *269*, 114049. [CrossRef]
40. Chang, Y.; Wen, W.; Xu, Y.; Cui, H.; Xu, Y. Quasi-static mechanical behavior of filament wound composite thin-walled tubes: Tension, torsion, and multi-axial loading. *Thin-Walled Struct.* **2022**, *177*, 109361. [CrossRef]
41. Tableau, N.; Aboura, Z.; Khellil, K.; Laurin, F.; Schneider, J. Multiaxial loading on a 3D woven carbon fiber reinforced plastic composite using tensile-torsion tests: Identification of the first damage envelope and associated damage mechanisms. *Compos. Struct.* **2019**, *227*, 111305. [CrossRef]
42. Hinton, M.; Kaddour, A. Triaxial test results for fibre-reinforced composites: The Second World-Wide Failure Exercise benchmark data. *J. Compos. Mater.* **2012**, *47*, 653–678. [CrossRef]
43. Kawakami, H.; Fujii, T.J.; Morita, Y. Fatigue Degradation and Life Prediction of Glass Fabric Polymer Composite under Tension/Torsion Biaxial Loadings. *J. Reinf. Plast. Compos.* **1996**, *15*, 183–195. [CrossRef]
44. Quaresimin, M.; Susmel, L.; Talreja, R. Fatigue behavior and live assessment of composite laminates under multiaxial loadings. *Int. J. Fatigue* **2010**, *32*, 2–16. [CrossRef]

45. Wil'deman, V.E.; Staroverov, O.A.; Lobanov, D.S. Diagram and parameters of fatigue sensitivity for evaluating the residual strength of layered GFRP composites after preliminary cyclic loadings. *Mech. Compos. Mater.* **2018**, *54*, 313–320. [CrossRef]
46. Wildemann, V.E.; Staroverov, O.A.; Yankin, A.S.; Mugatarov, A.I. Description of fatigue sensitivity curves and transition to critical states of polymer composites by cumulative distribution functions. *Frat. Integrita Strutt.* **2023**, *17*, 91–99. [CrossRef]
47. Wu, Z.; Fang, G.; Fu, M.; Chen, X.; Liang, J.; Lv, D. Random fatigue damage accumulation analysis of composite thin-wall structures based on residual stiffness method. *Compos. Struct.* **2019**, *211*, 546–556. [CrossRef]
48. Lu, W.; Gao, Z.; Adluru, H.K.; Hoos, K.H.; Seneviratne, W.P.; Mollenhauer, D.H.; Iarve, E.V. Fatigue damage modeling in laminated composite by using Rx-FEM and strength tracking method. *Compos. Part A Appl. Sci. Manuf.* **2022**, *163*, 107199. [CrossRef]
49. Guo, R.; Li, C.; Niu, Y.; Xian, G. The fatigue performances of carbon fiber reinforced polymer composites—A review. *J. Mater. Res. Technol.* **2022**, *21*, 4773–4789. [CrossRef]
50. Joosten, M.W.; Dávila, C.G.; Yang, Q. Predicting fatigue damage in composites subjected to general loading conditions. *Compos. Part A Appl. Sci. Manuf.* **2022**, *156*, 106862. [CrossRef]
51. Yadav, I.N.; Thapa, K.B. Strain-based theoretical fatigue damage model of woven glass-epoxy fabric composite material. *Compos. Part C Open Access* **2020**, *3*, 100067. [CrossRef]
52. Mao, H.; Mahadevan, S. Fatigue damage modelling of composite materials. *Compos. Struct.* **2002**, *58*, 405–410. [CrossRef]
53. Gu, Y.; Zhang, D.; Zhang, Z.; Sun, J.; Yue, S.; Li, G.; Qian, K. Torsion damage mechanisms analysis of two-dimensional braided composite tubes with digital image correction and X-ray micro-computed tomography. *Compos. Struct.* **2021**, *256*, 113020. [CrossRef]
54. Xu, D.H.; Cerbu, C.; Wang, H.W.; Rosca, I.C. Analysis of the hybrid composite materials reinforced with natural fibers considering digital image correlation (DIC) measurements. *Mech. Mater.* **2019**, *135*, 46–56. [CrossRef]
55. Azadi, M.; Saeedi, M.; Mokhtarishirazabad, M.; Lopez-Crespo, P. Effects of loading rate on crack growth behavior in carbon fiber reinforced polymer composites using digital image correlation technique. *Compos. Part B Eng.* **2019**, *175*, 107161. [CrossRef]
56. Wildemann, V.V.; Tretyakova, T.V.; Strungar, E.M.; Tretyakov, M.P. Deformation and failure of carbon fiber composite specimens with embedded defects during tension-torsion test. *Frat. Integrita Strutt.* **2018**, *46*, 295–305. [CrossRef]
57. Pankow, M.; Justusson, B.; Riosbaas, M.; Waas, A.M.; Yen, C.F. Effect of fiber architecture on tensile fracture of 3D woven textile composites. *Compos. Struct.* **2019**, *225*, 111139. [CrossRef]
58. Kovács, L.; Romhány, G. Derivation of ply specific stiffness parameters of fiber reinforced polymer laminates via inverse solution of classical laminate theory. *Period Polytech Mech. Eng.* **2018**, *62*, 158–164. [CrossRef]
59. Strungar, E.; Lobanov, D.; Wildemann, V. Evaluation of the Sensitivity of Various Reinforcement Patterns for Structural Carbon Fibers to Open Holes during Tensile Tests. *Polymers* **2021**, *13*, 4287. [CrossRef]
60. Harizi, W.; Chaki, S.; Bourse, G.; Ourak, M. Damage mechanisms assessment of Glass Fiber-Reinforced Polymer (GFRP) composites using multivariable analysis methods applied to acoustic emission data. *Compos. Struct.* **2022**, *289*, 115470. [CrossRef]
61. Rishikesan, V.; Bhagyesh, C.; Arunachalam, N. Characterisation of drilling-induced damage in GFRP Honeycomb Sandwich Composites using Acoustic Emission. *Procedia Manuf.* **2021**, *53*, 664–672. [CrossRef]
62. Kishore Kumar, P.; Kuppan, P. Online monitoring of delamination mechanisms in drilling of Mwcnts reinforced Gfrp nanocomposites by acoustic emission. *Mater. Today Proc.* **2018**, *5*, 13036–13047. [CrossRef]
63. Friedrich, L.; Colpo, A.; Maggi, A.; Becker, T.; Lacidogna, G.; Iturrioz, I. Damage process in glass fiber reinforced polymer specimens using acoustic emission technique with low frequency acquisition. *Compos. Struct.* **2021**, *256*, 113105. [CrossRef]
64. Ichenihi, A.; Li, W.; Gao, Y.; Rao, Y. Feature selection and clustering of damage for pseudo-ductile unidirectional carbon/glass hybrid composite using acoustic emission. *Appl. Acoust.* **2021**, *182*, 108184. [CrossRef]
65. Lobanov, D.S.; Zubova, E.M. Damage accumulation after temperature aging in structural GFRP in interlayer shear tests. *IOP Conf. Ser. Mater. Sci. Eng.* **2021**, *1093*, 012019. [CrossRef]
66. Wildemann, V.E.; Staroverov, O.A.; Strungar, E.M.; Lunegova, E.M.; Mugatarov, A.I. Stability of Postcritical Deformation of CFRP under Static $\pm 45^{\circ}$ Tension with Vibrations. *Polymers* **2022**, *14*, 4502. [CrossRef]
67. Taheri-Behrooz, F.; Moghaddam, H.S. Nonlinear numerical analysis of the V-notched rail shear test specimen. *Polym. Test.* **2018**, *65*, 44–53. [CrossRef]
68. Fallahi, H.; Taheri-Behrooz, F.; Asadi, A. Nonlinear Mechanical Response of Polymer Matrix Composites: A Review. *Polym. Rev.* **2020**, *60*, 42–85. [CrossRef]
69. Fallahi, H.; Taheri-Behrooz, F. Phenomenological Constitutive Modeling of the Non-Linear Loading-Unloading Response of UD Fiber-Reinforced Polymers. *Compos. Struct.* **2022**, *292*, 115671. [CrossRef]
70. Polilov, A.N.; Tatus, N.A. Experimental substantiation of strength criteria for fibrous composites exhibiting directional characteristics of failure. *PNRPU Mech. Bull.* **2012**, *2*, 140–166.
71. Hashin, Z. Failure Criteria for Unidirectional Fiber Composites. *J. Appl. Mech.* **1980**, *47*, 329–334. [CrossRef]
72. Samareh-Mousavi, S.S.; Mandegarian, S.; Taheri-Behrooz, F. A nonlinear FE analysis to model progressive fatigue damage of cross-ply laminates under pin-loaded conditions. *Int. J. Fatigue* **2019**, *119*, 290–301. [CrossRef]

73. Lobanov, D.S.; Lunegova, E.M.; Mugatarov, A.I. Influence of preliminary thermal aging on the residual interlayer strength and staging of damage accumulation in structural carbon plastic. *PNRPU Mech. Bull.* **2021**, *1*, 41–51. [CrossRef]
74. Lobanov, D.S.; Lunegova, E.M. Evaluation of the thermal and moisture aging influence in aggressive environments on the change in the of the mechanical behavior of fiberglass by a short beam bending tests based on acoustic emission technique. *PNRPU Mech. Bull.* **2022**, *4*, 42–53. [CrossRef]

polymers

MDPI

Article

Enhanced Open-Hole Strength and Toughness of Sandwich Carbon-Kevlar Woven Composite Laminates

Mohammad K. A. Khan [1], Harri Junaedi [2], Hassan Alshahrani [1], Ahmed Wagih [3,4,5,*], Gilles Lubineau [3,4] and Tamer A. Sebaey [2,5,*]

[1] Department of Mechanical Engineering, College of Engineering, Najran University, Najran 11001, Saudi Arabia; mkkhan@nu.edu.sa (M.K.A.K.); haalshahrani@nu.edu.sa (H.A.)
[2] Engineering Management Department, College of Engineering, Prince Sultan University, Riyadh 11586, Saudi Arabia; hlukman@psu.edu.sa
[3] Mechanical Engineering Program, Physical Science and Engineering Division, King Abdullah University of Science and Technology (KAUST), Thuwal 23955-6900, Saudi Arabia; gilles.lubineau@kaust.edu.sa
[4] Mechanics of Composites for Energy and Mobility Lab, King Abdullah University of Science and Technology (KAUST), Thuwal 23955-6900, Saudi Arabia
[5] Mechanical Design and Production Department, Faculty of Engineering, Zagazig University, Zagazig 44519, Egypt
* Correspondence: ahmed.abdelhady.1@kaust.edu.sa (A.W.); tsebaey@psu.edu.sa (T.A.S.)

Abstract: Fiber-reinforced plastic composites are sensitive to holes, as they cut the main load-carrying member in the composite (fibers) and they induce out-of-plane stresses. In this study, we demonstrated notch sensitivity enhancement in a hybrid carbon/epoxy (CFRP) composite with a Kevlar core sandwich compared to monotonic CFRP and Kevlar composites. Open-hole tensile samples were cut using waterjet cutting at different width to diameter ratios and tested under tensile loading. We performed an open-hole tension (OHT) test to characterize the notch sensitivity of the composites via the comparison of the open-hole tensile strength and strain as well as the damage propagation (as monitored via CT scan). The results showed that hybrid laminate has lower notch sensitivity than CFRP and KFRP laminates because the strength reduction rate with hole size was lower. Moreover, this laminate showed no reduction in the failure strain by increasing the hole size up to 12 mm. At w/d = 6, the lowest drop in strength showed by the hybrid laminate was 65.4%, followed by the CFRP and KFRP laminates with 63.5% and 56.1%, respectively. For the specific strength, the hybrid laminate showed a 7% and 9% higher value as compared with CFRP and KFRP laminates, respectively. The enhancement in notch sensitivity was due to its progressive damage mode, which was initiated via delamination at the Kevlar–carbon interface, followed by matrix cracking and fiber breakage in the core layers. Finally, matrix cracking and fiber breakage occurred in the CFRP face sheet layers. The specific strength (normalized strength and strain to density) and strain were larger for the hybrid than the CFRP and KFRP laminates due to the lower density of Kevlar fibers and the progressive damage modes which delayed the final failure of the hybrid composite.

Keywords: hybrid composites; mechanical properties; fractography; notch sensitivity

Citation: Khan, M.K.A.; Junaedi, H.; Alshahrani, H.; Wagih, A.; Lubineau, G.; Sebaey, T.A. Enhanced Open-Hole Strength and Toughness of Sandwich Carbon-Kevlar Woven Composite Laminates. *Polymers* **2023**, *15*, 2276. https://doi.org/10.3390/polym15102276

Academic Editors: Emilia P. Collar and Jesús-María García-Martínez

Received: 20 April 2023
Revised: 9 May 2023
Accepted: 10 May 2023
Published: 11 May 2023

1. Introduction

Composite laminates are ubiquitous in the aerospace, automotive, construction, and marine industries as structural components [1]. Moreover, the superior mechanical, fatigue, and durability performance of CFRP has supported their application in many primary structures [2–4]. Combining various types of materials and components into structures necessitates the existence of holes to facilitate the joining of these distinct components. Thus, it is usual practice and unavoidable in the design, manufacturing, and assembly of structures to introduce holes in composite laminates [5,6]. Hole opening creates a discontinuity in the fiber and the matrix and results in stress concentration during loading. These geometrical perturbations also come with large out-of-plane stresses, while laminates are designed to bear the in-plane load. Holes, as free edges, are systematic hot points in structures, as out-of-plane-stress-related mechanisms, such as delamination, are active in these areas. As a result, studies on opening and using a hole in composite structures have been conducted for many years and are still being conducted for the complicated nature of the damage mechanisms associated with such a problem [7–13].

The sensitivity of a laminate with the presence of a hole or a notch, in terms of its mechanical properties, is defined by several factors. These factors include the hole or notch dimensions, shape, laminate dimensions, fiber orientations, ply stacking sequence, hole-cutting quality, and materials [14]. Failure stress (strength) and damage propagation vary significantly by changing one or more of these factors, even for unnotched laminates [15,16]. For instance, Wysmulski [17,18] developed FE models and validated them experimentally to study the compression behavior and damage prediction in composite laminates with different cross-sections, demonstrating the sensitivity of these structures to stacking sequences and ply orientations. The most common damage mechanisms associated with the open-hole tensile test are brittle failure, fiber pull-out, delamination, and their combination [14]. Several experimental and numerical investigations have been conducted to determine how the composite laminate responds to a hole. Two types of holes are produced in laminates upon assembly: open hole and filled hole. An open hole is when the hole is left empty, and a filled hole is when the hole is occupied by a bolt or similar. The scenario with the open hole is the topic of this research. Research into the open hole of hybrid composite laminates has been one of the recent focuses [19–23].

Carbon fiber (CF) and Kevlar (aramid) are synthetic fibers known for possessing high strength, stiffness, and low density. The strength and stiffness of CF are higher compared to Kevlar. The superiority of aramid over CF is in its toughness, ability to absorb an impact load, and lower density (1.44 g/cm^3) [12–27]. Due to its toughness property, aramid fiber is frequently employed in military applications for ballistic body armor and cut-resistant safety equipment. However, aramid has poor compressive strength, is susceptible to moisture absorption, and is more expensive. As a result, hybrid laminates have been increasingly popular in recent decades due to their coupling properties.

The use of multiple types of fibers in a single laminate allows the designer to fine-tune the properties of the laminate to the design requirements [22,28–34]. Hybridization is expected to gain some enhancement from some properties. A study by Guled and Chittappa [30] into hybrid CF/Kevlar/Epoxy laminates concluded that using CF laminate on the outer skin enhances the interlaminar shear strength of the overall laminate. Shaari et al. [20,21,23] studied the open-hole tension (OHT) test for Kevlar/glass fiber (GF)/epoxy fiber hybrid laminate composites for different hybridization ratios, hole sizes, and stacking sequences. The OHT of GF laminates improved due to the existence of Kevlar fiber. Hybridization with glass and Kevlar fibers improved hybrid composite specimens' tensile strength and failure strain. The failure mechanism shifted from being dominated by the matrix to being dominated by the fibers as the hole size increased. De Medeiros et al. [22] investigated the CF/Kevlar hybrid composite laminates. The hybrid laminates that were studied were composed of CF and Kevlar with different orientations. They concluded that the hybrid laminate suffered greater deterioration in mechanical properties when the orientation of the Kevlar fibers was aligned with the direction of the load.

Sebaey and Wagih [26] examined the flexural properties of notched carbon–aramid hybrid laminates in two different forms: intra-ply and sandwich laminates. Sandwich laminates demonstrated progressive damage; meanwhile, catastrophic damage was indicated for intra-ply hybrid laminates. Moreover, sandwich specimens showed higher specific strength compared to intra-ply laminates. Wagih et al. [27] studied the impact and flexural properties of a hybrid carbon–aramid/epoxy composite in the form of a sandwich, in which carbon/epoxy plies were used as face sheets and aramid/epoxy plies were used as the core. They found out that the carbon fiber plies in the bottom part of the laminate were not ruptured after the impact and flexural test. Localized damage was found only at the top carbon fiber ply and the upper plies of the aramid core. Basha et al. [31] examined aramid plies sandwiched between two face sheets of CFRP. CFRP/aramid sandwich composite laminates significantly improved their damage resistance under impact and compression after impact loads and showed better damage tolerance compared to CFRP laminate only.

This study aims to determine how the mechanical performance of composite laminates is affected by the type of fiber and the size of the holes. The composite laminates were constructed out of carbon fiber, aramid fiber, and the hybridization of both. Pristine and open-hole samples were prepared and subjected to a tensile test. The effect of varying hole diameters on mechanical properties was investigated. To understand the failure mechanism, a computed tomography (CT) scan was used to analyze the laminates after failure.

2. Materials and Methods

As-received laminated panels used in this study were supplied from Dragonplate, ALLRed & Associates Inc. (Elbridge, NY, USA). Composite laminates with three different types of fiber-reinforced polymers were utilized, namely, carbon-fiber-reinforced polymer, Aramid (Kevlar)-fiber-reinforced polymer, and carbon-Kevlar-fiber-reinforced polymer (hybrid). The hybrid laminate consisted of a single ply of CFRP on the outer layers and Kevlar laminates in the core. The benefits of Kevlar are the higher toughness and resistance to impact. Although not as stiff as carbon fiber, carbon/Kevlar products can be more suitable than one solely comprising carbon fiber or fiberglass in certain applications (bulletproof). Additionally, since Kevlar is non-conductive, it is usually suitable for applications where carbon may cause electromagnetic interference issues, for example, in radomes and UAV components. In all such applications, connecting composite parts requires a hole, for which the notch sensitivity is considered as a key design factor.

After cutting the plies from the woven fiber mat, they were laid, and resin transfer molding was used. The laminate was left to reach its full strength for 20 days. After that, 30 mm was removed from all of the specimen edges to avoid any edge effects. The specimens, with the desired dimensions, and the holes were cut using waterjet cutting.

All of the laminates had a plain woven-roving nature, with $0/90^{\circ}$ orientation and a total thickness of 3 mm. The stacking sequences of the three laminates were $\left[(0/90)_4^C\right]_S$, $\left[(0/90)_8^K\right]_S$, and $\left[(0/90)_2^C/(0/90)_4^K\right]_S$ for CFRP, KFRP, and hybrid, respectively. The CFRP, KFRP, and hybrid densities were 1.45, 1.31, and 1.32 g/cm^3, respectively. The resin used in this study was NCT 304 with a 140 °C glass transition temperature. Other mechanical properties of the constituent materials are listed in Table 1.

Table 1. Mechanical properties of the fiber and matrix used.

	Carbon Fiber	Kevlar Fiber	Epoxy Matrix
Tensile strength (MPa)	3500	2920	66
Tensile modulus (GPa)	230	70.5	3.7

The pristine and OHT test samples were cut from the laminate panels using a water jet machine. The samples were cut to the dimensions of 200 mm × 18 mm for the pristine test samples according to the ASTM D3039M standard [33] and 200 mm × 36 mm according to the ASTM D5766M standard [34] for the OHT test samples. For the pristine samples, tabs of 2 mm thickness made of glass fiber composite with ±45 stacking sequence were glued to the sample edges, leaving a 90 mm gauge length to avoid undesirable damage modes as suggested by the ASTM standard. Meanwhile, the OHT samples were tested without tabs, as the hole presented a higher stress concentration that ensured failure in the hole section. The gauge length of all of the samples, open-hole and pristine, was kept constant at 90 mm. A total of 5 different holes with diameters of 4, 6, 9, 12, and 18 mm were considered to evaluate the notch sensitivity of the produced hybrid sandwich composites, resulting in 5 different width-to-hole-diameter (w/d) ratios of 9, 6, 4, 3, and 2, respectively. Figure 1 shows the samples and X-ray images of the samples, indicating that the samples were free of damages such as delamination and matrix cracking around the hole, which is a critical parameter in evaluating the notch sensitivity of structures.

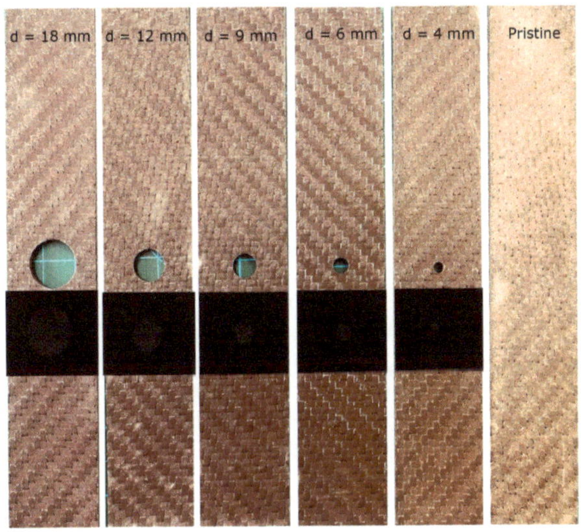

Figure 1. Photograph of the samples with different hole diameters and X-ray images around the hole.

Tensile tests were performed using an Instron 5882 universal testing machine (Buckinghamshire, UK) with a 100 KN load cell. A mechanical grip was used to clamp samples during loading, as shown in Figure 2. The test was conducted with a crosshead speed of 2 mm/min until there was complete failure for a batch of 3 samples for each configuration. An additional batch of samples with 6 mm hole diameters was tested and interrupted before the final failure to observe the damage progression modes. To avoid confusion, the sample used to characterize the damage mode at a certain loading level was not used again for mechanical characterization.

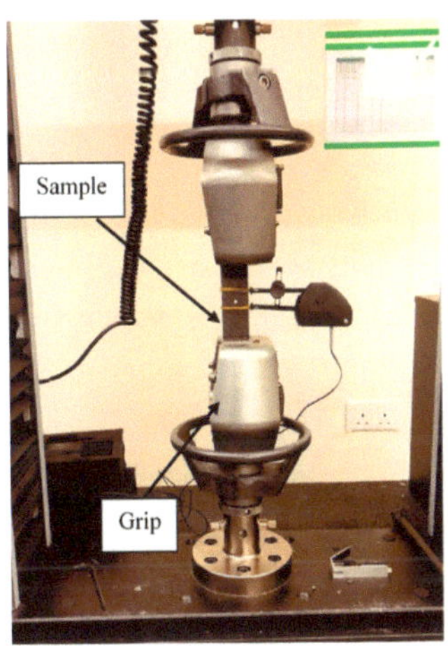

Figure 2. Experimental tensile test setup.

After testing, the hole area was inspected using an X-ray CT scan (Nikon XT H 225, Nikon Metrology NV, Derby, UK). The prepared samples were then placed inside the scanning machine for examination. The current was set to 140 μA, the voltage was set to 120 kV, the exposure time per frame was 375 ms, and a voxel size of 10.5 μm was set. A total of 2001 projections per scan were collected during the scanning process and the 3D CT Pro reconstruction software was used to reconstruct the scanned images. These images were then sliced and analyzed using Aviso 2.0 software (Redwood City, CA, USA) to inspect the different damage modes inside the samples. So, for CFRP, 3 samples were inspected, where one of them was inspected after final failure while two of them were inspected in different loading stages. For KFRP, a sample was inspected after final failure and another one was inspected before final failure. For the hybrid laminate, four samples were inspected: one after final failure and three in different loading stages.

3. Results

Figure 3 shows the sample results of the stress-strain curves for the pristine hybrid laminate and the KFRP laminate with a hole of 6 mm diameter. As is clear from Figure 3, the stress-strain profile shows good repeatability. It is worth remarking that the same repeatability level was monitored for all of the test conditions performed. The elastic regions of the curves are similar and the profiles after the ultimate stress also show similar behavior.

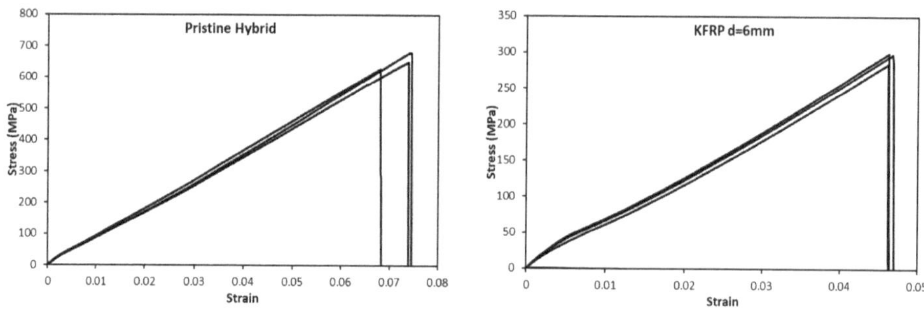

(a) (b)

Figure 3. Stress-strain curve repeatability of (**a**) pristine hybrid and (**b**) KFRP open-hole composite laminates, d = 6 mm.

Figure 4 compares the stress-strain curves of different hole diameters for different laminates. The CFRP laminates show a decrease in the ultimate stress and strain at failure by increasing the hole diameter. The stress-strain profiles for different hole diameters and pristine samples are similar. The stress rises linearly until reaching the maximum stress, and then a progressive decrease in the stress occurs, followed by a complete loss of its resistance. Unlike CFRP, KFRP laminates show a standard brittle behavior for the pristine and OHT samples, where the stress increases linearly with increasing strain until a sudden stress drop occurs, causing the complete loss of the load-bearing capacity of the laminate, indicating brittle failure. The hybrid laminates show a different profile between the pristine and OHT samples. The pristine hybrid sample shows a typical brittle stress–strain curve similar to the KFRP laminates. However, the OHT samples show progressive failure, indicating the failure of different plies of fibers (Kevlar plies and CF plies). Figure 5 shows the typical macroscopic failure modes in the samples with different hole sizes. It demonstrates that the KFRP laminate suffered from brittle localized fiber breakage in the high stress concentration area. On the other side, CFRP and hybrid laminates show dispersed damage modes around the hole area, which explains the progressive damage mode observed in the stress-strain behavior (see Figure 4a,c). Table 2 summarizes the modulus and the strength and strain at failure for the three configurations with different hole diameters. It is worth noting that the modulus here is calculated based on the crosshead displacement, as in most of the samples, the extensometer values are not available.

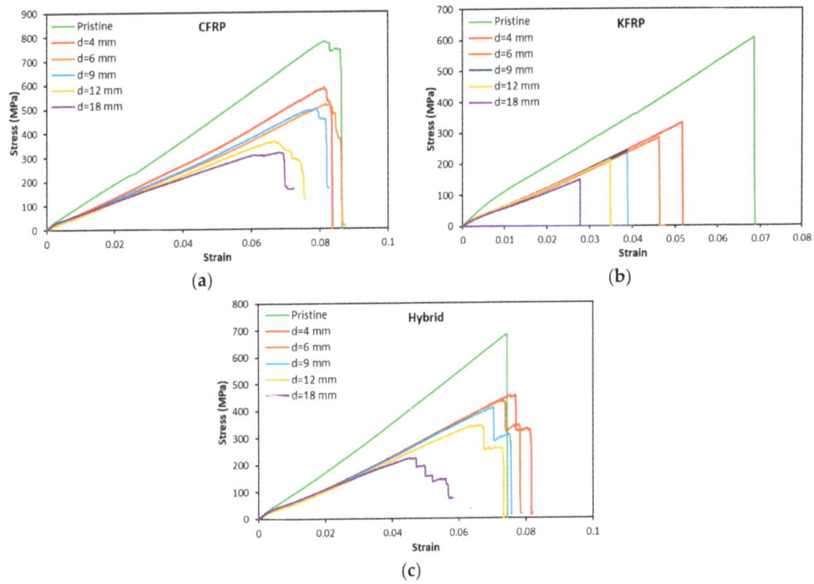

Figure 4. Stress-strain curves of OHT samples of different hole diameters for (**a**) CFRP, (**b**) KFRP, and (**c**) hybrid.

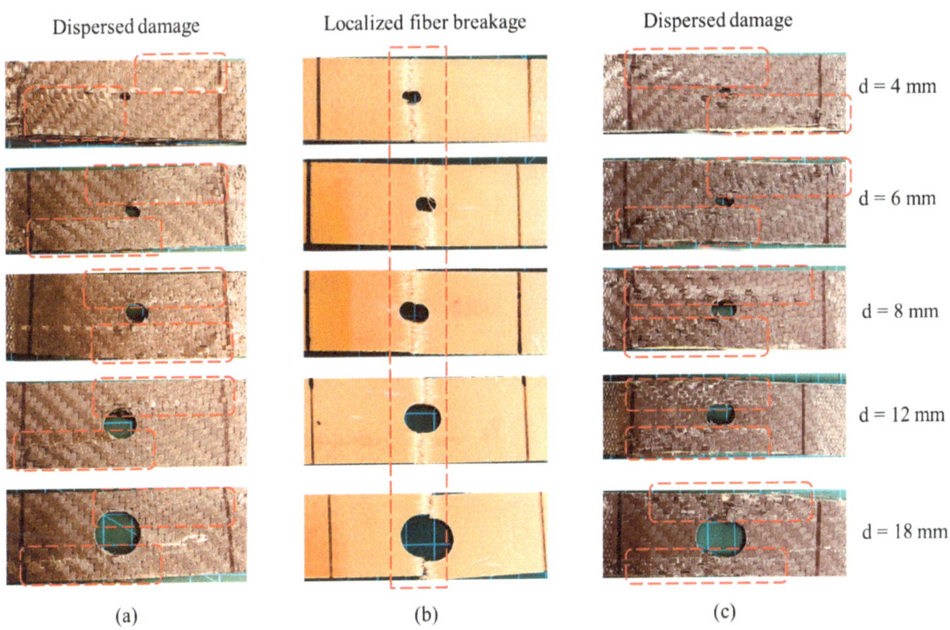

Figure 5. Open-hole samples after the tensile test: (**a**) CFRP, (**b**) KFRP, and (**c**) hybrid. (Red dotted frames show the damage area).

Table 2. Strain and stress at failure for the test campaign.

Hole Diameter (mm)	Apparent Modulus (MPa)			Strain at Failure			Tensile Strength (MPa)		
	CFRP	KFRP	Hybrid	CFRP	KFRP	Hybrid	CFRP	KFRP	Hybrid
0	9482 ± 220	8733 ± 399	8786 ± 369	0.093 ± 0.012	0.061 ± 0.008	0.074 ± 0.005	806 ± 48	524 ± 76	653 ± 27
4	6426 ± 139	6183 ± 109	5872 ± 171	0.085 ± 0.003	0.051 ± 0.005	0.081 ± 0.002	581 ± 8	322 ± 28	454 ± 19
6	6458 ± 093	6122 ± 215	5712 ± 489	0.081 ± 0.006	0.047 ± 0.001	0.076 ± 0.004	512 ± 11	294 ± 9	427 ± 14
9	6030 ± 347	6112 ± 302	5693 ± 162	0.082 ± 0.003	0.042 ± 0.002	0.076 ± 0.003	490 ± 12	246 ± 6	380 ± 31
12	5748 ± 562	5851 ± 334	5359 ± 078	0.075 ± 0.008	0.035 ± 0	0.071 ± 0.003	365 ± 2	209 ± 4	326 ± 18
18	5372 ± 011	5284 ± 331	5149 ± 392	0.074 ± 0.002	0.03 ± 0.001	0.06 ± 0.002	320 ± 14	154 ± 4	221 ± 9

4. Discussion

4.1. Damage Sequence

Figure 6 displays the CT scan image of samples with a 6 mm diameter hole after failure. It justifies the different apparent failure modes experienced by each sample type. Delamination dominates the CFRP's failure mode, as seen in Figure 6a. Fiber breakage is also observed near the hole. For KFRP in Figure 6b, the failure mode is governed by fiber breakage, followed by cracks propagating throughout the matrix between the layers and repeating until fully broken. Delaminations are completely inhibited for KFRP laminates due to the fact that Kevlar is much less anisotropic compared to carbon fiber. Therefore, the generated out-of-plane stresses are less. For the hybrid laminate, fiber breakages of Kevlar and the separation of the CF laminates from the Kevlar laminate are observed in Figure 6c.

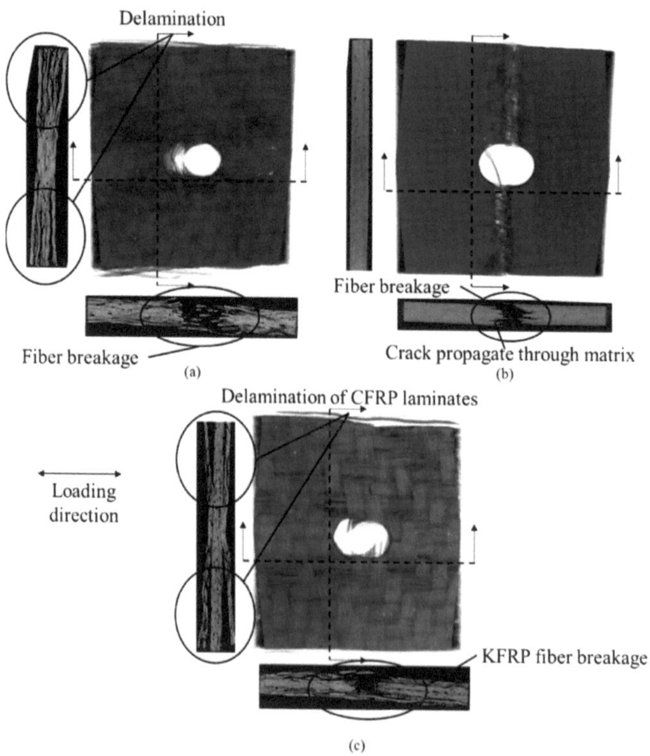

Figure 6. CT scan images of failed open-hole tension samples for 6 mm hole diameter: (**a**) CFRP, (**b**) KFRP, and (**c**) hybrid.

To better understand the different damage modes, the evolution of these damage modes, and how they interact with each other, some tests were interrupted before the final failure of the laminate and the damage modes were examined. Figure 7 shows the damage sequence of the CFRP laminate with a hole diameter of 6 mm in different stages of loading, before the laminates ultimately failed, and the corresponding stress-strain curves. The damage modes are shown using a 2D cross-sectional view of the laminate in 2 perpendicular planes at the hole center (Figure 7c) and at the edge of the hole (Figure 7d), as explained in the 3D micrograph in Figure 7b. Matrix cracks appeared first around the hole edge early during loading, where no significant change in the stress–strain profile could be observed. These matrix cracks initiated inside the tows are marked with yellow in Figure 7c,d due to the high interlaminar stress concentrations on the matrix and the free edge effect around the hole. The presence of the hole and the free edge induced out-of-plane stress that generated interlaminar shear stress [35]. Just before the final failure, "Point 2" in Figure 7a, delaminations were observed (see Figure 7c) that were initiated at the hole edge due to the interlaminar shear stress generated near the hole [36]. Similar phenomena have been reported elsewhere [14,15,37,38]. These delaminations were initiated and propagated around the hole in both directions, as shown in Figure 7d, due to the high shear stresses generated and the woven-roving nature of the plies. The propagation of delaminations reflects small stress drops in the stress–strain curve just before "Point 2". These stress drops were repeated before the final failure due to the propagation of more delaminations until the saturation of the composite with delamination. Once the laminate was saturated with delaminations, fiber kinking and a fiber cut occurred, as shown in Figure 7c,d, causing a large stress drop that reduced the laminate strength to almost zero, referred to as "Point 3". In this stage, due to the unstable growth of damage, extensive matrix cracks and delaminations propagated throughout the whole laminate width, as shown in Figure 7d.

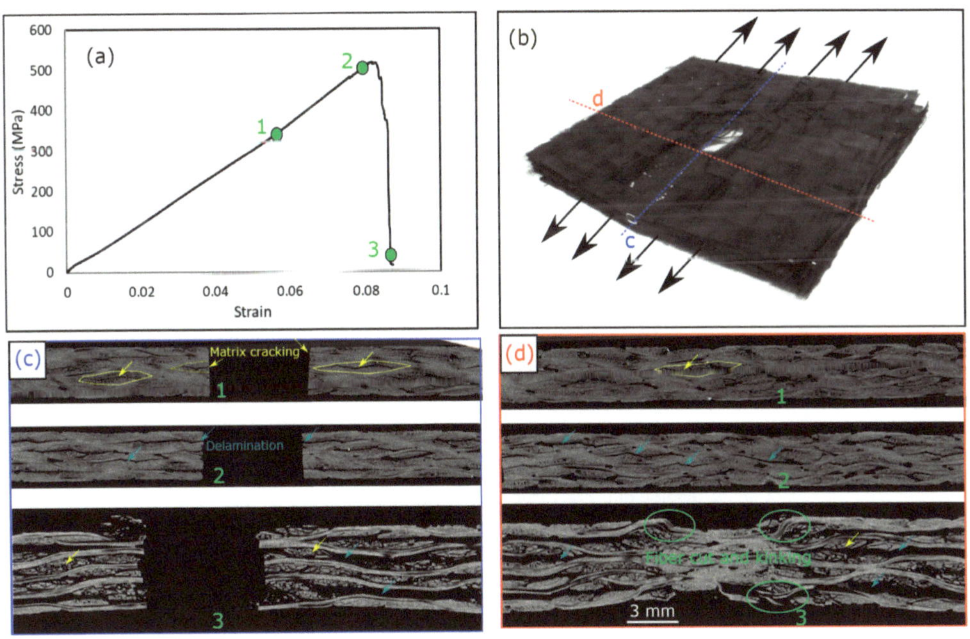

Figure 7. Damage sequence in CFRP OHT laminate with 6 mm hole diameter: (**a**) stress-strain curve, (**b**) 3D CT-micrograph of the damage after failure at "Point 3", (**c**) 2D cross-section at "Points 1, 2, and 3" in the loading direction at the sections marked in (**b**), (**d**) 2D cross-section at "Points 1, 2, and 3" in the perpendicular to the loading direction at the sections marked in (**b**).

Figure 8 shows the damage sequence in KFRP with a 6 mm hole diameter. The damage modes were recorded in two stages: during the load, and before and after the maximum stress (Point 1 and Point 2). As previously stated, the stress-strain response is monotonically linear until the final brittle failure. It is demonstrated that, in the loading direction, the hole edges do not suffer from any kind of damage even after the final failure, as shown in Figure 8c. The damage was very localized in the hole center in the perpendicular direction to the loading. At "Point 1", single delamination of the first layer was observed in the direction perpendicular to the loading direction (see Figure 8d), which can be justified by the cut of fibers during the machining of the holes that produced a larger stress concentration at this particular interface and hence delamination growth. Apart from this single delamination, unlike the CFRP laminate, there was no sign of any kind of damage throughout the laminate thickness. It is well known that KFRP laminates possess lower interlaminar shear stress (ILSS) compared to CF/epoxy composite laminates [28]. In contrast to CFRP, no delamination was shown on KFRP, even though matrix cracking throughout the thickness was observed. This can be justified by the lower tensile strength (lower stress level) of KFRP compared to CFRP, as shown in Figure 4. The damage mode, after failure, indicated at "Point 2" in Figure 8a, shows a brittle cut of fibers throughout the whole laminate thickness with few pulled out fibers. During the formation of the fiber bundle pull-out, a few local delaminations appeared just around the hole. In summary, the damage mode in KFRP was different than that of CFRP. In KFRP, localized damage in the hole center occurred, whereas CFRP showed distributed damage throughout the whole width of the sample with extensive matrix cracks and delaminations.

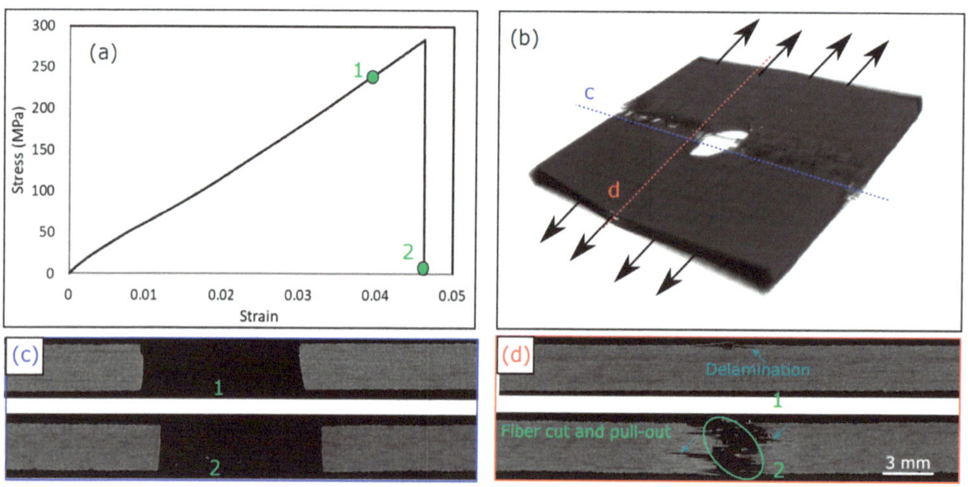

Figure 8. Damage sequence in KFRP OHT laminate with 6 mm hole diameter: (**a**) stress–strain curve, (**b**) 3D CT-micrograph of the damage after failure at "Point 2", (**c**) 2D cross-section at "Points 1 and 2" in the loading direction at the sections marked in (**b**), (**d**) 2D cross-section at "Points 1 and 2" in the perpendicular to the loading direction at the sections marked in (**b**).

The damage sequence in the hybrid laminate with a 6 mm hole diameter is shown in Figure 9. In this laminate, we tested 4 samples under different loading strains until complete failure, defined as "Points 1, 2, 3, and 4" in Figure 9a. The stress increased monotonically by increasing the strain until "Point 1", where local delamination at the CFRP/Kevlar interface was initiated around the hole (see Figure 9c,d) due to the low interlaminar toughness of this interface [26]. Owing to the free edges at the hole, out-of-plane stresses acted as opening stresses at the interface between the CFRP and Kevlar layers. Once these stresses reached the interlaminar shear strength of that interface, delaminations were initiated. These delaminations propagated until "Point 2", as shown in Figure 9c,d. At "Point 2", the initiation of the Kevlar fiber cut occurred at the lower Kevlar plies (see Figure 9(d2)) due to the loss of the support from the neighbor CFRP plies, as a result of the separation between the CFRP and Kevlar layers. Just after "Point 2", a large stress drop occurred due to the complete fiber cut in the Kevlar plies' core, as shown in "Point 3". Additionally, extensive delaminations at the CFRP/Kevlar interface occurred in this stage, as shown in Figure 9(c3). It is notable that two interesting phenomena happened with the KFRP core in the hybrid laminate that did not appear in the KFRP laminate. The first is that Kevlar fiber breakage (cut) occurred progressively due to the support of the Kevlar plies with CFRP plies, which increased their in situ strength. The second interesting phenomenon is that the failure strain of the Kevlar core was larger than the failure strain of the pure Kevlar laminate. For the pure Kevlar laminate, the failure strain was 0.046, whereas the initiation failure strain of the Kevlar core in the hybrid laminate was somewhere between "Points 1 and 2", corresponding to strains 0.06 and 0.072. Moreover, the strength of the Kevlar core plies in the hybrid laminate was increased compared to the pure Kevlar plies in the KFR laminate due to the in situ strength effect. Therefore, the CFRP face plies strengthened and toughened the Kevlar plies' core. After "Point 3", the CFRP face sheets were the main load carrier during loading, which supported the material until the failure strain of CFRP was reached and then complete failure occurred, see Figure 9(c4,d4). The failure strain of the CFRP plies in the hybrid laminate was equal to that of the pure carbon fiber laminate. Moreover, extensive matrix cracks in the CFRP and Kevlar core plies were observed after failure, as shown in Figure 9(c4,d4).

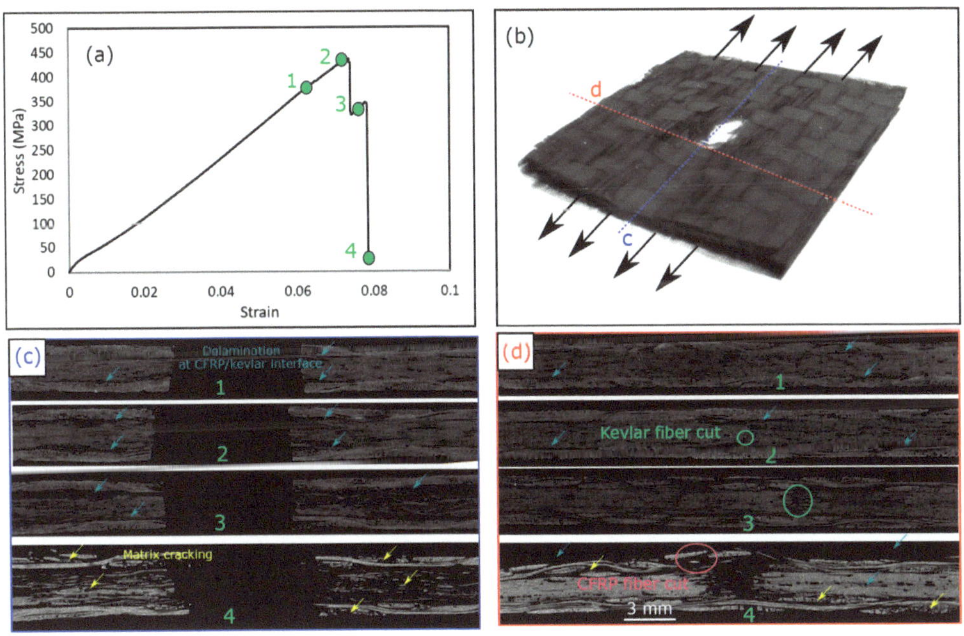

Figure 9. Damage sequence in hybrid OHT laminate with 6 mm hole diameter: (**a**) stress-strain curve, (**b**) 3D CT-micrograph of the damage after failure at "Point 4", (**c**) 2D cross-section at "Points 1, 2, 3 and 4" in the loading direction at the sections marked in (**b**), (**d**) 2D cross-section at "Points 1, 2, 3 and 4" in the perpendicular to the loading direction at the sections marked in (**b**).

Figure 10 shows the damage modes in samples with a 12 mm hole diameter to evaluate the effect of hole size on the damage modes. As shown in the figure, the damage modes observed after failure were similar to those observed for laminates with a 6 mm hole diameter, as shown in Figures 7–9. Therefore, we can summarize the damage modes in each of the three configurations considered. For the CFRP samples, progressive damage was observed as matrix cracking initiation and propagation occurred, followed by delamination at the free edges around the hole. Finally, fiber breakage and kinking at the hole edges and extensive delamination throughout the whole laminate width occurred. For the KFRP samples, the damage mode was a pure brittle failure with fiber cut and pull-out. For the hybrid laminate, a progressive damage mode was observed as delamination (separation) was initiated at the CFRP/Kevlar interface and then propagated, which weakened the Kevlar plies in the laminate core. After that, the initiation of the fiber cut in the Kevlar plies in the laminate was followed by the propagation of the fiber cut in the Kevlar plies. Finally, the fiber was cut in the CFRP plies, which resulted in complete failure.

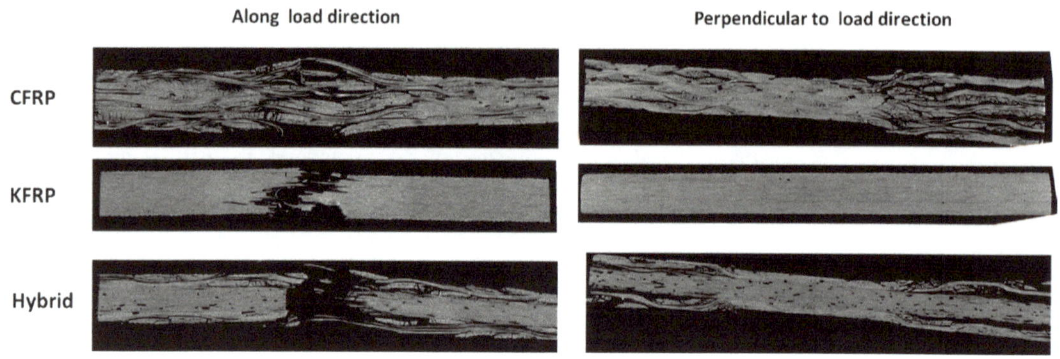

Figure 10. CT scan cross-section of failed open-hole tension samples with 12 mm hole diameters near the hole.

4.2. Notch Sensitivity

Figure 11a shows the tensile strength of the three configuration composites as a function of the hole diameter. The strength of the laminates decreased with the increase in the open-hole diameter of all types of laminates. A hole in a laminate produced stress concentration and out-of-plane stresses. These two factors altered the failure mechanisms and degraded the strength of the laminate, which was well established for the OHT test [39]. Other studies also confirmed that larger holes caused a decrement in the tensile strength of the composites. Stress could be disseminated to a larger region of the specimen cross-section with a small hole diameter, whereas with a larger hole, the damage zone extended to a greater portion of the specimen; therefore, less of a region remained to support the stress [15,21].

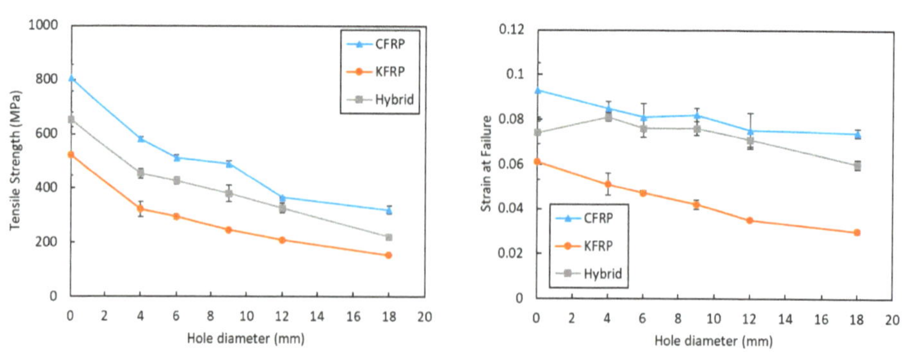

(a) (b)

Figure 11. (**a**) Tensile strength and (**b**) strain upon failure of the composite laminates with different hole diameters.

Figure 11b shows the evolution of the strain upon failure with an increase in different hole diameters. Similar to the tensile strength, strain at failure also showed a drop due to the increase in the open-hole diameter, with an exception being the hybrid laminate. The higher drop in the strain at failure was experienced by the KFRP composites. The drop in the strain due to the increase in the hole diameter is due to the stress concentration at the hole, which leads to the early failure of laminates. Meanwhile, a slight increase in the strain at failure was observed for hybrid laminates with 4, 6, and 8 mm hole diameters. The increase in strain at failure for the hybrid OHT sample, as seen from the failure mechanics (Figure 11b) and stress–strain curves (Figure 4), can be justified by the delamination of the CF plies. As seen in Figure 9, hybrid laminates showed two steps of failure. The first one occurred at the maximum stress due to the breakage of the Kevlar fiber plies, whereas the second one was due to the failure caused by the delamination between the CF plies and the Kevlar plies. The enhancement in the ductility of the OHT hybrid laminate samples could be one of the advantages of the laminate compared to the other two laminates.

Normalized strength was calculated by dividing the open-hole tensile strength of each composite system by the tensile strength of the pristine composite. The plot of normalized strength to the w/d is presented in Figure 12a. For a hole diameter of 4 mm (w/d = 9), a drop to 72.1% of the pristine sample for samples was observed for CFRP, whereas, for the same ratio, 69.5% and 61.5% were observed for the hybrid and KFRP laminates, respectively. Meanwhile, at w/d = 6, the lowest drop in strength was shown by the hybrid laminate at 65.4% of the pristine sample, followed by CFRP and KFRP with 63.5% and 56.1%, respectively. At the largest hole diameter of 18 mm (w/d = 2), the lowest strength drop was shown by CFRP with 39.7%, followed by the hybrid and KFRP laminates with 33.9% and 29.4%, respectively. In summary, the KFRP composite had the highest drop for all conditions compared to CFRP and the hybrid laminate. On the other hand, the hybrid composite laminate normalized OHT strength was very close to that of CFRP. A drop in the OHT strength for all of the samples was observed by the decrease in w/d. The same trends have also been shown by Salleh et al. [19] for fiberglass-reinforced epoxy composites. KFRP is more notch-sensitive compared to CFRP. The replacement of the outer layers of KFRP with face CFRP plies to produce the hybrid laminate reduces the notch sensitivity of the KFRP laminate.

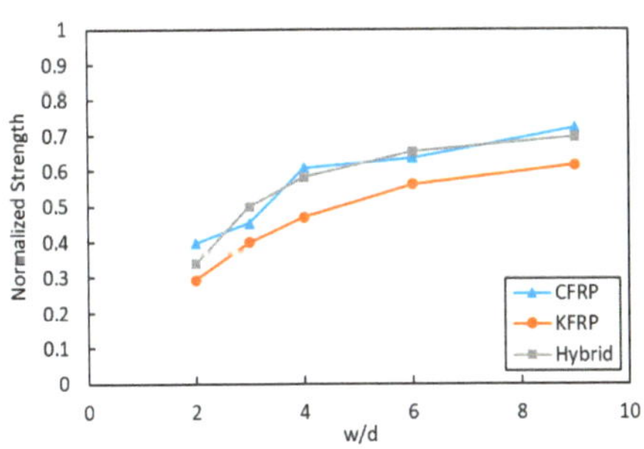

(a)

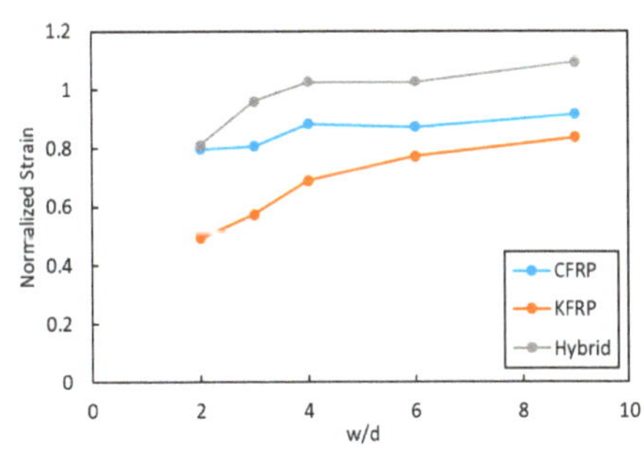

(b)

Figure 12. (**a**) Normalized strength and (**b**) normalized strain at different w/d.

The normalized strain was also calculated, and the result is presented in Figure 12b. The figure displays the normalized strain of the three laminates and the change in the w/d. It shows that KFRP had the highest drop in the strain at failure in comparison to its pristine sample. The hybrid laminates had higher normalized strain values compared to CFRP. The presence of CFRP plies in the outer layers of the KFRP laminates enhanced the OHT strength and the laminate's ductility simultaneously.

In comparison to CF, Kevlar has a lower density, which also results in composite laminates that have lower density. Thus, normalized strength and strain in regard to density (specific strength and strain, respectively) were also calculated, and the result is presented in Figure 13. It shows that hybrid laminate has higher normalized strength/density (Figure 13a) and normalized strain/density (Figure 13b) compared to CFRP and KFRP for the OHT samples. It showed almost around 7% and 9% larger specific strength than the CFRP and KFRP laminates. More interestingly, it showed a large enhancement in specific strain at a larger width-to-hole ratio reaching 20% and 18% compared to CFRP and KFRP laminates.

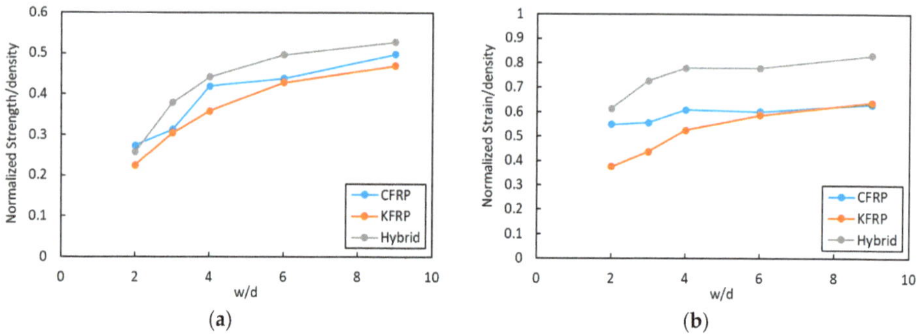

Figure 13. (**a**) Normalized strength/density and (**b**) normalized strain/density at different w/d.

5. Conclusions

Pristine and open-hole tension (OHT) composite laminates of pure CFRP and KFRP and a hybrid laminate with a core of KFRP surrounded by two face sheets of CFRP were tested. Micro-computed tomography was deployed to characterize the damage modes and their sequence during the tensile test. Based on the results and discussion, the following conclusions can be drawn:

- The progressive damage mode occurred for CFRP and the hybrid laminate under the OHT test, while KFRP showed a brittle damage mode.
- For CFRP, the damage was initiated as matrix cracks around the hole due to the interlaminar shear stresses at the hole edges, followed by delaminations at the same position. Then, fiber breakage occurred progressively after the saturation of the laminate with delaminations. For the hybrid laminate, the damage was initiated as delaminations at the CFRP/Kevlar interface that propagated by increasing the applied strain. Then, the Kevlar core fiber cut was initiated and propagated, causing a partial loss of laminate strength. Finally, CFRP face sheet fiber breakage and matrix cracking occurred, causing the final failure of the samples.
- The hybrid laminate showed better notch sensitivity than the CFRP and KFRP laminates due to the progressive damage mode, where the reduction in the strength of the hybrid laminate by increasing the hole size was the lowest.
- The hybrid laminate showed 7 and 9% higher specific strength compared to the CFRP and KFRP laminates. At the same time, it showed a 20% and 18% higher specific failure strain compared to the CFRP and KFRP laminates.

Author Contributions: Conceptualization, A.W. and T.A.S.; methodology, H.J.; validation, M.K.A.K. and H.A.; investigation, A.W. and H.J.; resources, G.L. and T.A.S.; writing—original draft preparation, A.W. and H.J.; writing—review and editing, A.W., T.A.S. and G.L. supervision, T.A.S.; project administration, M.K.A.K. and H.A.; funding acquisition, M.K.A.K. All authors have read and agreed to the published version of the manuscript.

Funding: This research was funded by the Deanship of Scientific Research at Najran University under the General Research Funding program, grant code (NU/DRP/SERC/12/22).

Institutional Review Board Statement: Not applicable.

Informed Consent Statement: Not applicable.

Data Availability Statement: The data will be made available upon request to the authors.

Acknowledgments: The authors are thankful to the Deanship of Scientific Research at Najran University for funding this work under the General Research Funding program, grant code (NU/DRP/SERC/12/22). Also, T.A.S. and H.J. would like to thank Prince Sultan University for their support.

Conflicts of Interest: The authors declare no conflict of interest.

References

1. Harris, C.E.; Morris, D.H. Fracture of Thick Laminated Composites. *Exp. Mech.* **1986**, *26*, 34–41. [CrossRef]
2. Xian, G.; Guo, R.; Li, C.; Hong, B. Mechanical properties of carbon/glass fiber reinforced polymer plates with sandwich structure exposed to freezing-thawing environment: Effects of water immersion, bending loading and fiber hybrid mode. *Mech. Adv. Mater. Struct.* **2023**, *30*, 814–834. [CrossRef]
3. Pan, Y.; Yan, D. Study on the durability of GFRP bars and carbon/glass hybrid fiber reinforced polymer (HFRP) bars aged in alkaline solution. *Compos. Struct.* **2021**, *261*, 113285. [CrossRef]
4. Guo, R.; Li, C.; Xian, G. Water absorption and long-term thermal and mechanical properties of carbon/glass hybrid rod for bridge cable. *Eng. Struct.* **2023**, *274*, 115176. [CrossRef]
5. Huang, C.; Ju, S.; He, M.; Zheng, Q.; He, Y.; Xiao, J.; Zhang, J.; Jiang, D. Identification of Failure Modes of Composite Thin-Ply Laminates Containing Circular Hole under Tension by Acoustic Emission Signals. *Compos. Struct.* **2018**, *206*, 70–79. [CrossRef]
6. Zheng, K.; Hu, H.; Cao, D.; Zhong, Y.; Li, S. Experimental and numerical studies on the tensile behaviors of thin-ply and thick-ply open-hole laminates. *Thin Walled Struct.* **2023**, *186*, 110649. [CrossRef]
7. Khashaba, U.A.; Sebaey, T.A.; Alnefaie, K.A. Failure and Reliability Analysis of Pinned-Joints Composite Laminates: Effects of Stacking Sequences. *Compos. B Eng.* **2013**, *45*, 1694–1703. [CrossRef]
8. Nguyen-Hoang, M.; Becker, W. Open Holes in Composite Laminates with Finite Dimensions: Structural Assessment by Analytical Methods. *Arch. Appl. Mech.* **2022**, *92*, 1101–1125. [CrossRef]
9. Feistle, M.; Pätzold, I.; Golle, R.; Volk, W. Open Hole Tensile Tests for the Determination of the Edge-Crack Sensitivity of Sheared Holes Dependent on Specimen Geometry, Cutting Parameters, and the Notch Factor. *Procedia Manuf.* **2019**, *29*, 412–419. [CrossRef]
10. Zhang, D.; Zheng, X.; Wu, T. Damage Characteristics of Open-Hole Laminated Composites Subjected to Longitudinal Loads. *Compos. Struct.* **2019**, *230*, 111474. [CrossRef]
11. Guo, Q.; Zhang, Y.; Li, D.; Lv, Q.; Sun, X.; Ma, M.; Chen, L. Experimental and numerical investigation of open-hole tensile properties and damage mechanisms of 3D woven composites under weft-loading. *Thin Walled Struct.* **2021**, *161*, 107455. [CrossRef]
12. Sivakumar, D.; Ng, L.F.; Zalani, N.F.M.; Selamat, M.Z.; Ab Ghani, A.F.; Fadzullah, S.H.S.M. Influence of kenaf fabric on the tensile performance of environmentally sustainable fibre metal laminates. *Alex. Eng. J.* **2018**, *57*, 4003–4008. [CrossRef]
13. Zappalorto, M.; Ricotta, M. Understanding the Effect of Notches in Orthotropic Solids Subjected to Static Loads. *Theor. Appl. Fract. Mech.* **2021**, *116*, 103110. [CrossRef]
14. Green, B.G.; Wisnom, M.R.; Hallett, S.R. An Experimental Investigation into the Tensile Strength Scaling of Notched Composites. *Compos. Part A Appl. Sci. Manuf.* **2007**, *38*, 867–878. [CrossRef]
15. Wisnom, M.R.; Hallett, S.R. The Role of Delamination in Strength, Failure Mechanism and Hole Size Effect in Open Hole Tensile Tests on Quasi-Isotropic Laminates. *Compos. Part A Appl. Sci. Manuf.* **2009**, *40*, 335–342. [CrossRef]
16. Aoki, R.; Higuchi, R.; Yokozeki, T.; Aoki, K.; Uchiyama, S.; Ogasawara, T. Effects of Ply Thickness and 0° Layer Ratio on Failure Mechanism of Open-Hole and Filled-Hole Tensile Tests of Thin-Ply Composite Laminates. *Compos. Struct.* **2022**, *280*, 114926. [CrossRef]
17. Wysmulski, P. Load Eccentricity of Compressed Composite Z-Columns in Non-Linear State. *Materials* **2022**, *15*, 7631. [CrossRef]
18. Wysmulski, P. Non-linear analysis of the postbuckling behaviour of eccentrically compressed composite channel-section columns. *Compos. Struct.* **2023**, *305*, 116446. [CrossRef]
19. Salleh, Z.; Berhan, M.N.; Hyie, K.M.; Taib, Y.M.; Kalam, A.; Roselina, N.R.N. Open Hole Tensile Properties of Kenaf Composite and Kenaf/Fibreglass Hybrid Composite Laminates. *Procedia Eng.* **2013**, *68*, 399–404. [CrossRef]
20. Shaari, N.; Jumahat, A.; Abdullah, S.A.; Hadderi, A.Z. Effect of Hybridization on Open-Hole Tension Properties of Woven Kevlar/Glass Fiber Hybrid Composite Laminates. *J. Teknol.* **2015**, *76*, 91–96. [CrossRef]
21. Shaari, N.; Jumahat, A. Hole Size Effects on the Open Hole Tensile Properties of Woven Kevlar-Glass Fibre Hybrid Composite Laminates. *Pertanika J. Sci. Technol.* **2017**, *25*, 309–318.
22. De Medeiros, R.J.; da Nóbrega, S.H.S.; de Aquino, E.M.F. Failure Theories on Carbon/Kevlar Hybrid Fabric Based Composite Laminate: Notch and Anisotropy Effects. *Mater. Res.* **2019**, *22*, 20180099. [CrossRef]
23. Shaari, N.; Abdul Wahab, M.F.; Shaari, N.S.; Jumahat, A. Unhole and Open Hole Tensile Properties of Hybrid Kevlar/Glass Fiber Polymer Composites with Different Stacking Sequence. *Mater. Today Proc.* **2021**, *46*, 1595–1599. [CrossRef]
24. Kowsari, E.; Haddadi-Asl, V.; Ajdari, F.B.; Hemmat, J. Aramid Fibers Composites to Innovative Sustainable Materials for Biomedical Applications. *Mater. Biomed. Eng. Biopolym. Fibers* **2019**, 173–204. [CrossRef]
25. Manocha, L.M. Carbon Fibers. *Encycl. Mater. Sci. Technol.* **2001**, 906–916. [CrossRef]

26. Sebaey, T.A.; Wagih, A. Flexural Properties of Notched Carbon–Aramid Hybrid Composite Laminates. *J. Compos. Mater.* **2019**, *53*, 4137–4148. [CrossRef]
27. Wagih, A.; Sebaey, T.A.; Yudhanto, A.; Lubineau, G. Post-Impact Flexural Behavior of Carbon-Aramid/Epoxy Hybrid Composites. *Compos. Struct.* **2020**, *239*, 112022. [CrossRef]
28. Monjon, A.; Santos, P.; Valvez, S.; Reis, P.N.B. Hybridization Effects on Bending and Interlaminar Shear Strength of Composite Laminates. *Materials* **2022**, *15*, 1302. [CrossRef]
29. Alsaadi, M.; Erkliğ, A.; Alrawi, H. Effect of S-Glass Fabric on the Mechanical Characteristics of a Hybrid Carbon/Aramid Fabric Reinforced Epoxy Composites. *Mater. Res. Express* **2017**, *4*, 055304. [CrossRef]
30. Guled, F.D.; Chittappa, H.C. Influence of Interply Arrangement on Inter-Laminar Shear Strength of Carbon-Kevlar/Epoxy Hybrid Composites. *AIP Conf. Proc.* **2019**, *2057*, 020045. [CrossRef]
31. Basha, M.; Wagih, A.; Melaibari, A.; Lubineau, G.; Eltaher, M.A. On the Impact Damage Resistance and Tolerance Improvement of Hybrid CFRP/Kevlar Sandwich Composites. *Microporous Mesoporous Mater.* **2022**, *333*, 111732. [CrossRef]
32. Basha, M.; Wagih, A.; Melaibari, A.; Lubineau, G.; Abdraboh, A.M.; Eltaher, M.A. Impact and Post-Impact Response of Lightweight CFRP/Wood Sandwich Composites. *Compos. Struct.* **2022**, *279*, 114766. [CrossRef]
33. *ASTM Standard D3039/D3039M*; Standard Test Method for Tensile Properties of Polymer Matrix Composite Materials. ASTM International: West Conshohocken, PA, USA, 2000. [CrossRef]
34. *ASTM Standard D5766/D5766M*; Standard Test Method for Open-Hole Tensile Strength of Polymer Matrix Composite Laminates. ASTM International: West Conshohocken, PA, USA, 2018. [CrossRef]
35. Katerelos, D.T.G. Investigation of the Free Edge Interlaminar Stresses Dependence on the Ply Thickness and Orientation. *Adv. Compos. Lett.* **2006**, *15*, 096369350601500102. [CrossRef]
36. Hu, F.Z.; Soutis, C.; Edge, E.C. Interlaminar Stresses in Composite Laminates with a Circular Hole. *Compos. Struct.* **1997**, *37*, 223–232. [CrossRef]
37. Chuaqui, T.R.C.; Nielsen, M.W.D.; Colton, J.; Butler, R.; Rhead, A.T. Effects of Ply Angle and Blocking on Open-Hole Tensile Strength of Composite Laminates: A Design and Certification Perspective. *Compos. B Eng.* **2021**, *207*, 108582. [CrossRef]
38. O'Higgins, R.M.; McCarthy, M.A.; McCarthy, C.T. Comparison of Open Hole Tension Characteristics of High Strength Glass and Carbon Fibre-Reinforced Composite Materials. *Compos. Sci. Technol.* **2008**, *68*, 2770–2778. [CrossRef]
39. Moure, M.M.; Otero, F.; García-Castillo, S.K.; Sánchez-Sáez, S.; Barbero, E.; Barbero, E.J. Damage Evolution in Open-Hole Laminated Composite Plates Subjected to in-Plane Loads. *Compos. Struct.* **2015**, *133*, 1048–1057. [CrossRef]

polymers

MDPI

Article

Effect of Number of Tests on the Mechanical Characteristics of *Agave sisalana* Yarns for Composites Structures: Statistical Approach

Mounir Gahgah [1], Ahmed Belaadi [1,*], Messaouda Boumaaza [2], Hassan Alshahrani [3,4] and Mohammad K. A. Khan [3,4]

[1] Department of Mechanical Engineering, Faculty of Technology, University 20 Août 1955-Skikda, El-Hadaiek Skikda 21000, Algeria
[2] Laboratory of Civil and Engineering Hydraulic (LGCH), University 8 Mai 1945 Guelma, Guelma 24000, Algeria
[3] Department of Mechanical Engineering, College of Engineering, Najran University, Najran P.O. Box 1988, Saudi Arabia
[4] Scientific and Engineering Research Centre, Deanship of Scientific Research, Najran University, Najran P.O. Box 1988, Saudi Arabia
* Correspondence: a.belaadi@univ-skikda.dz

Abstract: A designer of sustainable biocomposite structures and natural ropes needs to have a high confidence interval (95% CI) for mechanical characteristics data of performance materials, yet qualities for plant-based fibers are very diverse. A comprehensive study of the elements that enhance the performance of biocomposites or sustainable ropes created from vegetable fibers is necessary. The current study included five groups with varying numbers (N) of tests of 20, 40, 60, 80, and 100 on the mechanical characteristics at room temperatures. The purpose of this study was to determine how changing N affects the mechanical properties of sisal yarn. These properties include its strength, Young's modulus, and deformation at rupture. A significance testing program including more than 100 tests was performed. Owing to the heterogeneity of the plant yarn, each group received more than 20 samples at a gauge length (GL) of 100 mm. The tensile strength characteristics of sisal yarns produced a wide range of findings, as is common for natural fibers, necessitating a statistical analysis. Its dispersion was explored and measured using the statistical methods. The Weibull distribution with two parameters and a prediction model with a 95% confidence level for maximum likelihood (ML) and least squares (LS) were used to investigate and quantify its dispersion.

Keywords: sisal fiber; yarn; mechanical testing; hazard function; statistical methods; SEM

check for updates

Citation: Gahgah, M.; Belaadi, A.; Boumaaza, M.; Alshahrani, H.; Khan, M.K.A. Effect of Number of Tests on the Mechanical Characteristics of *Agave sisalana* Yarns for Composites Structures: Statistical Approach. *Polymers* **2023**, *15*, 2885. https://doi.org/10.3390/polym15132885

Academic Editors: Emilia P. Collar and Jesús-María García-Martínez

Received: 26 May 2023
Revised: 22 June 2023
Accepted: 27 June 2023
Published: 29 June 2023

1. Introduction

Due to their potential to be recycled and biodegraded, ecological composites reinforced with fibers are becoming more and more popular as substitutes for synthetic composite materials in a variety of applications [1]. Sisal fibers, the toughest natural fibers, are increasingly used in advanced materials such as composites to lessen the material's overall environmental effect [2]. High-performance composites made from renewable sources are becoming increasingly popular in the composites industry and among its consumers. Because of their exceptional formability, these composites are frequently manufactured using yarn-based textile engineering [3,4]. However, to use natural thread as a reinforcement system in structural composite applications that are highly loaded, its mechanical performance must be significantly improved [5,6].

Several studies are now being conducted to better understand the fundamental qualities of natural fibers, such as high-quality biocomposite raw materials [7,8]. These studies are critical for establishing their intended ideal usage. There is also a discussion about how to improve the longevity and commercial worth of natural fibers through the study and characterization of their qualities and potential applications. Natural fibers have a low density and high strength-to-weight ratio, making them good candidates for lightweight composites and reinforcing materials [9]. The chemical composition and microstructure of fibers have an impact on their mechanical properties, with the fiber cross section being the most influential variable on fiber strength [10]. Moreover, the anatomical qualities of fibers differ between species, influencing their density and mechanical properties. Various factors, such as fiber extraction, storage duration, and environmental conditions, affect the size and quality of natural fibers [7,11,12].

The mechanical characteristics using vegetable sisal yarns exposed to tension quasi-static stress with a 13° twist angle and 232 tex linear density were examined by Belaadi et al. [13]. A test procedure was performed on 150 samples. Because natural yarns vary in quality, more than 30 specimens were evaluated for each GL. Five distinct GLs (from 50 mm to 300 mm) were used. The findings obtained were then analyzed utilizing statistical Weibull distributions with two and three parameters for various probability indexes, and LS and ML estimates. The author showed that as the GL increases, the strain and tension on the yarn drop from 14.7% to 5% and 180 MPa to 122 MPa, respectively. Nevertheless, Young's modulus is unaffected by the GL. A recent study by Belaadi et al. [14] was taken as the basis to examine the behavior of 30 samples of jute yarn with a GL of 150 mm, an area angle torsion of 10° to 13°, a linearly distributed density of 27.021 tex (g/1000 m), and an average diameter of 500–1100 m. Based on the data of the tested fibers, we found the mean strength to be 91.69 ± 21.38 MPa, the average strain to be 2.18 ± 0.34%, and the Young's modulus to be roughly 3163 ± 994 MPa. These numbers (Young's modulus and strength) show an increase of almost 14.20% and 22.5%, respectively, compared to the results found by Codispoti et al. [15] for jute yarn with a gauge length of 100 mm. Another novel approach was employed by El-Geiheini et al. [16] to enhance the mechanical properties of cotton yarns. The resulting models were developed by applying artificial neural network techniques to processed images.

Yan et al. [17] used a 150 mm length soaked for 30 min in a solution of 5 wt% NaOH to test the tensile strength of three distinct types of processed bamboo, flax, and linen threads. Ten iterations of each test were conducted. The tensile failure stress of unprocessed flax single-strand yarn is 115.4% higher than that of unprocessed bamboo and 12.1% higher than that of unprocessed linen. Linen yarns have over 50% greater elongation at the break point than flax and bamboo yarns. All three fiber yarns' tensile strengths and moduli were lower compared to their alkali-treated counterparts. When compared to untreated specimens, potassium hydroxide-treated linen, flax, and bamboo strings had lower tensile strengths and tensile moduli of 16.4%, 18.5%, and 30.7% and 13.0%, 5.9%, and 27.8%, respectively.

On the other hand, alkali-treated linen and flax yarns had greater elongations at rupture. Gomes et al. [18] achieved a comparable outcome in the case of a curaua fiber with alkali treatment. Saaidia et al. [19] discussed the statistical characterization of the tensile characteristics of 13 series of 30 samples of raw and treated jute yarns with twisted surface angles ranging between 11° and 13°, with a linear density of 267 tex at GL = 50 mm. The yarns were treated with various NaOH concentrations and immersion periods. Weibull techniques with two and three parameters were used for the statistical analysis. The findings for uniaxial tensile yarns reveal a variance in terms of failure strain, stress, and Young's modulus, which is mostly determined by the period of immersion and the concentration of NaOH. An immersion duration of two hours with a NaOH dosage of 2% produced the best mechanical characteristics.

Abida et al. [20] conducted an experiment examination on the morphological and mechanical characteristics of yarns of flax, revealing the three characteristic distributions to be normal distributions. The suggested model implies that tissue is made up of many threads that work together like springs, and linen threads are fragile and flexible. A genetic algorithm with multiobjective analysis was utilized to solve the inverse issue to identify the yarn characteristics from experimental tensile testing. Abida et al. [21] conducted a new study on the coefficients of hygroscopic expansion of flax fiber- and flax thread-strengthened composites along three orientations using a combined experimental multiscale approach. This study found that flax yarn had a significant radial coefficient of expansion (β_r = 1.06), and the composite showed strongly anisotropic bulking behavior. In addition, the weft direction exhibited good dimensional stability: β_r = 0.13 ± 0.019. The radial expansion of the warp threads and matrix is balanced by the negative axial hygro-expansion factor in the weft threads, which can be explained by the longitudinal and out-of-plane swelling factors.

Wang et al. [22] developed a statistical model to examine the random tensile response of 30 natural jute fiber yarns at GL = 100 mm, taking fiber crimp and property distribution into account. The statistical properties of corrugated jute fibers are described as a distribution probability beta function fiber strain. Despite the fact that the effective modulus of elasticity and thread strength obey the law of normal distribution, the tension graphs of similar thread specimens exhibit identical form features to the beta dispersion in the crimped strain. This pattern may provide reasonable forecasting restrictions for the dispersion of the jute fiber nonlinear tensile response. Sohbatzadeh et al. [23] presented a new method for the low-cost modification of the synthetic aramid thread surface. In this study, plasma systems at atmospheric pressure were used to surface modify the threads with argon as the working gas and acetone as the precursor. These modifications resulted in improved tension strength and flotation characteristics, with the treated thread displaying superb waterproofing and buoyancy characteristics. In addition, the modifications by plasma enhanced the thread's mechanical strength and, thus, its suitability for reinforcing applications. Additionally, the study revealed that desirable carbon-based nano-structures were synthesizable on the yarn surface.

Consequently, an evaluation of the mechanical performance using the fibers and threads themselves is necessary for the application of sisal threads as load-bearing cables. The literature has covered a lot of research on synthetic yarn. However, the use of statistical methods to estimate the mechanical performance of natural yarns requires further investigation. In this study, the focus was on the mechanical characterization of yarns using various sample numbers. Two-parameter Weibull statistics for various estimators and techniques (ML and LS predictions) were used to assess the tensile properties. Additionally, a thorough analysis of the study's findings was conducted and contrasted with information from the literature. To the knowledge of the authors, this represents the first instance in which the yarns have been investigated at various sample sizes using a predictive model with a 95% confidence level.

2. Materials and Methods

The yarns made from the *Agave sisalana* utilized in this study were formed of carded sisal fibers, which are particularly suitable for producing yarns and ropes of varying widths. The Algerian packaging and cable company BLIDA provided the yarns. The sisal fiber was 0.76 to 1.4 m long, 248 μm in diameter, and had a cross-sectional area of 0.043 ± 0.008 mm^2. The mean mechanical parameters of individual separate-carded sisal fibers employed for this investigation, with a density of 1.43 g/cm^3, were previously published by Belaadi et al. [24] as = 463 MPa, ε = 7.84%, and E = 7.39 GPa. The sisal yarns were made up of about 70–80 sisal fibers with a diameter of approximately 2.41 ± 0.64 mm and a cross-sectional area of 4.52 ± 0.43 mm^2. In addition, the sisal yarns (Figure 1a,b) had a linear density of 223 ± 42 tex (g/1000 m) with a mean twisting angle of approximately 10–13° (Table 1). The diameter of each sisal thread was measured using a ZEISS microscope instrument fitted with a Moticam 2500 camera that was digitally controlled by the Motic-Images Pro V2.0 visualization software. Over its entire length, the diameter changed. Ten measurements were performed at various points along the yarn. Using the mean yarn diameter, the observable cross-sectional measurement for each yarn was computed. In addition, the fiber cross section was assumed to be circular with a constant diameter throughout its length [13,22].

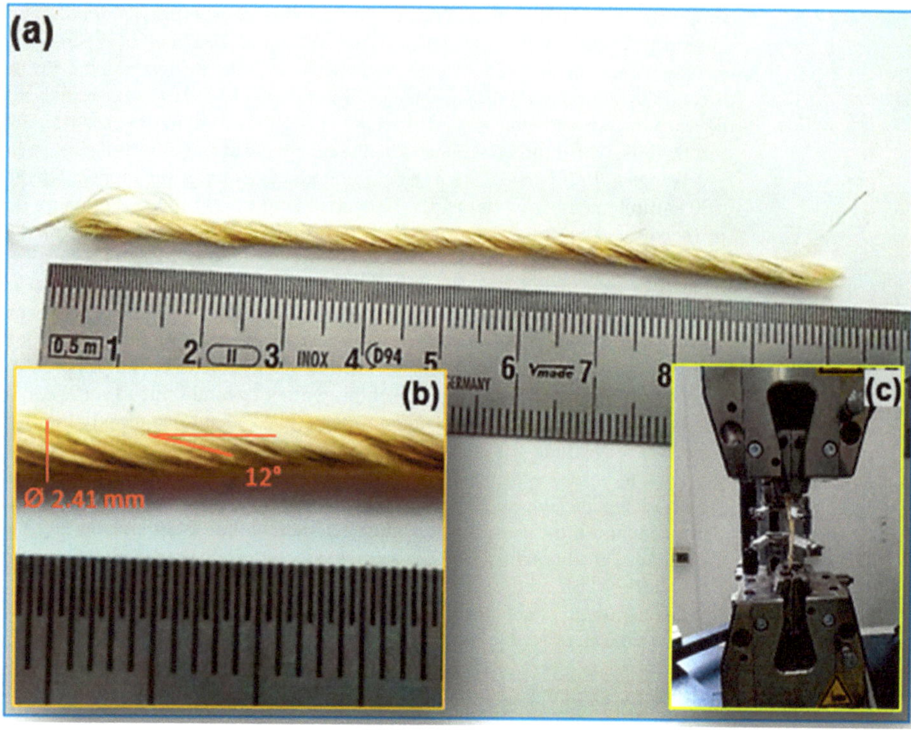

Figure 1. (**a**) Single specimen; (**b**) geometry of the sisal yarn used in this work; and (**c**) specimen clamped.

Table 1. Material characteristics of sisal yarn used in this work.

	The Number of Fibers	Density (g/cm^3)	Yarn Cross Section Area (mm^2)	Tex (g/1000 m)	Surface Twist Angle (○)
Sisal yarn	70–80	1.433 ± 0.012	4.52 ± 0.43	223 ± 42	10–13°

Minimum values of 20 basic sisal yarns were used to determine the tensile mechanical behavior characteristics (the modulus of elasticity, ultimate elongation, and tensile strength) for five groups (N) of 20, 40, 60, 80, and 100 in accordance with ASTM D2256-01 [25]. The sisal yarn samples were selected to be a mean length of 100 mm. The latter were carefully handled to avoid deterioration. The Zwick-Roell testing apparatus with a 50 kN load cell and mechanical grips was used for all of the experiments (Figure 1c) at GL = 100 mm. At a steady rate of 5 mm/min, tensile tests were carried out in the laboratory at a temperature of 25 °C and a relative humidity of 40%.

An individual sisal fiber, thread, and sections were examined using a JSM-7600F scanning electron microscopy (SEM). A thin layer of gold was applied to make the samples conductive. At an accelerating voltage, SEM micrographs were captured (Figure 2a–i).

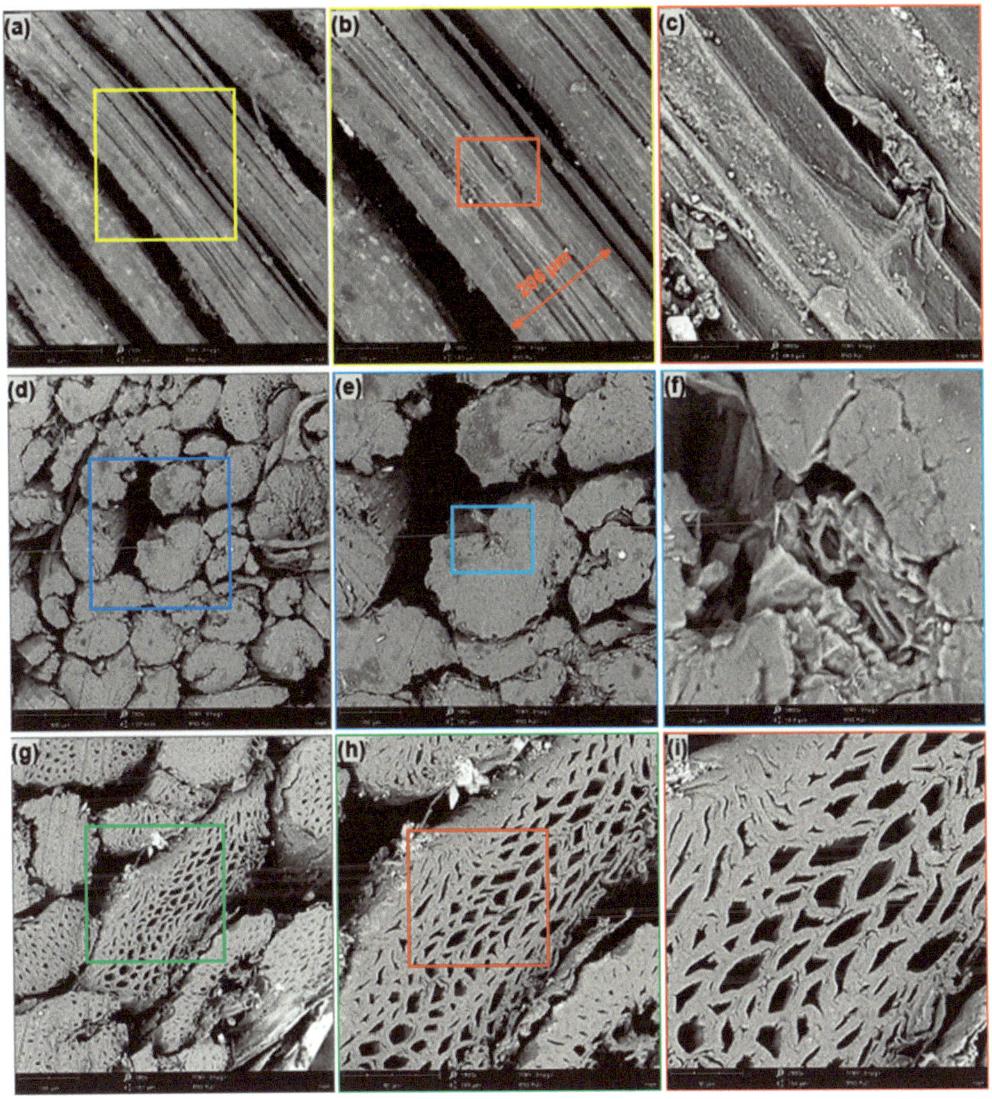

Figure 2. SEM micrographs of (**a**) longitudinal view of sisal yarn, (**b**) diameter measurements for single sisal fiber, (**c**) topographic surface of sisal fiber, (**d**) cross-sectional view of sisal yarn, (**e**,**f**) zoomed details of the selected zone in (**d**,**g**) another view of the cross section of the same yarn from (**d**), and (**h**,**i**) zoomed details of the selected zone in g showing the crushing of sisal fiber cells.

3. Weibull Statistics for Sisal Yarns Data

According to the literature, most natural fibers exhibit significant dispersion in their mechanical characteristics. Such scattering can be described by the concentration of imperfections on the surface and the microstructure of the fiber [26,27]. Two- and three-parameter models were used to statistically examine the values of the sisal yarns' tensile mechanical properties (σ, ε, and E).

To describe the degree of variable tensile properties, the Weibull distribution using different approaches was determined using ML and LS for a confidence level of 95%. Moreover, using mechanical characteristic data, Anderson–Darling (AD) testing with modified quality-of-fit estimations was employed to identify the best fit. Minitab V-16 was used to conduct the statistical analysis. The Laplace–Gauss normal distribution is a rule of absolutely continuous probability, which is determined by two parameters: the standard deviation (SD) and the mean (t and α^2 is the variance). The parameter (μ) provides information about the distribution center. The parameter s indicates the extent to which it has spread. Equation (1) [28] gives the probability density function (PDF). The variable $y = (x - \mu)/\alpha$ is the reduced-centred Gaussian probability density determined using Equation (2) for the variable $\mu = 0$ and $\alpha = 1$. The log-normal distribution, on the other hand, is characterized using two parameters: ξ and λ, with $\lambda > 0$, and one obtained for the PDF (Equation (3)). Equation (4) defines the PDF of the three-parameter Weibull distribution, often known as the failure distribution, or as the cumulative distribution function (CDF). In addition, Equation (5) yields the probability density for $s_0 = 0$ [22]. By simplifying Equation (5), we obtained a Weibull survival probability with two parameters by assuming $s_0 = 0$ (s_0 stands for a threshold or localization parameter), which denotes a mean parameter value of x (minimum survival time) that can be characterized by Equation (6) [22,29]. Where m and s represent real positive numbers that represent the shape's factor such as The Weibull modulus and the scale parameter or characteristic value, respectively.

$$P(x|\mu,\alpha) = \frac{1}{\sqrt{2\pi s}} e^{-\frac{1}{2}\left(\frac{x-\mu}{\alpha}\right)^2} \tag{1}$$

$$\varphi(y) = \frac{1}{\sqrt{2\pi}} e^{-\frac{y^2}{2}} \tag{2}$$

$$P(x|\xi,\lambda) = \frac{1}{x\xi\sqrt{2\pi}} e^{-\frac{1}{2}\left(\frac{\ln(x)\ -\ \lambda}{\xi}\right)^2} \tag{3}$$

$$F(x|s_0,s,m) = 1 - e^{\left[-\left(\frac{x\ -\ s_0}{s}\right)^m\right]},\quad x \geq s_0 \tag{4}$$

$$F(x|s,m) = \frac{m}{s}\left(\frac{x}{s}\right)^{m-1} e^{\left[-\left(\frac{x}{s}\right)^m\right]} \tag{5}$$

$$F(x|s,m) = 1 - e^{\left[-\left(\frac{x}{s}\right)^m\right]} \tag{6}$$

The ML technique is described by the formula in Equation (7) [30] and is a key approach for estimating the probability density parameters or Weibull probability. This strategy depends on the parameters being selected so as to maximize the probability for the specimen information. Since this method provides estimated parameters that have higher statistical properties, this approach was selected due to its statistical consistency. The advantage of this method is that the ambiguity of the distribution of Weibull parameters with confidence intervals of 95% can be easily and efficiently identified. Another approach, known as the LS estimate using linear regression, (Equation (8)) [31] was applied from member to member using the natural logarithm. The Weibull modulus m was determined from the slope of a straight line of the $\ln\left(\ln\left(\frac{1}{1-P}\right)\right)$ vr. $\ln(x)$. This line's intercept enabled us to determine s. In our case, distinctive mechanical parameters, such as the characteristic Young's modulus, strength, and strain at failure, i.e., E_0, σ_0, and ε_0, respectively, were obtained. The biggest issue with this methodology is the estimation of the survival probability (p). The p value was calculated using an estimator or probability index. $P_i = (i - \tau)/(n - \beta)$ where $\tau = 0.3$, and $\beta = 0.4$ are the generic forms of the estimator. The probability index was extensively used to evaluate the estimator, as shown in Equation (9) [13]. Additionally, numerous alternatives for estimating the probability index are available in the Minitab program, such as the median rank or Benard index (Equation (9)).

$$\text{Likelihood}(m,s|x_1,\ldots,x_2) = \prod_{i=1}^{n} F(x_i) = \prod_{i=1}^{n} \frac{m}{s}\left(\frac{x_i}{s}\right)^{m-1} e^{\left[-\left(\frac{x}{s}\right)^m\right]} \tag{7}$$

$$\ln\left(\ln\left(\frac{1}{1-P}\right)\right) = m\ \ln(x) - m\ \ln(s) \tag{8}$$

$$P_i = \frac{i-0.3}{n+0.4} \tag{9}$$

4. Results

4.1. Tensile Behavior of an Elementary Sisal Yarn

At room temperature, the sisal yarns subjected to tensile tests at a speed of 5 mm/min were divided into five series of 20 specimens, representing a total of 100 elementary yarns that were chosen at random via the specific lot and gauged at GL = 100 mm. The stress–strain graphs from 20 static tensile tests performed on the sisal yarns are displayed in Figure 3a. Clearly, the wide dispersion of results, a phenomenon specific to natural fibers, calls for an in-depth statistical analysis. The same behavior was observed in the case of jute yarn reported by Wang et al. [22] and Saidia et al. [19] at GL = 50 mm, and in the case of sisal yarn [13] and flax yarn [20] in the last 50 samples. As illustrated in Figure 3b, the typical tensile curve behavior of sisal yarns may be characterized into three zones: the crimp, nonlinear, and damage region. The nonlinear region (0–4.5% of strain-ε) is most likely related to the rearrangement of elementary fibers in the yarn. In addition, this phenomenon has been explained by several authors [13,28] since sisal yarns are produced by twisting the fibers in a spiral, creating intermediate spaces in the fibers. The Young's modulus of the threads was also determined using the linear-elastic region (next phase), which had a relatively high slope (5–10% of ε and σ between 30–170 MPa). The specimen did not completely fail until the stress rapidly decreased to a mean strength of 81 MPa, with 80% of the yarn fibers failing as a result of the third area, i.e., the damage zone.

As a result of 100 tests of sisal yarns at various test numbers (N), the standard deviation (SD) and the mean values of the mechanical characteristics, notably strain at break, stress, and Young's modulus, are summarized in Table 1. They were analyzed by calculating the percentage of the coefficient of variation (CoV%). The coefficient of variation in percent is defined as the relationship of the mean (μ) to the standard deviation (σ), as determined through the expression CoV (%) = [(σ/μ) $\times$ 100], whereas at a low percentage CoV, it was assumed that there would be little difference in the data. A very high and significant scatter in the values given for tensile properties with various tests N can be seen in Table 2. The tensile strength was determined from the cylindrical assumed cross section of individual threads, with the theoretical area provided by the mean diameter of the filaments. Nevertheless, Young's modulus was determined in the elastic portion of the stress–strain curve, which is the graph's initial slope, usually in the range of 0.5 to 1.5% of ε. The high degree of scattering observed in the findings is a natural characteristic of yarns related to products with vegetable fibers. In fact, there were several factors contributing to these scatters, which affected the fiber [32,33] and its yarn: (1) Environmental factors that depend on the particular conditions in which the plant fiber was grown, its type, the specific position in the plants, and whether it has an irregular geometry; and (2) experimental factors associated with both the choice of test parameters (speed and precision of deformation, type of machine clamping device, and the environmental conditions in the laboratory (the relative hygrometry influences the behavior of the saw fiber in relation to its hydrophilic character)), and the choice of the measurement geometry (the cross section of the yarn) used in the calculation of the resistance (porosity in percentage). Indeed, it is clear from Table 2 that the number of tests (N = 20, 40, 60, 80, and 100) has a significant impact on the performance of sisal yarns. Thus, in this work, the sisal yarn showed maximum stress values equivalent to 148, 146, 139, 138, and 135 MPa, with a progressive decrease according to the number of tests until N = 80, followed by stabilization until N = 100 tests. The same phenomenon was observed with respect to the values of the failure strain, which also decreased with increasing N: ε = 8.41%, 7.83%, 7.37%, 7.15%, and 6.70%. However, when N was increased to 100, a higher Young's modulus E was obtained, i.e., it went from

528 MPa for N = 20, to 660 MPa. Aside from the strength, there was a decrease in the stress of about 82% (148 for N = 20 and 135 MPa for N = 100).

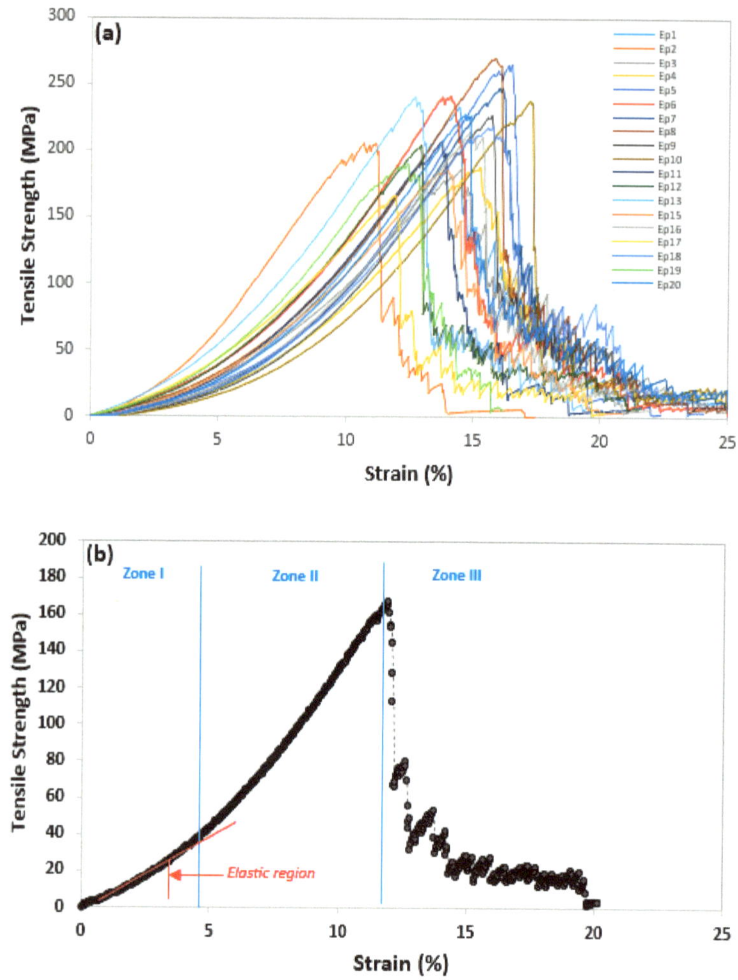

Figure 3. Stress-strain curves (**a**) for first 20 sisal yarns and (**b**) typical behavior for tensile test at GL = 100 mm.

Table 2. Mean, standard deviation, and covariance for different mechanical properties of sisal yarn tested in tensile static tension in this work.

N	Strength (MPa)			Strain (%)			Young Modulus (MPa)		
	Mean	SD	CoV	Mean	SD	CoV	Mean	SD	CoV
20	148	23.85	10.81	8.41	0.91	16.03	528	93	17.72
40	146	25.71	17.49	7.83	1.29	16.43	633	181	28.48
60	139	27.16	19.53	7.37	1.27	17.28	648	169	26.07
80	138	30.67	26.06	7.15	1.21	16.83	657	172	26.06
100	135	29.87	23.97	6.70	1.14	21.02	660	158	23.97

It is difficult to compare the experimental data obtained from the literature (Table 3) given the types of yarn used, its maturity, the environmental conditions in which these plants were grown, and the methods used to perform the tests, especially to determine the rate of stress. Nevertheless, it was possible from the literature [1,5,8,13,15,19,20,22,28,34–37] to synthesize the results on plant yarns in tensile static tests (Table 3). The strength value obtained from our experiments for sisal yarn equates to 148 MPa for GL = 100 mm in N = 20 tests, which is similar to the reference for N = 30 [13]. Additionally, it shows that, for testing with N = 20, a value of E = 528 MPa was obtained, whereas the Young's modulus for this reference for the same gauge length of 100 mm was 556 MPa. These results are near to those from this study for N = 100, which is equivalent to 660, for the same GL and N = 100. In contrast to the results of the present investigation, the researchers in [15] discovered lower values for σ and E for GL 100 mm (σ = 31.5 MPa and E = 85.2 MPa) for N = 10 to 15 tests. With GL = 100 mm and N = 20 trials, the strain rate of sisal yarn during breakage in this investigation was 8.41%, which is nearly identical to the 8.37% for the same GL with N = 30 found in [13]. On the other hand, a lower result for strain rupture equal to 6.7% was obtained by raising the number of trials to 100.

Table 3. Summarized mechanical properties data with GL and number of tests of different vegetal yarns from the literature.

Yarn	Strength (MPa)	Strain (%)	Young's Modulus (MPa)	GL (mm)	Number of Tests	Reference
Jute	52.8 ± 15.23	3.77 ± 0.77	2.28 ± 0.63 GPa	50	25	[34]
Jute	43	7.5	310	50	50	[5]
Jute	117	-	-	100	30	[22]
Jute	248	3.5	-	80	10	[8]
Jute	117	4.39	1978	50	30	[19]
Jute	53.03	3.84	2150	50	10	[35]
Jut	74.8	0.03	2769.7	100	10–15	[15]
Flax	224 ± 45.5	1–3	11,400 ± 2110	500	95	[28]
Flax	271	2.69	10,800	80	30	[20]
Flax	198.1	3.22	5913.6	100	10–15	[15]
Flax	32.7 (N)	1.93	1351 (N)	250	20	[36]
Hemp	124.1	2.94	4236.9	100	10–15	[15]
Hemp	74.86± 6.42	0.277 ± 0.031	1.046 ± 0.172	30	-	[1]
Hemp	449	4.28	11,910	50	20	[37]
Coir	51.8	0.16	322.1	100	10–15	[15]
Sisal	31.5	0.38	85.2	100	10–15	[15]
Sisal	180 ± 25	14.76 ± 1.59	336 ± 184	50	30	[13]
Sisal	148 ± 31	8.37 ± 0.97	556 ± 106	100	30	[13]
Sisal	141 ± 28	6.39 ± 0.59	792 ± 232	150	30	[13]
Sisal	127 ± 24	5.70 ± 0.59	688 ± 175	200	30	[13]
Sisal	122 ± 26	5.00 ± 0.46	616 ± 124	300	30	[13]

4.2. Statistical Distribution of Sisal Yarn Data

The wide variability of mechanical characteristics of cellulosic fibers of plant origin is a challenge for designers of composite structures. To gain a better understanding of the biocomposites benefits or strings produced from plant-based fibers, it is essential to have a good knowledge of the constituents that enhance their properties. Figure 4a,b show the relationship variation between tensile stress and strain at rupture, Young's modulus, and power fit, with a power prediction model for a 95% CI for different test groups. The Young's modulus dropped as the strength increased, as shown in the figure, and the relationship between strain at rupture and Young's modulus also followed this pattern [13].

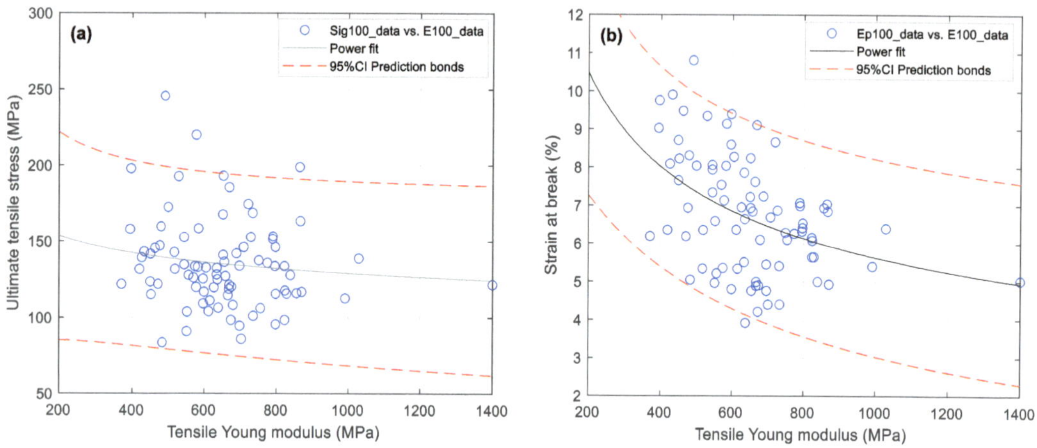

Figure 4. Scatter plots for all 100 tests of sisal yarns for (**a**) tensile strain as function of Young's modulus and (**b**) tensile strength as a function of Young's modulus.

Figure 5 illustrates the experimental histograms for the mechanical properties, in particular, σ, ε, and E in the case of N = 100 tests, according to the various distribution methods. To select the cells required for the histogram, we followed the standard square root rule based on the amount of data to be considered. In the field of materials science, various statistical distributions are available, with the Weibull distribution, normal distribution, and lognormal distribution being the ones that are employed the most. Therefore, in materials science, lognormal, Weibull, and normal distributions are frequently used to describe various properties of materials. Indeed, in contrast to the lognormal distribution, which is used to represent the distribution of the size of particles and cracks for materials, the distribution known as the Weibull distribution is frequently employed for strength forecasting and to assess the brittle fracture and reliability of materials. On the other hand, the normal distribution is a versatile distribution that can be used to model a wide range of material properties and describe random phenomena. For mechanical property information, the lognormal distribution or Weibull distribution is typically the most appropriate solution. In this case, to justify this choice, the normality Kolmogorov–Smirnov test (Table 4) was performed and the estimates of the goodness-of-fit with four distinct distributions via the Anderson–Darling test (Table 5) and the AD with *p*-values (Table 6) are presented.

Table 4. Summary table of test for normality using Kolmogorov–Smirnov test criteria for N100 tests.

	Failure Strength (MPa)	Strain (%)	Young's Modulus (MPa)
p-value	0.025	0.043	0.015
Skewness parameter	1.00585	0.75178	1.04974

Table 5. AD and *p*-values estimates for different distributions for N100 tests.

	Strength (MPa)		Strain (%)		Young Modulus (MPa)	
Distribution	**AD**	***p***	**AD**	***p***	**AD**	***p***
Normal	1.474	0.005	0.751	0.049	0.640	0.092
Lognormal	0.374	0.081	0.442	0.083	0.516	0.087
2P-Weibull	2.606	0.110	1.302	0.132	1.694	0.212
3P-Weibull	0.587	0.133	0.434	0.307	0.620	0.109

Table 6. AD goodness-of-fit estimates for different distributions.

N	Strength (MPa)				Strain (%)				Young Modulus (MPa)			
	Normal	**Weibull**	**Log-Normal**	**3P-Weibull**	**Normal**	**Weibull**	**Log-Normal**	**3P-Weibull**	**Normal**	**Weibull**	**Log-Normal**	**3P-Weibull**
20	1.529	0.921	3.499	1.036	1.255	0.989	2.869	0.835	1.090	0.837	1.923	0.920
40	1.316	0.756	3.402	0.765	0.474	0.484	1.662	0.584	1.389	0.537	2.883	0.628
60	1.085	0.475	2.784	0.860	0.864	0.543	2.358	0.660	1.895	0.482	1.938	0.452
80	1.375	0.503	3.597	0.946	1.721	0.989	4.917	1.252	1.866	0.664	1.814	0.690
100	1.618	0.523	4.400	0.826	0.890	0.551	2.545	0.604	1.771	0.626	1.821	0.637

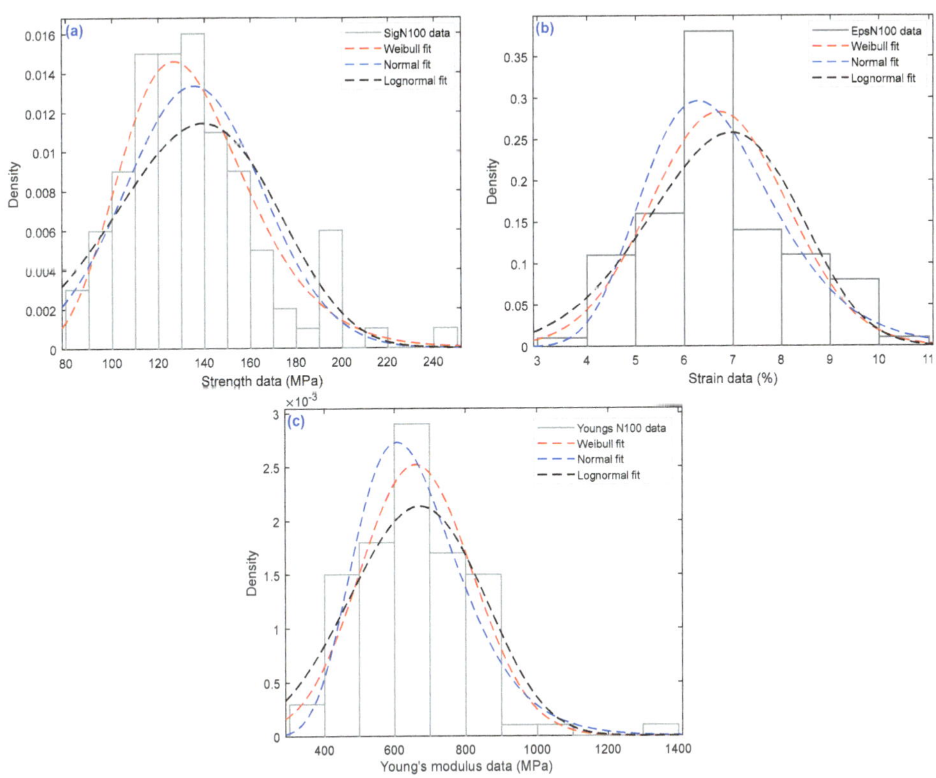

Figure 5. Histograms of the tensile strength data of the sisal yarns at different numbers of tests with the estimation of density functions Weibull, normal, and lognormal for N = 100 (**a**) strength data, (**b**) strain data and (**c**) Young's modulus data.

As a function of the number of specimens, Tukey or box plots were used to display the variance in the average mechanical characteristics of sisal yarns, such as tensile strength, elastic modulus, and strain. (see Figure 6). Box plots were used to show the general trends in the responses of a group. These diagrams are useful for visualizing the distribution and other characteristics of the responses of a large group, such as the mechanical properties. The diagram illustrated in Figure 6 shows a variety of box-plot shapes and positions. This study presents the distribution of data between the number of samples (N) and the mechanical characteristics of sisal yarns with a 95% confidence interval of the forecast. Variable data with maximum, median, minimum, and quartiles (Q1 and Q3) characterize this representation. The extremes are then extended with segments, resulting in extreme values for the first and ninth deciles. For instance, for N = 100, the median, first and third quartiles, and both maximum and minimum numbers of the samples were 131, 118, 148, 245, and 83 MPa for stress, 6.85, 5.85, 7.41, 10.7, and 3.8% for strain, and 651, 551, 763, 1230, and 371 MPa for Young's modulus. Additionally, it appears that the yarn's tensile stress (Figure 6a), which followed a power tendency line, decreased as N increased (20–100 tests). Similarly, Figure 5b illustrates the number of samples, N, as a function of elongation at break. It is evident that when N increased, the elongation reduced. On the other hand, as N increased, the mean Young's modulus rose (Figure 6b). This behavior is similar to that of single Washingtonia fibers, which Dembri et al. [38] observed from 30 to 90 tests before stabilizing at N = 120 and 150 tests.

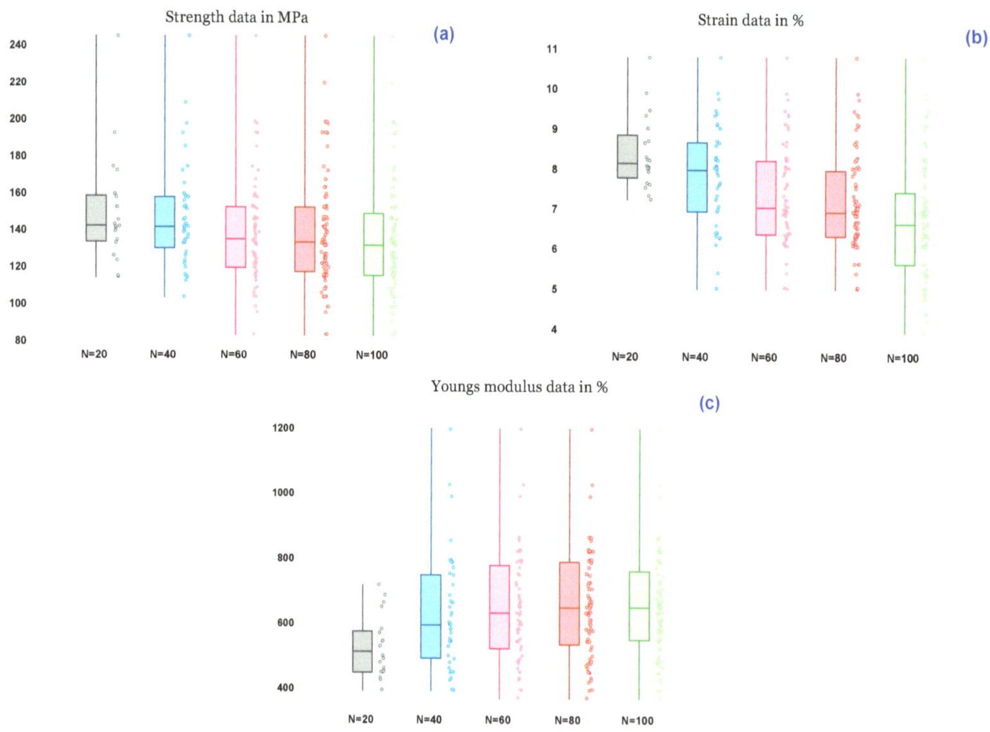

Figure 6. Average tensile properties with the number of tests N for (**a**) strengths data, (**b**) strain at break data, and (**c**) Young's modulus data.

4.3. Normality and Kolmogorov–Smirnov Tests

The test results grouped in Table 4 show that the *p*-value of normality using the Kolmogorov–Smirnov test criteria was >0.15 for elastic modulus data only. Thus, the property dispersion can be described by a 2P Weibull, 3P Weibull, and lognormal distribution. Therefore, it is necessary to determine which of these three laws best describes the experimental results. Additionally, Table 5's test results demonstrate that the *p*-values for the 2P and 3P Weibull distributions were above 0.1, while the *p*-value for the lognormal distribution was less than 0.1. Although the *p*-value was less than 0.1, in order to exclude a distributional law, the *p*-value recorded for the 3P Weibull law indicates that it was not as effective as the other 2P Weibull laws in describing our data. Thus, the dispersion of the properties can be described using a 2P Weibull distribution.

4.4. AD Goodness-of-Fit of Normality of Sisal Yarn Data

Following the Kolmogorov–Smirnov test for normality, we can conclude that the Weibull distribution is the best in terms of describing the behavior of our results. To determine which one of the four proposed distributions (Table 6)—normal, lognormal, 3P-Weibull, or 2P-Weibull—best fits the series (five groups) on which the experimental data was collected, the Minitab software was used to compute and perform an Anderson–Darling (AD) good fit test on every distribution in order to determine the smallest values of AD. This test produces the K-S (Kolmogorov-Smirnov) adjustment. For different numbers of trials (N = 20, 40, 60, 80, and 100 trials), the 2P-Weibull distribution was used to estimate the AD quantities for the sisal yarns Young's modulus, i.e., AD = 0.837, 0.537, 0.482, 0.664, and 0.626, respectively. These were lower compared to the 3P-Weibull distribution estimates for AD = 0.920, 0.628, 0.452, 0.690, and 0.637. In addition, with respect to strain and stress, the smallest amounts were found for the two-parameter Weibull distribution (strain at break: AD = 0.989, 0.484, 0.543, 0.989, and 0.551 and stress: AD = 0.921, 0.756, 0.475, 0.503, and 0.523). However, the highest values were for the logarithmic distribution, which are given as follows: AD = 3.499, 3.402, 2.784, 3.597, and 4.400 for σ; AD = 2.869, 1.662, 2.358, 4.917, and 2.545 for ε; and finally, AD = 1.923, 2.883, 1.938, 1.814, and 1.821 for E.

4.5. Weibull Analysis of Sisal Yarn Data

In this study, the two-parameter Weibull law was used to investigate the data of sisal yarns' mechanical properties, including the stress, Young's modulus, and strain, which revealed a significant amount of variation. Thus, Figure 7 well represents the distribution curves according to Weibull-LS and ML, corresponding to the stress, strain, and Young's modulus of the experimental data. In addition, using the Minitab software, the associated parameters are listed in Table 7. From Figure 7, it can be seen that for different numbers of tests (N), the behavior of the straight lines from the Weibull diagram of two-parameter LS and ML concerning the sisal yarn was almost linear with a superposition and a slight difference between them. However, there was an inflection of the line with respect to the LS and LM Weibull distributions when N = 20 trials with respect to E (Figure 7c,f), which resulted in the lowest values. This nonlinearity and almost overlapping behavior was observed at the level of *Agave americana* plant fibers with different GLs, and at the level of sisal elemental fibers [39] and yarn [13] (depending on GL). The R^2 coefficient is the primary control for evaluating the variation in the Weibull modulus (m). Furthermore, it is worth noting the satisfactory linearity of the fit between all datasets (Figure 7). Indeed, for each estimator, we found a correlation factor (R^2) of 0.900, 0.930, 0.958, 0.957, and 0.955 for σ, 0.910, 0.990, 0.964, 0.946, and 0.972 for ε, and 0.944, 0.930, 0.958, 0.964, and 0.969 for the Young's modulus. In addition, the two-parameter LS-Weibull model provides higher correlations (R^2 = 0.970–0.990 (Table 7)) with respect to the strain at failure compared to the other features (σ and E).

Table 7 lists the Weibull distribution's parameters, forms, and localities (which are its defining values) for each of the mechanical properties. The corresponding Weibull moduli (2P-Weibull-LS) least squares (LS) concerning strain (m_ε), stress (m_σ), and Young's modulus (m_E) relative to various N at 20, 40, 60, 80, and 100 mm were m_ε = 15.4, 8.7, 6.6, 5.9, 5.2, m_σ = 8.3, 6.4, 5.9, 6.2, and 5.4, and m_E = 1.7, 6.6, 4.8, 4.8, and 5.4, respectively. Consistent with reference [13], the length and amount of testing had a significant impact on the Weibull modulus of the tensile properties of elementary sisal yarns. Thus, for example, at 2P Weibull-LS, the stress moduli (m_σ) were 7.29, 7.21, 6.42, 6.18, and 6.11, respectively, for N = 20 to 100 tests. A similar behavior was observed with respect to all values of the 2P-Weibull modulus (ML) for different values of N (m_σ = 4.91, 4.69, 4.66, 4.52, and 4.47). For comparison (Table 7), the m_σ and σ_0 of the sisal yarn for N = 100 of 2P-Weibull-LS are 6.11 and 145 MPa, respectively. In contrast, in the 2P-Weibull-ML case, we found that m_σ = 4.47 and σ_0 = 147 MPa. The experimental value obtained in this case was 135 MPa (N = 100 tests).

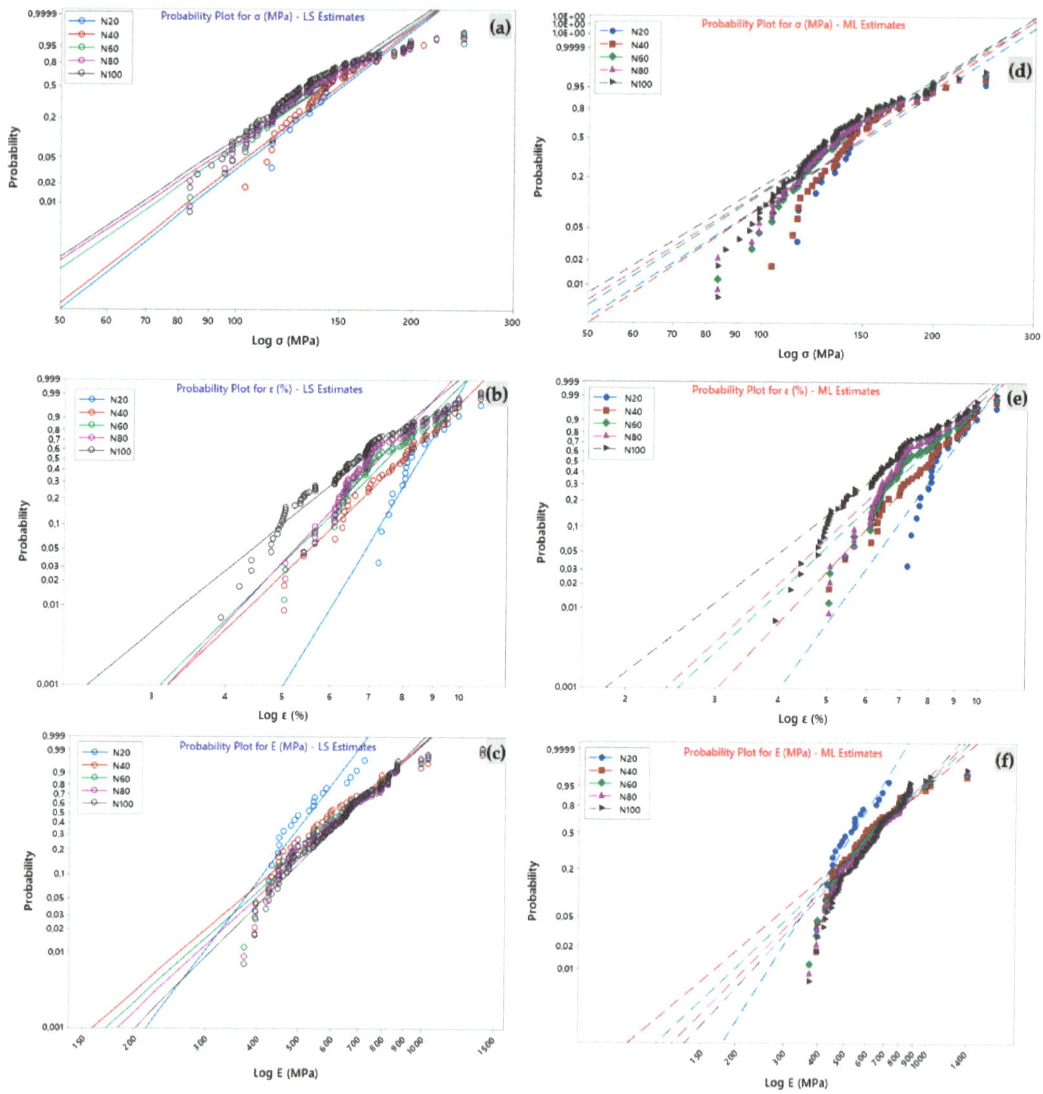

Figure 7. Probability plot for tensile properties with the number of tests N for (**a**–**c**) least square-LS estimation and (**d**–**f**) maximum likelihood-ML estimations.

Table 7. Summarized mechanical properties with statistical data of sisal yarns.

N	Strength, σ (MPa)					Strain, ε (%)					Young Modulus, E (GPa)				
	Weibull Statistic-LS														
	Scale	95% CI (Lower, Upper)	Shape	95% CI (Lower, Upper)	R^2	Scale	95% CI (Lower, Upper)	Shape	95% CI (Lower, Upper)	R^2	Scale	95% CI (Lower, Upper)	Shape	95% CI (Lower, Upper)	R^2
20	159	145, 164	7.29	6.17, 9.77	0.900	8.75	8.31, 9.03	12.42	9.83, 14.48	0.910	561	515, 591	7.20	5.42, 9.10	0.944
40	157	147, 162	7.21	6.25, 8.61	0.930	7.34	6.89, 8.66	7.21	5.71, 8.81	0.990	691	630, 729	4.69	4.03, 5.43	0.930
60	149	141, 154	6.42	5.60, 7.41	0.958	7.84	6.51, 8.10	7.45	6.32, 8.41	0.964	706	659, 738	4.86	4.23, 5.52	0.958
80	149	140, 155	6.18	5.32, 7.09	0.957	7.57	6.91, 8.28	8.01	6.98, 9.28	0.946	713	631, 722	5.06	4.61, 6.12	0.964
100	145	142, 150	6.11	5.37, 7.22	0.955	7.20	6.43, 8.54	6.10	5.22, 8.01	0.972	712	629, 725	5.49	4.92, 6.80	0.969
	Weibull Statistic-ML														
20	163	147, 179	4.91	3.50, 6.29		8.83	8.37, 9.32	8.77	6.46,11.90		568	525, 613	5.98	4.32, 8.28	
40	160	149, 171	4.69	3.97, 6.06		8.37	7.98, 8.79	6.78	5.36, 8.57		709	639, 786	3.19	2.59, 3.94	
60	151	142, 160	4.66	3.91, 5.54		7.92	7.57, 8.28	6.04	5.02, 7.27		719	667, 775	3.55	2.99, 4.22	
80	151	144, 159	4.52	3.88, 5.27		7.67	7.37, 7.97	5.93	5.07, 6.94		723	680, 769	3.80	3.28, 4.41	
100	147	140, 154	4.47	3.92, 5.12		7.28	6.98, 7.59	4.99	4.32, 5.77		721	685, 759	4.04	3.55, 4.61	

Using the Minitab software for Weibull-LS and ML analyses, the probability of survival at multiple estimates corresponding to the mechanical properties, including strain, stress, and Young's modulus, were plotted in Figure 8. For example, the graph in Figure 8a was constructed with the LS approximation of the probability index (Equation (9)) for five pairs of Weibull stress parameters (m_σ = 7.29 and σ_0 = 159, m_σ = 7.2, m_σ = 6.42 and σ_0 = 149, m_σ = 6.18 and σ_0 = 149, m_σ = 6.11 and σ_0 = 145, respectively, for N = 20, 40, 60, 80, and 100 mm). Thus, it can be observed that, when the likelihood $P(\sigma)$ = 0.4 in the case of 2P-Weibull-LS, corresponding to a 40% population survival for the wire samples with N = 80 and 100 trials, the stress was evaluated simultaneously as 147 and 143 MPa. However, when the survival rate reached approximately 50%, the lowest values were 137 and 140 MPa for the same number of trials. For tensile strain and stress, $P(\varepsilon) = P(\sigma) = 0.4$, we obtained 7.2% and 143 MPa, respectively. In addition, for N = 100 tests and for N = 80, we obtained ε = 7.6%, and σ = 147.

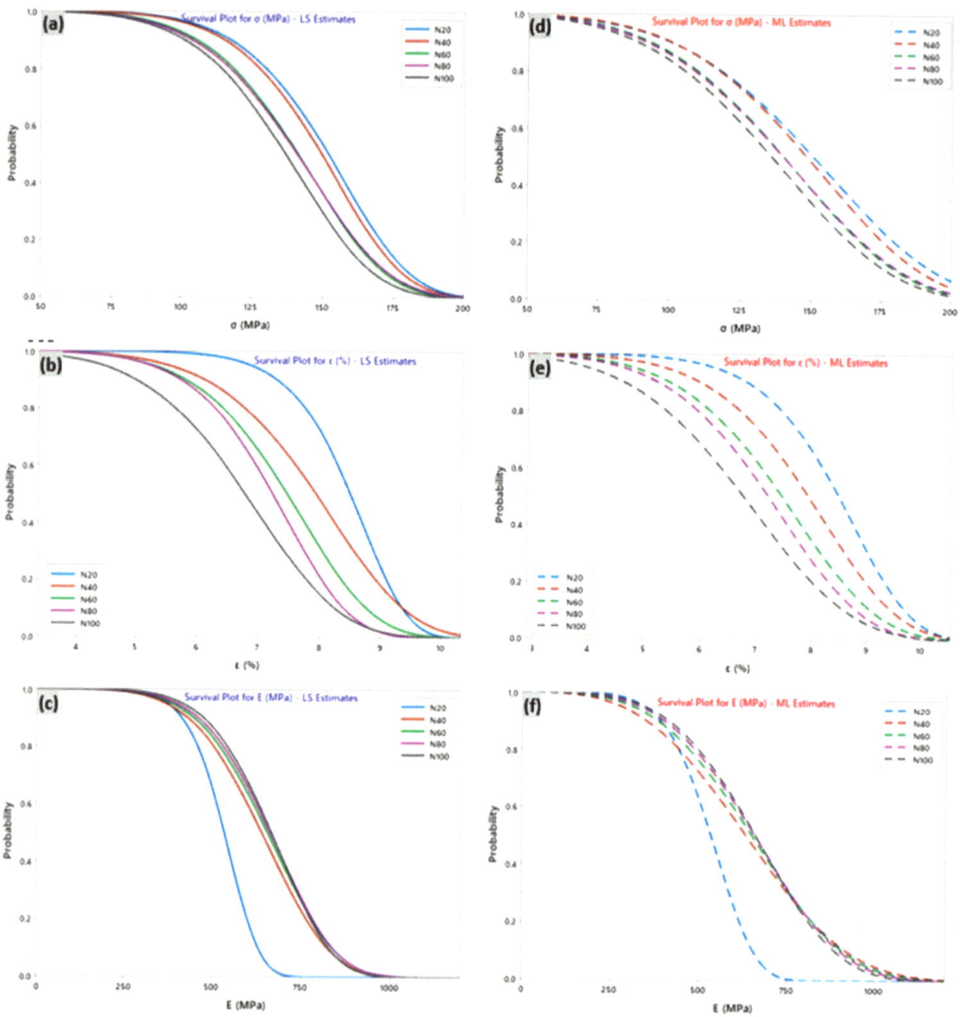

Figure 8. Survival plot for tensile properties with the number of tests N for (**a**–**c**) least square estimation-LS and (**d**–**f**) maximum likelihood-ML estimations.

The cumulative failure plot (Figure 9) allows us to determine the cumulative probability in % for a yarn element that meets a failure at load levels less than or equal to the specified load level, and thus, to evaluate the reliability of the sisal yarn from its failure. Furthermore, the cumulative failure function is the difference between 1 and the survival function. As an example, based on the data (N = 100 tests) related to the breaking stresses of the yarns, it appears that the probability of a yarn breaking when the stress reaches 165 MPa is about 0.90. This means that we can be 90% sure that the yarn will break at a tensile stress of 165 MPa for the LS estimate. This was similar for the other properties: E = 846 MPa and ε = 8.3% for the Young's modulus and strain, respectively. For the ML estimate, the stress, strain, and Young's modulus values were 172 MPa, 8.5%, and 905 MPa, respectively.

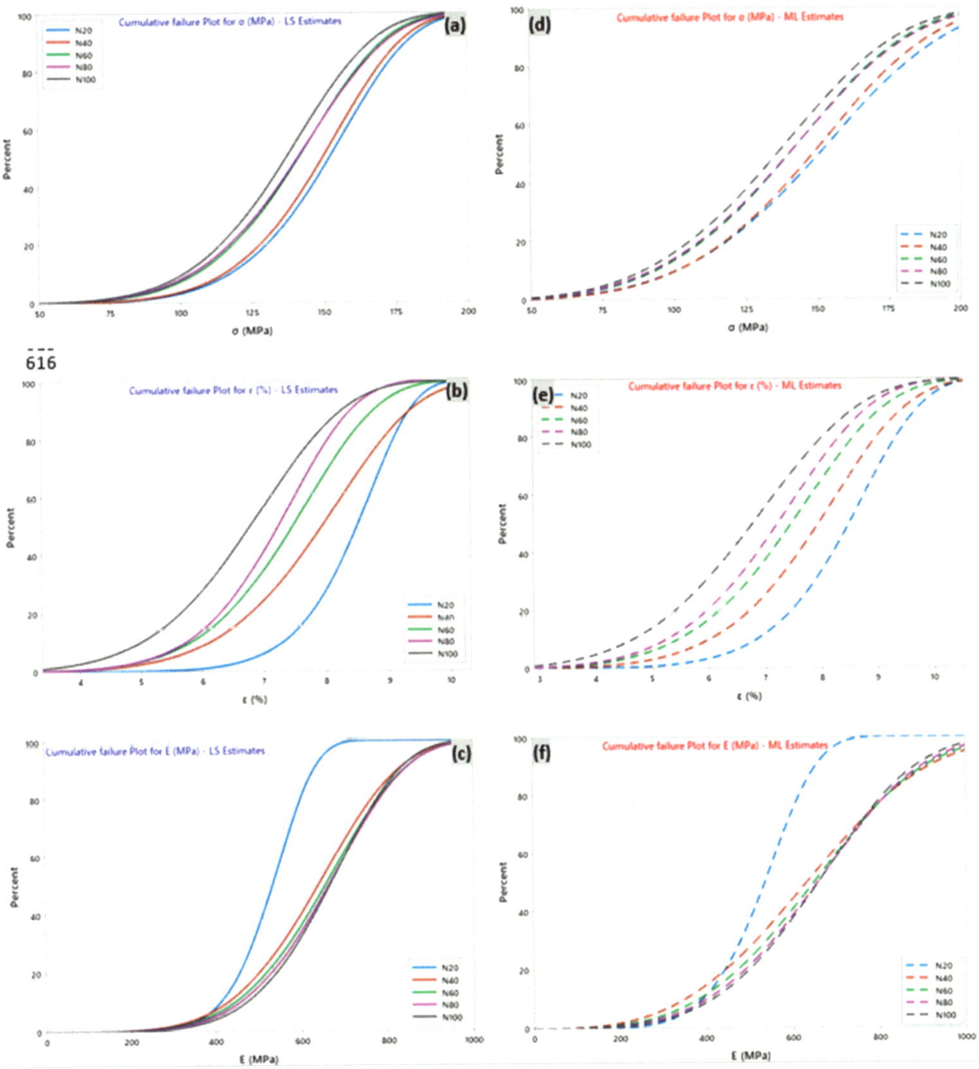

Figure 9. Cumulative failure plot for tensile properties with the number of tests N for (**a**–**c**) least square estimation-LS and (**d**–**f**) maximum likelihood-ML estimations.

There are several methods to determine the hazard function, which represents the probability of a defect as a function of the survival time of the sisal yarn. In the case of sisal yarn, the probability of a defect is a function of the survival time of the yarn. It can be observed that all the curves in the figures (Figure 10) have an increasing exponential trend, which means that the elements have a higher hazard of breaking as the load increases. In general, an increase in hazard occurs at the end of yarn breakage, especially when many fibers break simultaneously, resulting in sudden yarn breakage. The profile of the curve depended on the data, and the model was chosen for the analysis. As an example, Figure 10a is plotted according to the Weibull-LS function with five pairs of shape and scale parameters of the data (m_σ = 7.29 and σ_0 = 159, m_σ = 7.21 and σ_0 = 157, m_σ = 6.42 and σ_0 = 149, m_σ = 6.18 and σ_0 = 149, m_σ = 6.11 and σ_0 = 145 for N = 20, 40, 60, 80, and 100 mm, respectively). This translates into a hazard plot; therefore, the hazard rate increases with increasing loading. Furthermore, it is clear that for the same equivalent load rate of 195 MPa, the hazard rate was reduced when the number of tests N increased (rate = 0.17 for N = 20 up to 0.22 for N = 100). Moreover, for Weibull-ML (Figure 10d), the estimated rate was from 0.09 for N = 20 to 0.13 for N = 100, which is significantly lower than the rate estimated by the Weibull-LS model. The same interpretation applies to the other figures (Figure 10b,c,e,f).

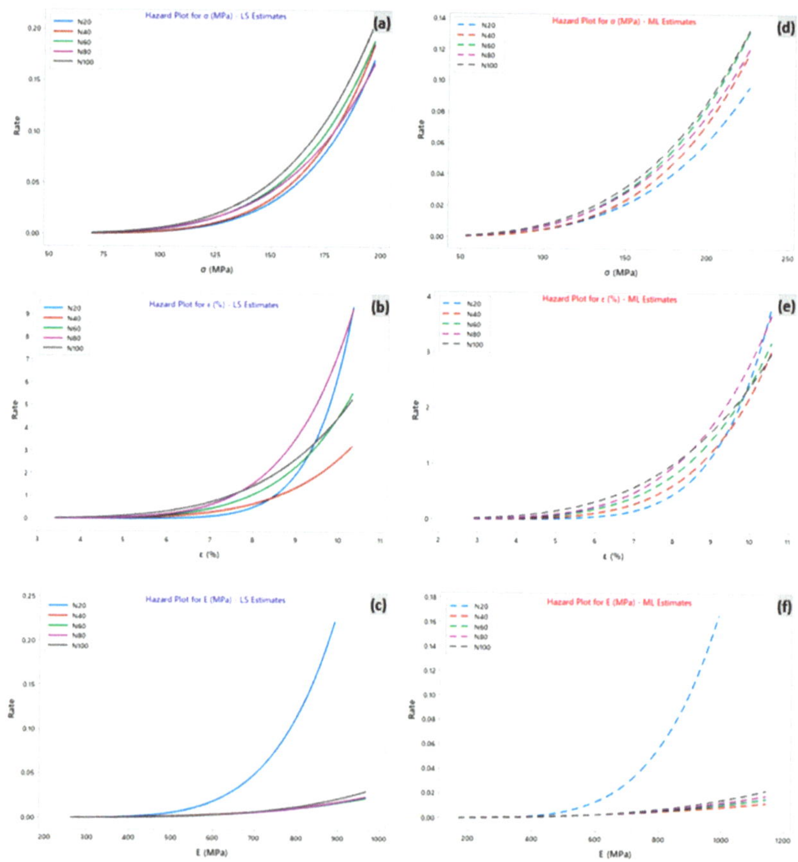

Figure 10. Hazard plot for tensile properties with the number of tests N for (**a–c**) least square estimation-LS and (**d–f**) maximum likelihood-ML estimations.

4.6. ANOVA Analysis of the Mechanical Properties for Yarn Data

The statistical treatment applied to the data is a method to better understand and analyze the significance of the experiment's findings. In the current study, the population samples were examined using one-way analysis of variance (ANOVA) based on the mean and distribution of data. This is because there was a large difference between the averages of multiple groups. The two potential hypotheses for the ANOVA technique's kind of hypothesis test are as follows: The first hypothesis states that all sample means are identical or not statistically distinct from one another. In addition, the number of yarns is a determining factor in the choice of sample. Thus, to best define any mechanical property parameters, Fisher's test, P, CI, MS, and SS were utilized for ANOVA to establish the impact of the number of tests on the answers.

Table 8 lists all of the results of the one-way analysis of variance (ANOVA) test with a 95% confidence level for the sisal yarn mechanical characteristics (ε, σ, and E) for the various research groups (20, 40, 60, 80, and 100). Due to the p value = 0.000 ($p < 0.001$) being below the significance limit (0.05), it is, therefore, not possible to retain the null hypothesis, indicating that averages would be identical.

Table 8. ANOVA test for strength, strain, and Young's modulus data of the sisal yarns for 95% CI.

Source	DF	Seq SS	Adj MS	Adj SS	Contribution	F-Value	*p*-Value
A. ANOVA test for ultimate tensile stress (for N = 20 to 100 samples)							
BG	4	6860	1714.9	6860	2.55%	1.93	0.105
WG	295	262,000	888.1	262,000	97.45%		
Total	299	268,860	-	-	100.00%		
S = 29.8016; R-sq = 2.55%; R-sq(adj) = 1.23%; PREESS = 270,897; R-sq(pred) = 0.00%							
B. ANOVA test for strain at failure data (for N = 20 to 100 samples)							
BG	4	71.82	17.955	71.82	12.82%	10.85	0.000
WG	295	488.19	1.655	488.19	87.18%		
Total	299	560.01	-	-	100.00%		
S = 1.28642; R-sq = 12.82%; R-sq(adj) = 11.64%; PREESS = 50.461; R-sq(pred) = 10.10%							
C. ANOVA test for young's modulus (for N = 20 to 100 samples)							
BG	4	312,564	78,141	312,564	3.57%	2.73	0.029
WG	295	8,434,610	28,592	8,434,610	96.43%		
Total	299	8,747,175	-	-	100.00%		
S = 169.091, R-sq = 3.57%; R-sq(adj) = 2.27%; PREESS = 8,707,691; R-sq(pred) = 0.45%							

BG: between group; WG: within group; DF: degree of freedom; SS: sum of squares; MS: mean square; F: F-test for ANOVA-one way.

5. Conclusions

In this work, it was evaluated to what extent the number of tests (N) could influence the mechanical characteristics of elementary sisal yarns (tensile stress (σ), Young's modulus (E), and strain at break (ε)). For this purpose, it was necessary to conduct a series of experimental static tensile tests, which allowed us to determine the mechanical properties for five series of N tests (20, 40, 60, 80, and 100 tests) to then identify and deduce the most efficient number of tests. The main conclusions drawn by this study from the experimental results and Weibull's law analysis can be summarized as follows:

- From the tensile tests applied to sisal yarn, it was found that σ and ε of the yarn decreased with the increase in N from 20 to 80 mm and stabilized from 148 MPa to 138 MPa and from 8.41% to 7.15%. This was followed by a slight decrease in values for N = 100 tests, which produced 135 MPa for stress and 6.70% for strain at break;

- ○ According to the experimental results, as far as the sisal yarn is concerned, it appears that the best mechanical performance was obtained for N = 100 tests;
- ○ Moreover, the mechanical properties of the yarns were more consistent with the 2P-Weibull-LS distribution than with the other ML method;
- ○ Finally, a one-way ANOVA analysis was also employed and revealed that N strongly influenced the sisal yarn mechanical characteristics.

Research results on the tensile properties of sisal yarns have significant practical implications for improving the manufacture of durable ropes and composite structures. They provide essential information for optimizing design, improving mechanical performance, reducing risk, and developing new materials and manufacturing techniques in these fields.

Author Contributions: Conceptualization, A.B. and M.G.; methodology, M.G. and A.B.; software, M.G.; validation, M.G., A.B., M.B. and H.A.; formal analysis, A.B. and M.B.; investigation, M.G., A.B. and M.B.; resources, A.B.; data curation, M.G. and A.B.; writing—original draft preparation, M.G., A.B., M.B., H.A. and M.K.A.K.; writing—review and editing, M.G., A.B., M.B., H.A. and M.K.A.K.; visualization, A.B.; supervision, A.B.; funding acquisition, H.A. All authors have read and agreed to the published version of the manuscript.

Funding: This project was funded by the Deanship of Scientific Research and under the supervision of the Scientific and Engineering Research Centre at Najran University grant code (NU/RCP/SERC/12/19).

Institutional Review Board Statement: Not applicable.

Data Availability Statement: The data presented in this study are available on request from the corresponding author.

Acknowledgments: The authors are thankful to the Deanship of Scientific Research and the supervision of the Scientific and Engineering Research Centre at Najran University for funding this work under the Research centers funding program grant code (NU/RCP/SERC/12/19).

Conflicts of Interest: The authors declare no conflict of interest.

References

1. Antony, S.; Cherouat, A.; Montay, G. Experimental, analytical and numerical analysis to investigate the tensile behaviour of hemp fibre yarns. *Compos. Struct.* **2018**, *202*, 482–490. [CrossRef]
2. Das, O.; Babu, K.; Shanmugam, V.; Sykam, K.; Tebyetekerwa, M.; Neisiany, R.E.; Försth, M.; Sas, G.; Gonzalez-Libreros, J.; Capezza, A.J.; et al. Natural and industrial wastes for sustainable and renewable polymer composites. *Renew. Sustain. Energy Rev.* **2022**, *158*, 112054. [CrossRef]
3. de Carvalho Bello, C.B.; Boem, I.; Cecchi, A.; Gattesco, N.; Oliveira, D.V. Experimental tests for the characterization of sisal fiber reinforced cementitious matrix for strengthening masonry structures. *Constr. Build. Mater.* **2019**, *219*, 44–55. [CrossRef]
4. Shi, J.; Yuan, S.; Zhang, W.; Wang, G.; Zhang, J.; Chen, H.; Cheng, H. Jute yarn-wound composites: Optimization of methods for evaluating mechanical properties and improvement of mechanical properties. *J. Mater. Res. Technol.* **2022**, *21*, 827–840. [CrossRef]
5. Ashadujjaman, M.; Saifullah, A.; Shah, D.U.; Zhang, M.; Akonda, M.; Karim, N.; Sarker, F. Enhancing the mechanical properties of natural jute yarn suitable for structural applications. *Mater. Res. Express* **2021**, *8*, 55503. [CrossRef]
6. Farhad, M.; Chowdhury, M.; Nayemul, M. Heliyon Qualitative and statistical analysis of cotton- fl ax blend yarn. *Heliyon* **2022**, *8*, e10161. [CrossRef]
7. Karimah, A.; Ridho, M.R.; Munawar, S.S.; Adi, D.S.; Ismadi; Damayanti, R.; Subiyanto, B.; Fatriasari, W.; Fudholi, A. A review on natural fibers for development of eco-friendly bio-composite: Characteristics, and utilizations. *J. Mater. Res. Technol.* **2021**, *13*, 2442–2458. [CrossRef]
8. Cottrell, J.A.; Ali, M.; Tatari, A.; Martinson, D.B. Effects of Fibre Moisture Content on the Mechanical Properties of Jute Reinforced Compressed Earth Composites. *Constr. Build. Mater.* **2023**, *373*, 130848. [CrossRef]
9. Shaker, K.; Waseem Ullah Khan, R.M.; Jabbar, M.; Umair, M.; Tariq, A.; Kashif, M.; Nawab, Y. Extraction and characterization of novel fibers from Vernonia elaeagnifolia as a potential textile fiber. *Ind. Crops Prod.* **2020**, *152*, 112518. [CrossRef]
10. Placet, V.; Cissé, O.; Lamine Boubakar, M. Nonlinear tensile behaviour of elementary hemp fibres. Part I: Investigation of the possible origins using repeated progressive loading with in situ microscopic observations. *Compos. Part A Appl. Sci. Manuf.* **2014**, *56*, 319–327. [CrossRef]
11. Vijay, R.; Vinod, A.; Lenin Singaravelu, D.; Sanjay, M.R.; Siengchin, S. Characterization of chemical treated and untreated natural fibers from Pennisetum orientale grass- A potential reinforcement for lightweight polymeric applications. *Int. J. Lightweight Mater. Manuf.* **2021**, *4*, 43–49. [CrossRef]

12. Porras, A.; Maranon, A.; Ashcroft, I.A. Characterization of a novel natural cellulose fabric from Manicaria saccifera palm as possible reinforcement of composite materials. *Compos. Part B Eng.* **2015**. [CrossRef]
13. Belaadi, A.; Bourchak, M.; Aouici, H. Mechanical properties of vegetal yarn: Statistical approach. *Compos. Part B Eng.* **2016**, *106*, 139–153. [CrossRef]
14. Belaadi, A.; Boumaaza, M.; Amroune, S.; Bourchak, M. Mechanical characterization and optimization of delamination factor in drilling bidirectional jute fibre-reinforced polymer biocomposites. *Int. J. Adv. Manuf. Technol.* **2020**, *111*, 2073–2094. [CrossRef]
15. Codispoti, R.; Oliveira, D.V.; Olivito, R.S.; Lourenço, P.B.; Fangueiro, R. Mechanical performance of natural fiber-reinforced composites for the strengthening of masonry. *Compos. Part B Eng.* **2015**, *77*, 74–83. [CrossRef]
16. El-Geiheini, A.; ElKateb, S.; Abd-Elhamied, M.R. Yarn Tensile Properties Modeling Using Artificial Intelligence. *Alex. Eng. J.* **2020**, *59*, 4435–4440. [CrossRef]
17. Yan, L.; Chouw, N.; Yuan, X. Improving the mechanical properties of natural fibre fabric reinforced epoxy composites by alkali treatment. *J. Reinf. Plast. Compos.* **2012**, *31*, 425–437. [CrossRef]
18. GOMES, A.; GODA, K.; OHGI, J. Effects of Alkali Treatment to Reinforcement on Tensile Properties of Curaua Fiber Green Composites. *JSME Int. J. Ser. A Solid Mech. Mater. Eng.* **2004**, *47*, 541–546. [CrossRef]
19. Saaidia, A.; Bezazi, A.; Belbah, A.; Bouchelaghem, H.; Scarpa, F.; Amirouche, S. Mechano-physical properties and statistical design of jute yarns. *Meas. J. Int. Meas. Confed.* **2017**, *111*, 284–294. [CrossRef]
20. Abida, M.; Baklouti, A.; Gehring, F.; Vivet, A.; Bouvet, C. Mechanics of Materials Inverse approach for flax yarns mechanical properties identification from statistical mechanical characterization of the fabric. *Mech. Mater.* **2020**, *151*, 103638. [CrossRef]
21. Abida, M.; Gehring, F.; Mars, J.; Vivet, A.; Dammak, F.; Haddar, M. Hygro-mechanical coupling and multiscale swelling coefficients assessment of flax yarns and flax/epoxy composites. *Compos. Part A Appl. Sci. Manuf.* **2020**, *136*, 105914. [CrossRef]
22. Wang, J.; Zhou, H.; Liu, Z.; Peng, X.; Zhou, H. Statistical modelling of tensile properties of natural fiber yarns considering probability distributions of fiber crimping and effective yarn elastic modulus. *Compos. Sci. Technol.* **2022**, *218*, 109142. [CrossRef]
23. Sohbatzadeh, F.; Shakerinasab, E.; Mirzanejhad, S. Surface modification of aramid yarn by atmospheric pressure plasma: Reinforcement and floating properties. *Polym. Test.* **2023**, *117*, 107836. [CrossRef]
24. Belaadi, A.; Bezazi, A.; Bourchak, M.; Scarpa, F. Tensile static and fatigue behaviour of sisal fibres. *Mater. Des.* **2013**, *46*, 76–83. [CrossRef]
25. *D2256/D2256M—1081*; Standard Test Method for Tensile Properties of Yarns by the Single—Strand Method. ASTM International: West Conshohocken, PA, USA, 2010. [CrossRef]
26. Silva, F.D.A.; Mobasher, B.; Soranakom, C.; Filho, R.D.T. Effect of fiber shape and morphology on interfacial bond and cracking behaviors of sisal fiber cement based composites. *Cem. Concr. Compos.* **2011**, *33*, 814–823. [CrossRef]
27. Silva, F.d.A.; Chawla, N.; Filho, R.D.d.T. Tensile behavior of high performance natural (sisal) fibers. *Compos. Sci. Technol.* **2008**, *68*, 3438–3443. [CrossRef]
28. Blanchard, J.; Sobey, A.J.; Blake, J.I.R. Multi-scale investigation into the mechanical behaviour of flax in yarn, cloth and laminate form. *Compos. Part B Eng.* **2016**, *84*, 228–235. [CrossRef]
29. Katragadda, S.C.; Madhu, S.; Raju, J.S.N.; Md, J.S. Characterization of novel natural cellulosic fiber extracted from the stem of Cissus vitiginea plant. *Int. J. Biol. Macromol.* **2020**, *161*, 1358–1370. [CrossRef]
30. Ahmadi, M.V.; Doostparast, M.; Ahmadi, J. Estimating the lifetime performance index with Weibull distribution based on progressive first-failure censoring scheme. *J. Comput. Appl. Math.* **2013**, *239*, 93–102. [CrossRef]
31. Gorjan, L.; Ambrožič, M. Bend strength of alumina ceramics: A comparison of Weibull statistics with other statistics based on very large experimental data set. *J. Eur. Ceram. Soc.* **2012**, *32*, 1221–1227. [CrossRef]
32. Belaadi, A.; Boumaaza, M.; Alshahrani, H.; Bourchak, M. Mechanical Properties of Natural Cellulosic Yucca treculeana L. Fiber for Biocomposites Applications: Statistical Analysis. *J. Nat. Fibers* **2022**, *19*, 15501–15518. [CrossRef]
33. Belaadi, A.; Amroune, S.; Seki, Y.; Keskin, O.Y.; Köktaş, S.; Bourchak, M.; Dufresne, A.; Fouad, H.; Jawaid, M. Extraction and Characterization of a New Lignocellulosic Fiber from Yucca Treculeana L. Leaf as Potential Reinforcement for Industrial Biocomposites. *J. Nat. Fibers* **2022**, *19*, 12235–12250. [CrossRef]
34. Ben Smail, Y.; El Moumen, A.; Imad, A.; Lmai, F.; Elminor, H. The effects of environmental conditions on the mechanical properties of jute yarns. *Mater. Today Proc.* **2020**, *30*, 860–864. [CrossRef]
35. Ben Smail, Y.; El Moumen, A.; Imad, A.; Lmai, F.; Elminor, H. Loading rate effect on mechanical properties of the jute yarns. *Mater. Today Proc.* **2021**, *37*, 3827–3833. [CrossRef]
36. Omrani, F.; Wang, P.; Soulat, D.; Ferreira, M. Mechanical properties of flax-fibre-reinforced preforms and composites: Influence of the type of yarns on multi-scale characterisations. *Compos. Part A Appl. Sci. Manuf.* **2017**, *93*, 72–81. [CrossRef]
37. Kim, J.T.; Netravali, A.N. Development of aligned-hemp yarn-reinforced green composites with soy protein resin: Effect of pH on mechanical and interfacial properties. *Compos. Sci. Technol.* **2011**, *71*, 541–547. [CrossRef]

38. Dembri, I.; Belaadi, A.; Boumaaza, M.; Bourchak, M. Tensile Behavior and Statistical Analysis of Washingtonia Filifera Fibers as Potential Reinforcement for Industrial Polymer Biocomposites. *J. Nat. Fibers* **2022**, *19*, 14839–14854. [CrossRef]
39. Belaadi, A.; Bezazi, A.; Bourchak, M.; Scarpa, F.; Zhu, C. Thermochemical and statistical mechanical properties of natural sisal fibres. *Compos. Part B Eng.* **2014**, *67*, 481–489. [CrossRef]

polymers

MDPI

Article

Study on the Tensile Behavior of Woven Non-Woven PLA/OLA/MgO Electrospun Fibers

Adrián Leonés, Laura Peponi *, Jesús-María García-Martínez † and Emilia P. Collar †

Instituto de Ciencia y Tecnología de Polímeros (ICTP-CSIC), Juan de la Cierva 3, 28006 Madrid, Spain; aleones@ictp.csic.es (A.L.); jesus.maria@ictp.csic.es (J.-M.G.-M.); ecollar@ictp.csic.es (E.P.C.)

* Correspondence: lpeponi@ictp.csic.es

† Grupo de Ingeniería de Polimeros.

Abstract: The present work deeply studied the mechanical behavior of woven non-woven PLA/OLA/MgO electrospun fibers, efibers, by using Box–Wilson surface response methodology. This work follows up a previous one where both the diameters and the thermal response of such efibers were discussed in terms of both the different amounts of magnesium oxide nanoparticles, MgO, as well as of the oligomer (lactic acid), OLA, used as plasticizer. The results of both works, in term of diameters, degree of crystallinity, and mechanical response, can be strongly correlated to each other, as reported here. In particular, the strain mechanism of PLA/OLA/MgO efibers was studied, showing an orientation of efibers parallel to the applied stress and identifying the mechanically weakest points that yielded the start of the breakage of efibers. Moreover, we identified 1.5 wt% as the critical amount of MgO, above which the plasticizing effect of OLA was weaker as the amount of both components increased. Moreover, the minimum elastic modulus value took place at 15 wt% of OLA, in agreement with the previously reported convergence point in the evolution of the degree of crystallinity. Regarding the yield point, a concentration of OLA between 20 and 30 wt% led to a slight improvement in the yielding capability in terms of tensile strength in comparison with neat PLA efibers. Therefore, the approach presented here permits the design of tailor-made electrospun nanocomposites with specific mechanical requirements.

Keywords: electrospinning; poly(lactic acid); magnesium oxide; oligomer(lactic acid); design of experiments; Box–Wilson response surface methodology

Citation: Leonés, A.; Peponi, L.; García-Martínez, J.-M.; Collar, E.P. Study on the Tensile Behavior of Woven Non-Woven PLA/OLA/MgO Electrospun Fibers. *Polymers* **2023**, *15*, 3973. https://doi.org/10.3390/polym15193973

Academic Editor: In Chul Um

Received: 21 August 2023
Revised: 26 September 2023
Accepted: 28 September 2023
Published: 3 October 2023

1. Introduction

Nowadays, poly(lactic acid), PLA, has emerged as a prominent polymer substitute for various petrochemical-based polymers due to its remarkable properties, such as biodegradability [1,2] or compostability [3,4]. However, from the mechanical point of view, its inherent brittleness and poor elongation at break has limited its applications [5]. This limitation is particularly significant in fields such as packaging [6,7] or tissues engineering [8]. Certainly, the brittleness of PLA constitutes a major problem in the main industrial fields, where there is no tolerance for cracking or tearing when subjected to force during manufacturing [9,10].

Efforts have been carried out to enhance these mechanical properties, including blending with other biodegradable polymers [11,12], copolymerization with other polymers [13,14], reinforcing with nanoparticles, NPs [15,16], or the incorporation of plasticizers agents [17–20]. Particularly, the design of PLA-based nanocomposites by adding NPs represents one of the most used strategies to increase and to enhance the mechanical properties of PLA [12,21].

Regarding the different processing techniques used to obtain PLA-based nanocomposites, the electrospinning technique has emerged as a suitable technique to fabricate PLA-based electrospun nanocomposites. This versatile technique enables the easy fabrication of woven non-woven electrospun fibers (from now on, referred to as efibers) reinforced with NPs through the application of a high electric voltage to a dispersion of NPs within a polymer solution in a volatile solvent under ambient conditions [8,22]. In particular, for biomedical applications, searching to improve the mechanical properties of pristine PLA efibers to mimic those of human tissues, the use of organic NPs is reported in the scientific literature, such as cellulose nanocrystals, CNC, or chitosan NPs with enhanced elongation at break and higher degree of crystallinity [23]. On the other hand, also, inorganic NPs such as magnesium oxide, MgO, NPs or silver, Ag, NPs can be used as reinforcements for PLA-based efibers due to their biocompatibility and antimicrobial properties [8,24,25].

Therefore, electrospun PLA-based nanocomposites represent the combination of, at least, two different components—in particular, the combination of the PLA polymeric matrix with other materials on the nanometric scale, such as NPs, in order to significantly improve their mechanical properties [12]. The improvement in the mechanical properties of efibers is related to the existence of interactions between NP surfaces and the macromolecular chains of the PLA matrix that surround them. Specifically, these interactions take place at the molecular level, where physical–chemical interactions between NPs and PLA chains play a crucial role in the overall behavior of the material as a whole [26]. Such a level of molecular interactions take place at the nanoscale and are related to the mobility of polymer chains adjacent to NPs and can be affected by some factors, such as the length of the polymer chains [27,28], the NP surface morphology [29,30], the NP concentration [31], or the interactions between polymer NPs and other components [32].

Furthermore, to properly tailor the thermal properties of electrospun PLA-based nanocomposites, such as its glass transition temperature, T_g, to approach the human body, the incorporation of plasticizers has been studied in the scientific literature [17,18]. The ideal plasticizer should exhibit a similar chemical structure, showing comparable intermolecular forces capable of strongly interacting with the PLA matrix. Additionally, oligomeric lactic acid, OLA, presents itself as a promising alternative to conventional plasticizers for PLA due to its similar chemical structure, relatively high molar mass, and renewable origin [33–35]. In fact, looking at the increase of both the mechanical and the thermal properties, preliminary studies of PLA-based nanocomposites plasticized with OLA can be found in the literature [17,36]. From both the mechanical and thermal points of view, Arrieta et al. reported a significant increase of the elongation at break of 20% with 15 wt% of OLA in comparison with neat PLA-based efibers [36], and Leonés et al. reported an increase in the elongation at break with 10–30 wt% OLA content, along with effective modulation of the glass transition temperature of PLA-based efibers [17].

Additionally, a certain concentration of plasticizer will be considered as the maximum, named the critical concentration [37], and, consequently, different amounts of plasticizer are expected to cause increasing improvements of the molecular interactions between PLA chains and NPs up to a limit close to the critical concentration. Once this critical concentration is overpassed, no further enhancements in the molecular interactions will occur between the components. In fact, above this concentration, the structural integrity of the polymer matrix could be compromised, leading to the possible segregation of the plasticizer, which could form its own phase decreasing the thermal and mechanical properties of the nanocomposites [38].

Therefore, understanding the interactions between NPs, plasticizer, and the surrounding polymeric chains implies an important aspect to be considered in the study of nanocomposites and their final mechanical performance. In this work, we describe the fitting of the mechanical behavior of PLA/OLA/MgO electrospun fibers to a mathematical model that permits the design of tailor-made electrospun nanocomposites with specific mechanical requirements. With this aim, and based on our previous results [18], the purpose of the present study is to find a correlation between each one of the tensile parameters and the composition of the PLA/OLA/MgO electrospun PLA-based nanocomposites. Thus, a Box–Wilson surface response methodology statistical design was used to model the system behavior in all the experimental range scanned. Once the PLA-based efibers were obtained, tensile tests were carried out, and the main mechanical properties were calculated. From these, a series of remarks concerning the tensile mechanism between MgO NPs, OLA plasticizer, and the PLA matrix was proposed and properly discussed in this work.

2. Materials and Methods

Polylactic acid (PLA3051D, 3% of D-lactic acid monomer, molecular weight 14.2 × 10^4 g·mol^{-1}, density 1.24 g·cm^{-3}) was supplied by NatureWorks®, Minneapolis, MN, USA. Lactic acid oligomer (Glyplast OLA8, ester content >99%, density 1.11 g·cm^{-3}, viscosity 22.5 mPa·s, molecular weight 1100 g·mol^{-1}) was kindly supplied by Condensia Quimica SA, Barcelona, Spain.

Before starting the electrospinning process, each solution was prepared following the next process. Firstly, the corresponding amounts of PLA and OLA were dissolved separately in $CHCl_3$ and stirred overnight at room temperature. Secondly, the amount of MgO NPs was weighed and dispersed in 20 mL of $CHCl_3$; after 30 min, the OLA solution was added to the MgO NPs solution and dispersed over 60 min. Then, the PLA solution was added to the MgO NPs and OLA solution and dispersed simultaneously for another 60 min. Finally, the necessary volume of DMF was added to assure the proportion of solvents ($CHCl_3$:DMF, 4:1). The dispersion process was carried out with a sonicator tip (Sonic Vibra-Cell VCX 750) of 750 watts and amplitude of 20%. Then, electrospun fiber mats were obtained in an Electrospinner Y-flow 2.2.D-XXX (Nanotechnology Solutions, Malaga, Spain). The electrospinning parameters were set as a voltage of 20 kV, distance of 17 cm between the tip and collector, and a polymer solution flux of 3.5 mL·h^{-1} in order to obtain randomly oriented efiber mats [18].

The amounts of OLA (wt%) and MgO (wt%) were determined in agreement with the specifications of the Box–Wilson experimental worksheet (Table 1). Box–Wilson methodology is based on a factorial design involving $2k + 2k + 1$ experiments with $2 + K$ additional replicated central runs, where k represents the number of independent variables [39,40]. The central point variables are coded as (0, 0), while the independent variable interval (OLA and MgO) is in the range between 6.00 and 30.00 wt% for OLA and between 0.60 and 3.00 wt% for MgO due to the factorial component being coded as (−1, 1). Additionally, the Box–Wilson experimental model considers $\alpha = \sqrt{2}$ as the coded variable for the star points of the worksheet [39,40]. Therefore, all coded and controlled factors are listed in Table 1 together with each run number. The trials were carried out in a randomized way, and the variables studied were statistically analyzed by one-way analysis of variance (ANOVA) using the statistical computer package Statgraphics Centurion XVII (Statpoint Technologies, Inc., Warrenton, VA, USA) [18].

Table 1. Worksheet for the Box–Wilson experimental design used.

Run	Coded Factors		Controlled Factors	
	OLA	MgO	OLA (wt%)	MgO (wt%)
I	−1	−1	6.00	0.60
II	1	−1	30.00	0.60
III	−1	1	6.00	3.00
IV	1	1	30.00	3.00
V	$-\sqrt{2}$	0	1.03	1.80
VI	$\sqrt{2}$	0	34.97	1.80
VII	0	$-\sqrt{2}$	18.00	0.10
VIII	0	$\sqrt{2}$	18.00	3.49
IX	0	0	18.00	1.80
X	0	0	18.00	1.80
XI	0	0	18.00	1.80
XII	0	0	18.00	1.80
XIII	0	0	18.00	1.80

Once the different polynomials fitted to quadratic models of each mechanical property studied by following the Box–Wilson surface response methodology were obtained, the data were then plotted using the 3D response surface plot, as well as the contour plots. In particular, the polynomial models corresponding to each mechanical property were plotted as a function of the OLA and MgO NP contents in all the experimental range scanned using OriginPro v8.5 software.

Once the efibers mat of each experimental run was obtained, the morphological and mechanical characterization was carried out. Scanning Electron Microscopy, SEM (PHILIPS XL30 Scanning Electron Microscope, Phillips, Eindhoven, The Netherlands), and Field Emission Scanning Electron Microscopy, FESEM (Hitachi S8000), were used in order to study the morphology of the efibers. All the samples were previously gold-coated (~5 nm thickness) in a Polaron SC7640 Auto/Manual Sputter (Polaron, Newhaven, East Sussex, UK).

The mechanical properties of the PLA-based electrospun nanocomposite mats were studied in the tensile test mode at room temperature. For this purpose, an Instron dynamometer (model 3366) equipped with a 100 N load cell at a crosshead speed of 10 mm·min^{-1} and initial length between clamps of 30 mm was used. At least five specimens of 10 mm, 6 mm width, and 100 μm of average thickness cut from the electrospun mats were measured. The mechanical properties identified over the stress–strain tests were the elastic modulus, E, as the slope of the curve between 0% and 2% of elongation, the tensile strength, and the elongation at the yield (σ_y and ε_y) and at break points (σ_b and ε_b), each one of them reported as the average value from each test. Nevertheless, since the Box–Wilson methodology does not contemplate the processing step but only a series of independent terms, the results obtained will be related to the interactions between components. In particular, in our system, the variables that can influence the final properties will be the amount of the components of the electrospun nanocomposites (PLA, MgO, and OLA). We used the same dispersion and electrospinning process parameters for all the experimental runs in order to reasonably control its effect in the final mechanical performance of the electrospun nanocomposites.

3. Results and Discussion

Once the different woven non-woven electrospun nanocomposites mats were obtained, the mechanical characterization of each run was carried out. Table 2 compiles all the results obtained from the tensile test experiments according to the Box–Wilson two independent variables experimental worksheet—in particular, the mechanical properties measured in the main regions at the stress–strain curves: the elastic zone (elastic modulus), and the yield and break points (strength and elongation in both). In Figure 1, the stress–strain curves for each run of the PLA-based efibers is reported.

As can be observed, the stress–strain curves are widely affected by the different compositions of each experimental run according to the Box–Wilson two independent variables experimental worksheet (Table 1). However, the central points—that is, from run IX to run XIII—showed similar tensile behaviors due to their similar compositions in terms of both OLA (wt%) and MgO (wt%) contents.

Table 2. Average mechanical properties obtained from the tensile test of each run.

Run	E (MPa)	σ_y (MPa)	ε_y (%)	σ_b (MPa)	ε_b (%)
I	70.8 ± 9.5	2.5 ± 0.8	5.3 ± 0.6	2.9 ± 1.1	37.2 ± 1.6
II	64.7 ± 8.9	1.3 ± 0.7	2.3 ± 0.2	2.7 ± 1.8	39.0 ± 1.3
III	36.3 ± 4.5	0.9 ± 0.3	2.7 ± 0.6	1.6 ± 0.4	22.4 ± 2.1
IV	169.4 ± 6.0	2.5 ± 0.6	2.1 ± 0.6	4.6 ± 1.8	4.2 ± 1.3
V	37.1 ± 7.9	1.0 ± 0.3	3.6 ± 0.4	2.9 ± 0.7	66.8 ± 1.7
VI	156.4 ± 9.3	3.2 ± 0.5	2.5 ± 0.4	6.0 ± 1.2	32.7 ± 1.7
VII	85.6 ± 2.7	2.5 ± 0.5	3.8 ± 0.7	3.8 ± 0.8	54.3 ± 1.4
VIII	56.6 ± 1.3	1.6 ± 0.4	3.9 ± 0.6	2.6 ± 0.6	32.4 ± 1.2
IX	40.9 ± 1.7	1.0 ± 0.4	3.1 ± 0.2	1.8 ± 0.4	27.6 ± 1.5
X	30.8 ± 1.8	1.2 ± 0.3	3.5 ± 0.1	1.2 ± 0.6	15.7 ± 1.7
XI	52.5 ± 9.9	1.6 ± 0.3	3.9 ± 0.5	2.2 ± 0.4	27.7 ± 1.7
XII	91.9 ± 2.4	1.9 ± 0.2	4.5 ± 0.5	4.0 ± 0.8	54.4 ± 1.4
XIII	31.8± 3.1	1.2 ± 0.4	4.4 ± 0.7	1.4 ± 0.8	23.8 ± 1.7

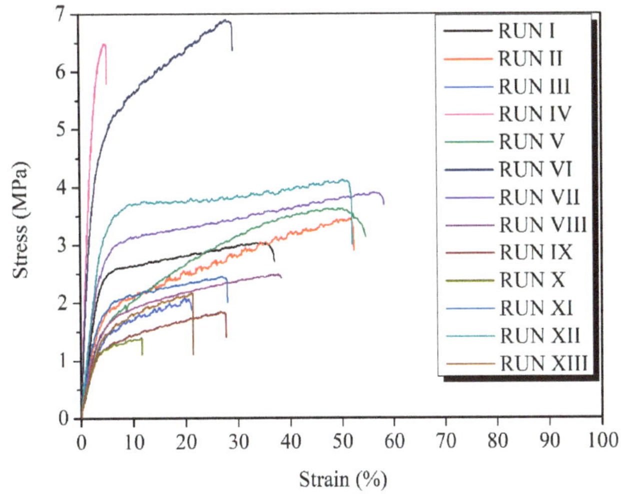

Figure 1. Stress–strain curve for PLA-based efibers obtained in each run studied.

The results obtained for each one of the studied parameters were fitted to quadratic models by following the Box–Wilson surface response methodology [39]. Five different polynomials describing the evolution of E (MPa), σ_y (MPa), ε_y (%), σ_b (MPa), and ε_b (%) were properly obtained, having the general form

$$Y = a_0 + a_1 \cdot x_1 + a_2 \cdot x_2 + a_3 \cdot x_1 \cdot x_2 + a_4 \cdot x_1^2 + a_5\, x_2^2$$

The coefficients obtained for each polynomial fit are summarized in Table 3, as well as its percentual confidence values for $\langle r^2 \rangle$ (%), the lack of fit, and the confidence factors coefficients, obtained from the ANOVA.

Table 3. Statistical parameters and coefficients of the polynomial equations from the Box–Wilson experimental design used ($Y = a_0 + a_1 \cdot x_1 + a_2 \cdot x_2 + a_3 \cdot x_1 \cdot x_2 + a_4 \cdot x_1^2 + a_5 \cdot x_2^2$).

	$\langle r^2 \rangle$ (%)	L. F. * (%)	C. F. * (%)	Ind. T. * a_0	L. T. * a_1	a_2	Int. T. * a_3	Q. T. * a_4	a_5
E (MPa)	82.47	51.0	97.8	145.7	−7.248	−68.16	2.416	0.1661	7.699
σ_y (MPa)	73.52	16.1	93.9	3.833	−0.1225	−1.695	0.04705	0.002058	0.1850
ε_y (%)	69.12	31.4	90.8	5.402	−0.007468	−0.8061	0.04288	−0.003415	−0.06892
σ_b (MPa)	67.29	62.2	89.3	5.648	−0.2686	−1.893	0.05469	0.006783	0.2217
ε_b (%)	45.79	32.2	59.1	59.91	−1.228	−5.980	−0.3477	0.03279	1.067

* L. F. (lack of fit), C. F. (confident factor), Ind. T. (independent term), L. T. (linear terms), Int. T (interaction term), and Q. T. (quadratic terms).

As reported in the literature [39,40], $\langle r^2 \rangle$ (%) is the statistical parameter that shows the quality of the fit of the model to a dataset and its predictive capability for the studied variables. In addition, $\langle r^2 \rangle$ (%) values higher than 70.00% are considered indicative of good fitted quadratic models. In our case, the $\langle r^2 \rangle$ (%) values obtained were 82.47% (E), 73.52% (σ_y), 69.12% (ε_y), 67.29% (σ_b), and 45.79% (ε_b), respectively. Therefore, in our case, for E and $\sigma_{y,}$ we can remark on the significance of the independent variables chosen, which are, in this case, the contents of both OLA (wt%) and MgO (wt%), to model the mechanical behavior of the PLA-based efibers in the experimental range studied. On the other hand, the $\langle r^2 \rangle$ (%) values for ε at yield and σ at break, 69.12% and 67.29% respectively, are in the limit and particularly low for ε at yield (45.79%) due to the wide variability of the experimental elongation values at the break point.

Table 3 also shows the lack of fit values, which are the percentage of pure error, related to possible factors not considered by the model but significantly influencing the response. High values of lack of fit suggest that this parameter is more susceptible to the noise effects present in the experiment process. Additionally, high values for the confident factor indicate the full significance of the independent variables chosen in the study. As described above, in our case, the OLA (wt%) and MgO (wt%) contents properly described the mechanical behavior of efibers. In particular, lack of fit values for the elastic modulus, 51.0%, and tensile strength at break, 62.2%, are the highest, which indicates that these are the parameters in which pure error is easier to accumulate. This is in agreement with the fact that both of them can be considered as the most sensitive parameters to the stress transmission between the PLA matrix and the MgO NPs, being, respectively, the starting and the end points of the mechanical test.

Likewise, very high values of the "confidence factor" indicate that almost all the factors considered to build the model play a prime role in the mechanical behavior of the system, except the elongation at break. As described above, this parameter showed a wide variability of the experimental values measured, leading to the lowest confidence factor value of 59.1%.

Additionally, in order to check the limitations of the model, Figure 2 shows the predicted versus the measured properties for the different mechanical properties. A good correlation between the predicted and measured values is observed but with some variability. As can be seen, homoscedastic distribution of the predicted values with respect to the measured one can be observed for all the mechanical properties studied.

Otherwise, Table 4 shows the confidence coefficient (%) and *t*-value related to the different terms of the Box–Wilson model obtained for the studied properties. As described in the literature, *t*-values higher than 2 are considered statistically significant, and the higher the *t*-value for a term, the greater the confidence in that particular term [39,40]. Among all the studied mechanical properties, E and σ at yield are the only ones able to show significant *t*-values higher than 2 for the linear terms of each independent variable, OLA (wt%) and MgO (wt%), as well as for the interaction term. Additionally, the elastic modulus showed a significant *t*-value of 2.54 for the quadratic term related to the OLA (wt%) content, which indicated the complexity of the interaction between the PLA matrix and the plasticizer. For the mechanical properties measured at the break point—that is, σ_b and ε_b—some considerations can be done. In particular, σ_b showed significant *t*-values higher than 2 for the linear and the quadratic terms related to OLA (wt%), while ε_b showed significant *t*-values higher than 2 only for the independent terms, which is in agreement with the previously described variability of the experimental elongation values measured at the break point, suggesting that, for this specific mechanical property, ε_b, the choice of both the OLA and MgO contents as independent variables is not the more adequate one.

3.1. Influence of the OLA and MgO Contents in the Elastic Modulus of Efibers

Figure 3a shows the 3D response surface plot and Figure 3b the isolines map of the elastic modulus, E, as a function of the OLA (wt%) and MgO (wt%) contents. It is shown as a convex saddle-shaped function not centered in the experimental space scanned and well defined in the region around the 15–22 wt% amount of OLA and above the 1 wt% amount of MgO.

In order to identify the critical points, Figure 3c,d show the parametric evolutions of the elastic modulus, E, with the OLA (wt%) and the MgO (wt%) contents, respectively, while the other one remains constant. In addition, the E values of neat PLA efibers have been included in these plots in order to compare with those of the different PLA-based efibers (purple dots in Figure 3c,d). It is important to remark that only the PLA-based efibers with OLA contents above 15 wt% and MgO contents above 1.4 wt% approach or overpass the E value of neat PLA efibers (E = 91 ± 8 MPa [17]). In fact, we can define a critical point at 15 wt% of OLA and below 15 wt% of OLA; all the PLA-based efibers show E values lower than the E value of neat PLA efibers and increase when the amount of MgO (wt%) decreases. On the other hand, above 15 wt% OLA, the PLA-based efibers with the lowest contents of MgO (lower than 1.4 wt% MgO) showed E values lower than neat electrospun PLA. However, once the amount of MgO (wt%) reaches and passes the 1 wt% content of MgO, the PLA-based efibers show the higher E values, overpassing the E value of neat PLA efibers. In particular, from 2 wt% MgO, the highest E values were observed.

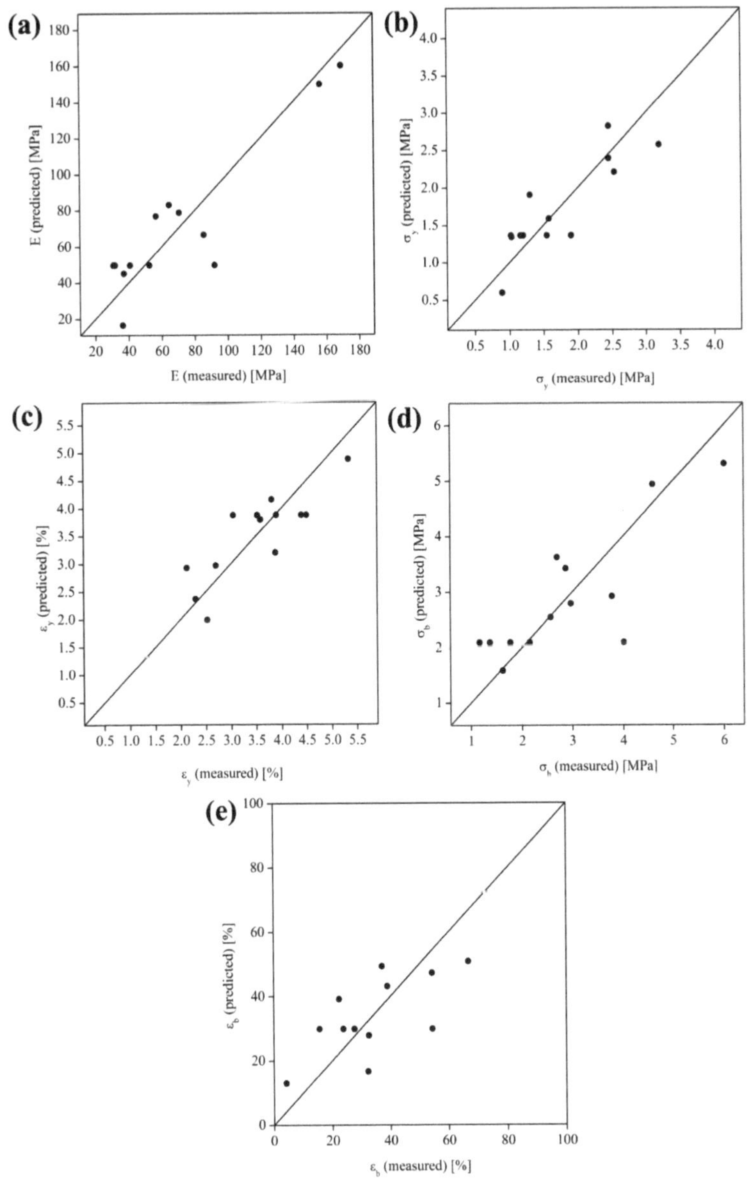

Figure 2. Predicted versus measured values for (**a**) E, (**b**) σ_y, (**c**) ε_y, (**d**) σ_b, and (**e**) ε_b.

Table 4. Confidence coefficients (%) and *t*-values for the different terms of the Box–Wilson model obtained for the mechanical properties (Polynomial Equation: $a_0 + a_1 \cdot x_1 + a_2 \cdot x_2 + a_3 \cdot x_1 \cdot x_2 + a_4 \cdot x_1^2 + a_5 \cdot x_2^2$).

	Ind. T. *	L. T. *		Int. T. *	Q. T. *	
		x_1	x_2	$x_1 \cdot x_2$	x_1^2	x_2^2
E (MPa)	3.52 (98.9%)	2.49 (95.9%)	2.34 (95.0%)	2.80 (97.4%)	2.54 (96.2%)	1.18 (71.0%)
σ_y (MPa)	4.68 (99.6%)	2.12 (93.0%)	2.94 (97.8%)	2.75 (97.2%)	1.59 (84.1%)	1.43 (79.7%)
ε_y (%)	4.72 (99.6%)	0.09 (19.1%)	1.00 (63.3%)	1.78 (88.4%)	1.89 (90.0%)	0.38 (31.7%)
σ_b (MPa)	3.23 (98.5%)	2.18 (93.6%)	1.54 (82.8%)	1.50 (81.9%)	2.46 (95.8%)	0.80 (53.6%)
ε_b (%)	2.19 (93.7%)	0.64 (44.9%)	0.31 (28.3%)	0.61 (43.5%)	0.76 (51.2%)	0.25 (25.4%)

* Ind. T. (independent term), L. T. (linear terms), Int. T (interaction term), and Q. T. (quadratic terms).

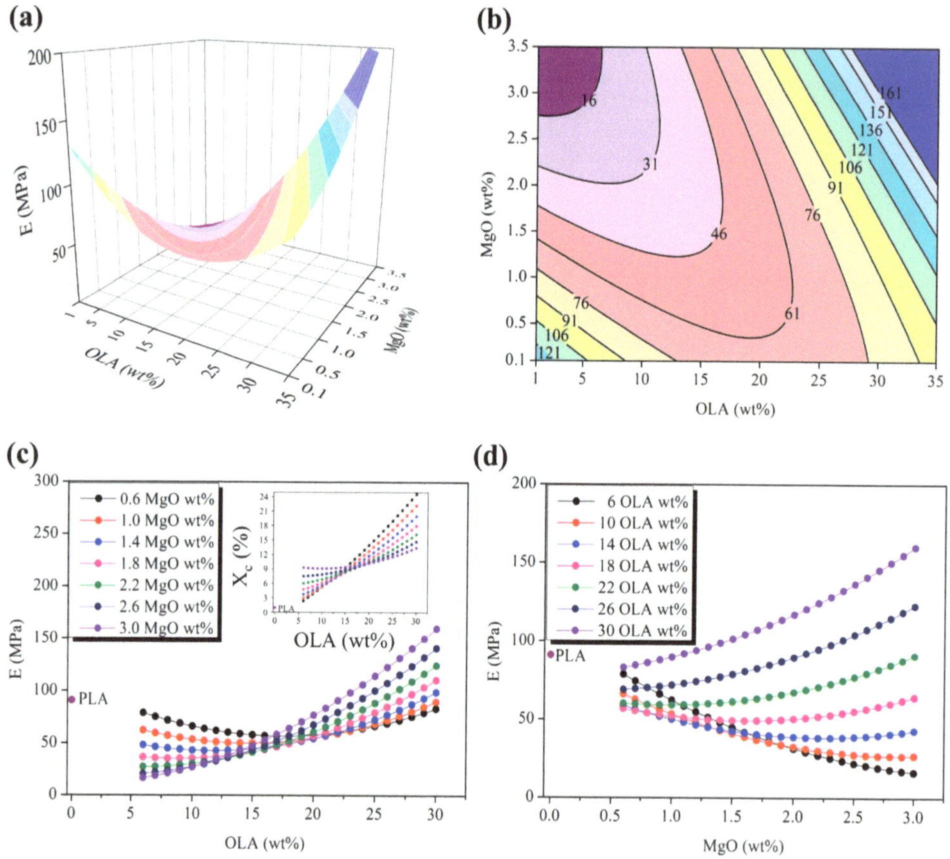

Figure 3. (**a**) Three-dimensional response surface plot and (**b**) contour plot of the elastic modulus as a function of the OLA and MgO NPs contents; colors changes are attributed to an increment of 15 MPa in the elastic modulus of efibers. (**c**) Parametric evolution of the elastic modulus with the OLA content remaining constant at the MgO NP level. (**d**) Parametric evolution of the elastic modulus with the MgO NP contents remaining constant at the OLA amounts.

Therefore, it is important to point out that the minimum in the evolution of E values takes place at 15 wt% of OLA, in agreement with the convergence point described in the evolution of the degree of crystallinity, X_c, values widely discussed in our previous work [18]. In order to compare the region that identifies the minimum stationary points placed at the surface responses of the electrospun PLA-based nanocomposites, both in terms of E as well as of X_c, an insert of the parametric evolutions of X_c with the OLA (wt%) content while the MgO (wt%) content remains constant has been included in Figure 3c.

As can be seen in Figure 3c, the highest E values, as well as the highest X_c values, are observed at the similar OLA (wt%) content regions—that is, above 15 wt%—but with inverse behavior related to the content of MgO (wt%). For instance, at the highest amount of OLA, 30 wt%, the highest E values were observed for the highest amount of MgO, 3 wt%, while, for the same amounts of OLA (wt%) and MgO (wt%), the X_c values was the lowest observed (purple isolines in Figure 3c).

The inversion between the E and the X_c evolutions on each corresponding parametric MgO (wt%) plots was discussed in a previous work [18] and, in this work, will be correlated with the mechanical response of the nanocomposites.

In fact, the decrease in the E values of the PLA-based nanocomposites with respect to the E values of neat PLA efibers and increasing the X_c (%) values always higher than the X_c (%) values of neat PLA efibers confirm the poor contribution of the PLA matrix to the stiffness of PLA-based nanocomposites. In terms of mechanical behavior, it means that PLA evolves as an almost amorphous polymeric material, which otherwise agrees with the T_g evolution of these PLA-based nanocomposites, as previously discussed [18].

From this perspective, the parametric evolutions of the elastic modulus, E, with the MgO (wt%) content for different contents of OLA (wt%) displayed in Figure 3d allows to identify 1.5 wt% as the critical amount of MgO; above which, the plasticizing effect of OLA that tends to decrease the E values is weaker as the amount of both OLA (wt%) and MgO (wt%) increase, until completely disappearing above the 15 wt% amount of OLA, where PLA-based efibers give the highest E values, even overcoming the E values of neat PLA efibers.

3.2. Mechanical Behavior at the Yield Point of the Woven Non-Woven PLA-Based Efibers

Figures 4 and 5 show the stress and strain responses for the woven non-woven PLA-based efibers at the yield point, respectively. As can be seen, the expected behavior for a tensile test carried out at room temperature is shown. Indeed, from the early stages of the strain processes under normal stress conditions, the matrix polymer chains tend to minimize the external force effects by flowing and then orienting parallel to the applied load direction. Then, for neat PLA efibers, the yield point indicates where the polymer chains of the PLA matrix are able to drag between them, continuing the strain of the material while the stress level remains constant. However, for PLA-based efibers, the other components such as the OLA chains and MgO NPs, would be unable to participate in the drag mechanism—the OLA chains because of their short lengths and the MgO NPs due to their stiffness. In both cases, the loss of interactions at nanoscale level between the NPs and the surrounding polymeric chains in those regions under strain processing at constant stress is expected, which provokes a decrease in the yield point values with respect to those of neat PLA efibers.

Accordingly, the stress reached at the yield point, defined as the yield strength, σ_y, is identified as the point where the slope on a stress–strain curve becomes zero. Figure 4a,b show the 3D response surface plot and the isolines map of the σ_y, respectively, in the experimental space scanned as a function of both the OLA (wt%) and the MgO (wt%) contents, respectively.

As previously discussed for the evolution of the elastic modulus, E, the σ_y response surface evolves as the convex side of a saddle-shaped function, with the minimum stationary points placed almost symmetrical in both sides of the experimental space scanned, as observed in the isolines plots in Figure 4b. The minimum distances between isolines appear at the lowest and the highest limits of the experimental space scanned for the OLA (wt%)

and the MgO (wt%) contents, respectively, which allow to identify the higher sensitivity σ_y response regions at the minimum variations in the amounts of OLA (wt%) and MgO (wt%).

Furthermore, the parametric plots shown in Figure 4c,d include the σ_y value of neat woven non-woven PLA efibers, 2.6 MPa. Only the PLA-based nanocomposites with a content of OLA of 30 wt% and a content of MgO higher than 2.6 wt%, respectively, exceed this value. To note again the excellent agreement with the elastic modulus evolution, it is important to remark that the convergence point for the σ_y curves is placed above 20–25 wt% OLA and slightly above E evolution 15 wt%. In particular, below this point, OLA 20–25 wt%, the σ_y values slightly decrease for the lowest MgO (wt%) content by increasing the OLA (wt%) content, remaining always below the σ_y value for PLA efibers. From the convergence point at OLA 20–25 wt%, increasing the MgO content up to 2.2 wt% allows to overpass the σ_y value of neat PLAs, just as it was found for the E evolution in Figure 3c. Above the convergence point in σ_y evolution, better interactions between NPs and the PLA matrix due to the presence of OLA was evidenced by the fact that the σ_y increased as the MgO content increased from 2.2 to 3.0 wt%. On the other hand, below the 2.2 wt% MgO content, the σ_y evolution isolines almost overlapping until the lowest MgO content, 0.6 wt%.

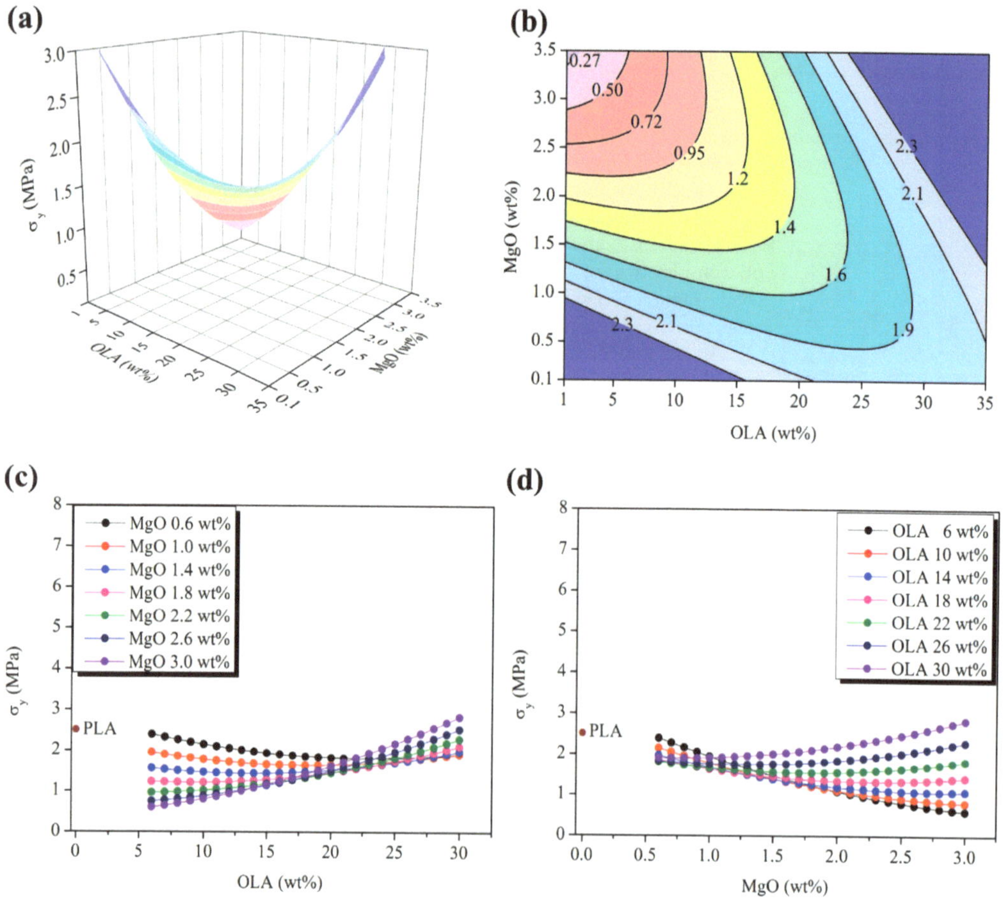

Figure 4. (**a**) Three-dimensional response surface plot and (**b**) contour plot of the σ_y as a function of OLA and MgO NPs contents; color changes are attributed to an increment of 0.2 MPa in the σ_y of efibers. (**c**) Parametric evolution of σ_y with the OLA content keeping constant the MgO NP levels. (**d**) Parametric evolution of σ_y with the MgO NPs content keeping constant the OLA amounts.

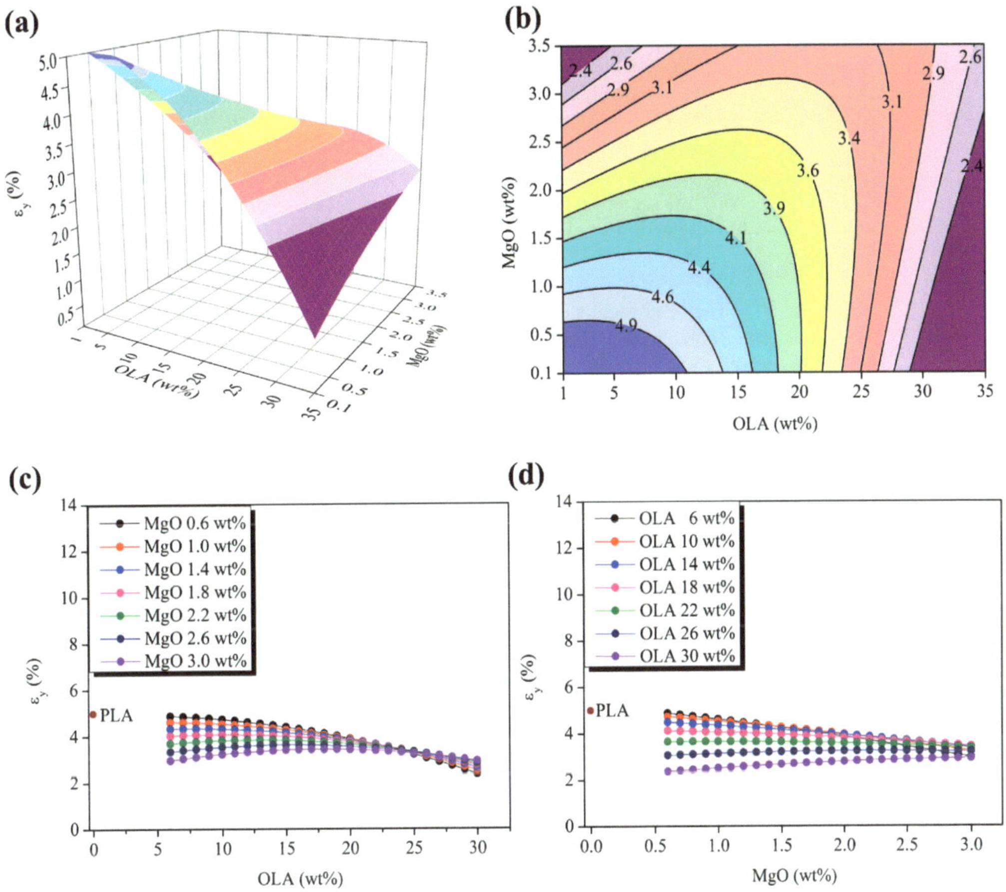

Figure 5. (**a**) Three=dimensional response surface plot and (**b**) contour plot of the ε_y as a function of the OLA and MgO NPs content; color changes are attributed to an increment of 0.3% in the ε_y of efibers. (**c**) Parametric evolution of ε_y with the OLA content keeping constant the MgO NPs level. (**d**) Parametric evolution of ε_y with the MgO NPs content keeping constant the OLA amounts.

Moreover, the response surface and isoline maps of the strain evolution at the yield point, ε_y (%), as a function of the OLA (wt%) and MgO (wt%) contents is shown in Figure 5a,b.

In this case, the response surface evolves as a saddle-shaped function inverting its curvature with respect to those previously discussed for both the E and the σ_y evolution. As can be seen in the isolines map in Figure 5d, the isoline distances decrease in the limits of the experimental space scanned for the OLA (wt%) and the MgO (wt%) contents, which is in total agreement with the previously described results for the E and the σ_y response surfaces.

In order to compare the evolution of the elongation at yield for the different PLA-based efibers, the parametric plots in Figure 5c,d show the ε_y values for the neat PLA efibers. In Figure 5c, the strain at yield values always decrease in the PLA-based efibers with respect to the PLA matrix, and tend to converge in a wide range of OLA contents between 20 and 25 wt%, which agrees with the previous results describing the σ_y evolution in Figure 4c. Otherwise, Figure 5d shows an almost lineal decrease of ε_y as the MgO (wt%) content increases, with smoothing and decreasing slopes up to zero for the OLA 22 wt% content. From this amount of OLA, 22 wt% (green isoline in Figure 5d), the ε_y isolines raised a constant value of ε_y in the experimental space scanned for MgO (wt%). However, once overpassing the OLA 22 wt% content, the ε_y values show slightly positive slopes for OLA contents of 26 and 30 wt%, respectively (dark blue and purple isolines in Figure 5d).

From the mechanical point of view at the yield point, in our PLA-based efibers, the presence of OLA in a range of concentration from 20 up to 30 wt% leads the system to approach a slight improvement in the yielding capability in terms of σ_y in comparison with neat woven non-woven PLA efibers. In the range of concentration of OLA between 20 and 30 wt%, the highest σ_y values were observed, which can be attributed to an improvement in the interaction level between the surrounding PLA polymeric chains and the NPs.

3.3. Influence of OLA and MgO Contents in the Mechanical Behavior at the Break Point of Efibers

The final breakage of PLA-based efibers is observed at the end of the tensile tests. This breakage, will occur at the mechanically weakest points of the highly strained system, which could be the points where randomly oriented efibers contacted between them during the electrospinning process (yellow circles in Figure 6a,b). However, other weak points would be those points close to the shortest polymer chains involved in the drag mechanism or the points on the surfaces of NPs embedded in PLA chains undergoing the strain process.

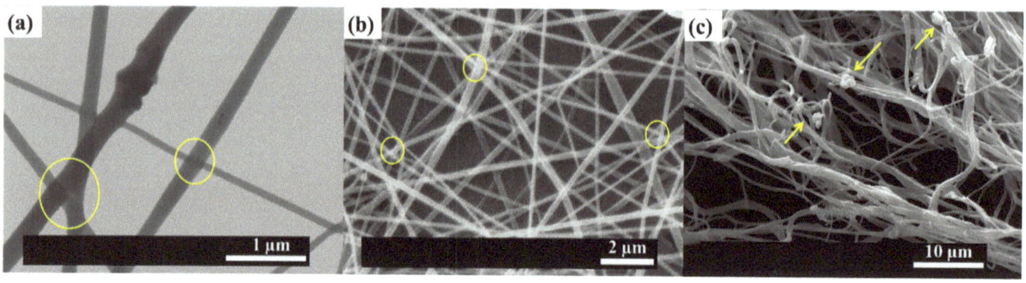

Figure 6. (**a**) FESEM images of PLA-based efibers from run X. (**b**) SEM images of tensile specimens of PLA-based efibers from run X before the tensile test, and (**c**) fracture surfaces of the same tensile specimens of PLA-based efibers from run X after the tensile test. Yellow circles are attributed to points where randomly oriented efibers contacted between them during the electrospinning process.

In order to verify the previous proposed tensile mechanism, the break point evolution of the woven non-woven efibers was studied in terms of σ_b (MPa) and ε_b (%). First of all, Figure 7a,b show both the 3D response surface plot and the isolines map of the σ_b evolution as a function of the OLA (wt%) and MgO (wt%) contents.

From the σ_b evolution point of view, in Figure 7b is observed a convex response surface well characterized by the presence of a minimum stationary point outside the experimental plane scanned moving towards the highest amount of MgO (wt%) and the lowest amount of OLA (wt%). As can be seen in Figure 7c,d, the σ_b values of PLA-based efibers are always lower than those of neat PLA efibers; only the electrospun nanocomposites with the highest content of OLA 30 wt% overcome the σ_b value of neat PLA when the content of MgO overpasses 1.5 wt%. As previously observed for the mechanical properties at the yield point, a convergence point located at OLA 20 wt% was observed, from which, by increasing the MgO (wt%) content yields, increases the σ_b values for the PLA-based efibers, which is in agreement with the previously described evolution for the elastic modulus, as well as the tensile strength at yield.

The described results obtained for the σ_b are correlated with the evolution of ε_b for the woven non-woven PLA-based efibers. Therefore, Figure 8 shows both the 3D response surface plot and the isolines map evolution of the ε_b as a function of the OLA (wt%) and the MgO (wt%) contents. In Figure 8a,b are observed a convex response surface with a very low decreasing slope moving towards the limits of the experimental space scanned for both the OLA (wt%) and the MgO (wt%) contents. Moreover, in Figure 8b, the isolines distance remains almost constant all along the contour map, suggesting an almost ε_b linear evolution in good agreement with the breakage of a highly strained material once it ended its yield capabilities.

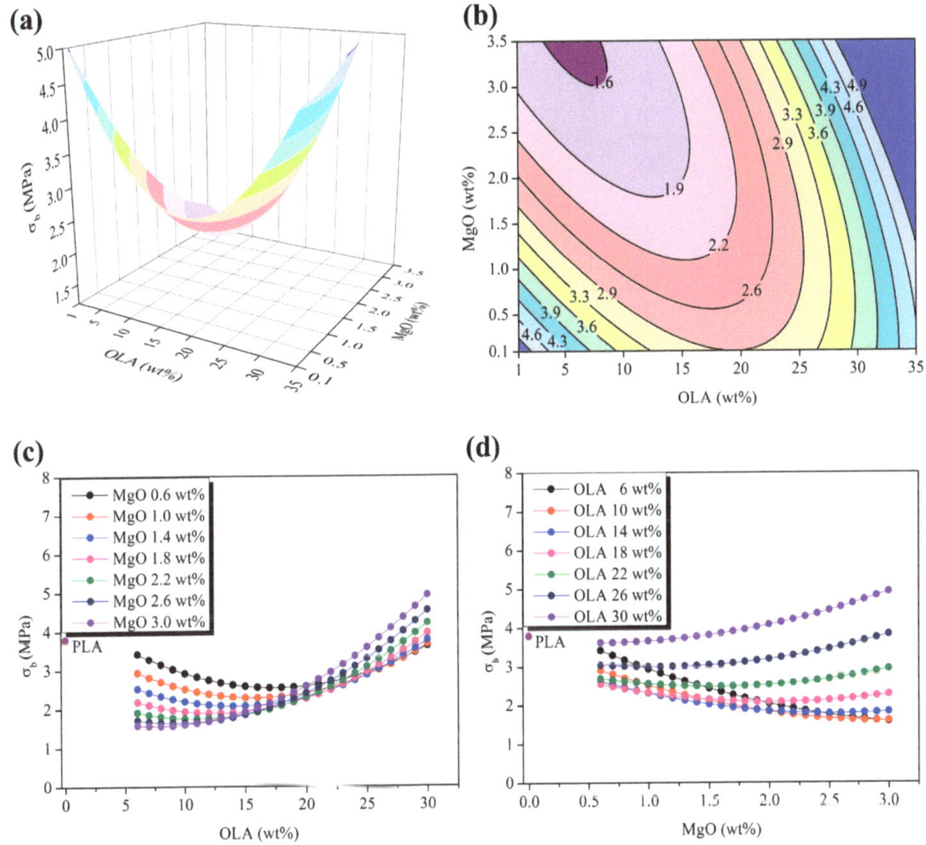

Figure 7. (**a**) Three-dimensional response surface plot and (**b**) contour plot of the σ_b as a function of the OLA and MgO NPs contents, color changes are attributed to an increment of 0.4 MPa in the σ_b of efibers. (**c**) Parametric evolution of σ_b with the OLA content keeping constant the MgO NPs level. (**d**) Parametric evolution of σ_b with the MgO NPs content keeping constant the OLA amounts.

In contrast with the previously discussed results for other mechanical properties, as it can be seen from the parametric plots shown in Figure 8c,d, the ε_b values are about 3 to 14 times lower than the ε_b values for neat PLA efibers in the experimental space scanned. Moreover, it is important to remark on the disappearance of the convergence point previously observed at OLA 20 wt%. In addition, in Figure 8c,d, both the parametric evolution of the ε_b versus OLA (wt%) and MgO (wt%) contents, respectively, showed almost linear and negative slopes of ε_b evolution as the contents of OLA (wt%) or MgO (wt%) increased, which is in excellent agreement with the previously proposed tensile mechanism.

In order to support the previously discussed tensile mechanism of the woven non-woven PLA-based efibers, SEM and FESEM images of PLA-based efibers are shown in Figure 6. These images correspond to a specimen of PLA-based efibers of the central point of the Box–Wilson worksheet—in particular, run X (18 wt% OLA and 1.80 wt% MgO). As previously discussed, these amounts of OLA (wt%) and MgO (wt%) are in the region of the critical point where the NPs are able to participate in the overall stress transmission mechanism in an efficient manner.

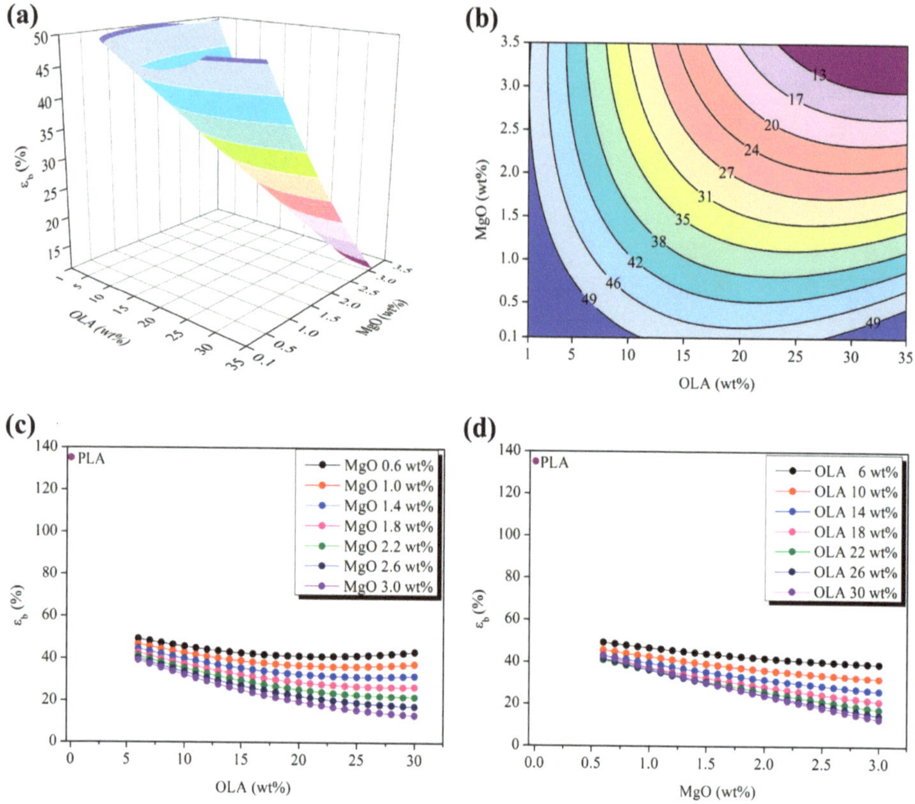

Figure 8. (**a**) Three-dimensional response surface plot and (**b**) contour plot of the ε_b as a function of the OLA and MgO NPs contents; color changes are attributed to an increment of 4% in the ε_b of efibers. (**c**) Parametric evolution of ε_b with the OLA content keeping constant the MgO NPs level. (**d**) Parametric evolution of ε_b with the MgO NPs content keeping constant the OLA amounts.

First of all, some considerations have to be taken into account. Electrospun woven non-woven mats are a kind of material constituted by fibers with homogeneous morphology and free spaces between them [35]. In addition, the framework of efibers usually shows some contact points between fibers (yellow circles), as observed in Figure 6a,b.

Due to the structure of electrospun woven non-woven mats, it is impossible to observe the whole tensile fracture surface of continuous materials. Thus, when electrospun woven non-woven mats are exposed to an external stress until the break point is reached, each efiber in the woven non-woven mat will break individually. As previously proposed, once past the yield point, polymer chains will start to align parallel to the applied force direction, and therefore, the embedded NPs will be dragged by them. Thus, the MgO NPs dispersed in the PLA matrix may be considered as stress concentrators where the breakage of the materials will start. It can be assumed that, accordingly with those previously exposed, the fractures will initiate around the stress concentration sites and lead to the final breakage of the fibers, as is schematically represented in Figure 9. Moreover, this behavior is similarly described in the literature [41]. For instance, Curgul et al. reported how the overall deformation of a fiber is determined by mobility of the polymer chains at the surface of the fiber and the number of oriented fragments present once past the yield point [42].

In addition, Arienstein et al. concluded that the orientation of the amorphous chains in the supramolecular region of the fibers influences the mechanical deformation process of the fibers [43]. On the other hand, Kim et al. proposed a similar model representation for the mechanical deformation process of electrospun polymer nanocomposite fibers in good agreement with our proposed mechanism shown in Figure 9a. In particular, they described how the presence of nanofillers within the fibers caused the stress concentration and acted as effective stress concentrators under the tensile load, because they have quite different elastic properties from those of the polymer matrix [44].

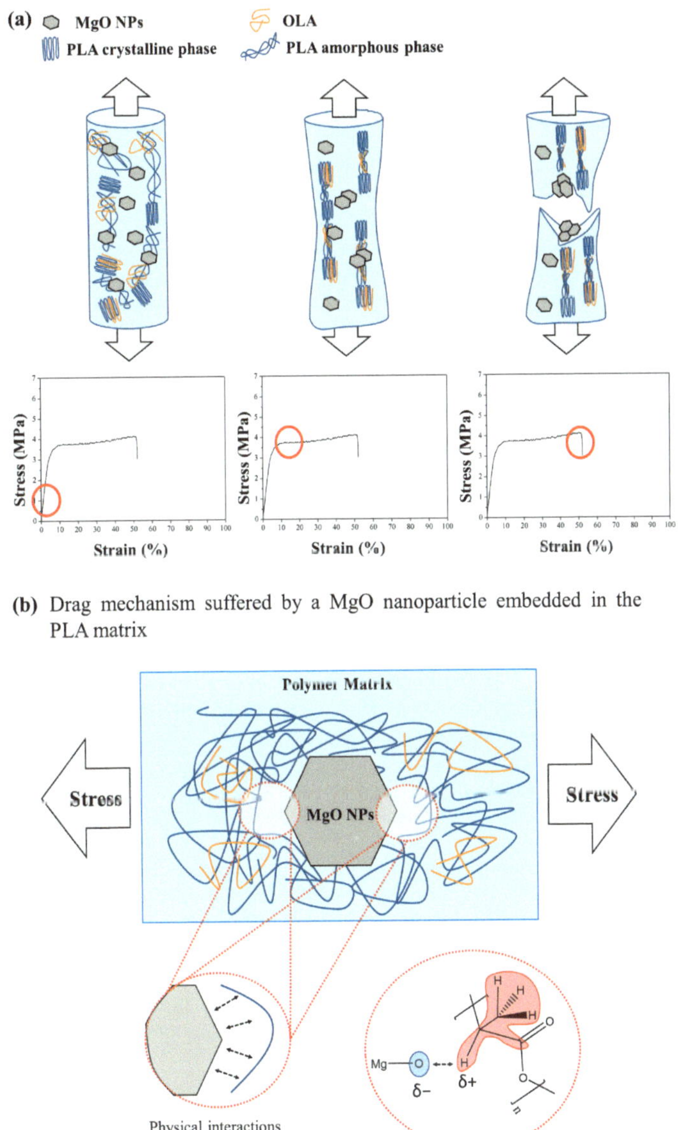

Figure 9. (**a**) Schematic representation of the mechanical behavior of the PLA-based efibers during the different stages of the tensile test. Red circles are attributed to the elastic region, yield point and break point (**b**) Proposed strain mechanism of PLA/OLA/MgO efibers.

In order to support the described strain mechanism, SEM images both before and after the tensile test are reported in Figure 8. Firstly, Figure 8a shows the MgO NPs embedded through the efibers, confirming the good dispersion of NPs by the electrospinning process carried out. Moreover, randomly oriented and smooth efibers can be observed in Figure 8b before the tensile test. After carrying out the tensile test, the fracture surface of the specimen was observed by SEM and is shown in Figure 8c. Firstly, the efibers showed an orientation parallel to the applied stress. In addition, as can be seen and it is indicated by the arrows, the broken efibers showed MgO NPs in the broken surface, corroborating the tensile mechanism. Once the PLA-based efibers are highly strained, the surfaces of MgO NPs in contact with PLA chains are the mechanically weakest points, yielding to the breakage of the efibers at these points.

4. Conclusions

The mechanical behaviors of woven non-woven PLA/OLA/MgO efibers were determined by studying their mechanical properties measured in the main regions at the stress–strain curves: the elastic zone (elastic modulus) and the yield and break points (strength and elongation in both). A Box–Wilson model was used in order to identify the level of interactions and look for the optimal compositional ratios. In our case, the $<r^2>$ (%) values obtained were 82.47% (E), 73.52% (σ_y), 69.12% (ε_y), 67.29% (σ_b), and 45.79% (ε_b), respectively. In addition, the predicted versus the measured values for E, σ_y, ε_y, σ_b, and ε_b showed very good correlations, evolving as homoscedastic distributions.

The parametric evolutions of the elastic modulus, E, with the MgO (wt%) contents for different contents of OLA (wt%) allowed to identify 1.5 wt% as the critical amount of MgO; above that, the plasticizing effect of OLA is weaker as the amount of both OLA (wt%) and MgO (wt%) increase. Moreover, the minimum E values take place at 15 wt% of OLA, in agreement with the previously reported convergence point in the evolution of the degree of crystallinity, $X_{c.}$

On the other hand, from the mechanical point of view, at the yield point, the presence of OLA in a range of concentrations from 20 to 30 wt% leads the system to approach a slight improvement in the yielding capability in terms of σ_y in comparison with neat woven non-woven PLA efibers. In the range of concentrations of OLA between 20 and 30 wt%, the highest σ_y values were observed, which could be attributed to an improvement in the interaction level between the surrounding PLA polymeric chains and the NPs.

Moreover, the strain mechanism of PLA/OLA/MgO efibers was described. Firstly, the efibers showed an orientation parallel to the applied stress. Then, once the material reached the yield point, polymer chains started to align parallel to the applied force direction, and therefore, the embedded NPs were dragged by them. Finally, when the PLA-based efibers were highly strained, the surfaces of the MgO NPs in contact with PLA chains were the mechanically weakest points, yielding to the breakage of the efibers at these points.

Author Contributions: Conceptualization, L.P., J.-M.G.-M. and E.P.C.; methodology, J.-M.G.-M. and E.P.C.; software, J.-M.G.-M. and A.L.; validation, L.P.; investigation, A.L.; resources, L.P. data curation, J.-M.G.-M.; writing—original draft preparation, A.L.; writing—review and editing, A.L., L.P., J.-M.G.-M., and E.P.C.; supervision, L.P.; project administration, L.P.; and funding acquisition, L.P. All authors have read and agreed to the published version of the manuscript.

Funding: This research was funded by the Agencia Estatal de Investigación (AEI, MICINN, Spain) and Fondo Europeo de Desarrollo Regional (FEDER, EU), PID2021-123753NB-C31 and TED2021-129335B-C21.

Institutional Review Board Statement: Not applicable.

Data Availability Statement: Not applicable.

Acknowledgments: The authors want to thank Condensia Quimica S.A. for providing Glyplast OLA8.

Conflicts of Interest: The authors declare no conflict of interest.

References

1. Garrison, T.; Murawski, A.; Quirino, R. Bio-Based Polymers with Potential for Biodegradability. *Polymers* **2016**, *8*, 262. [CrossRef] [PubMed]
2. Swetha, T.A.; Ananthi, V.; Bora, A.; Sengottuvelan, N.; Ponnuchamy, K.; Muthusamy, G.; Arun, A. A Review on Biodegradable Polylactic Acid (PLA) Production from Fermentative Food Waste—Its Applications and Degradation. *Int. J. Biol. Macromol.* **2023**, *234*, 123703. [CrossRef] [PubMed]
3. Pradhan, R.; Misra, M.; Erickson, L.; Mohanty, A. Compostability and Biodegradation Study of PLA–Wheat Straw and PLA–Soy Straw Based Green Composites in Simulated Composting Bioreactor. *Bioresour. Technol.* **2010**, *101*, 8489–8491. [CrossRef] [PubMed]
4. Jandas, P.J.; Mohanty, S.; Nayak, S.K. Sustainability, Compostability, and Specific Microbial Activity on Agricultural Mulch Films Prepared from Poly(Lactic Acid). *Ind. Eng. Chem. Res.* **2013**, *52*, 17714–17724. [CrossRef]
5. Farah, S.; Anderson, D.G.; Langer, R. Physical and Mechanical Properties of PLA, and Their Functions in Widespread Applications—A Comprehensive Review. *Adv. Drug Deliv. Rev.* **2016**, *107*, 367–392. [CrossRef]
6. Ahmed, J.; Varshney, S.K. Polylactides—Chemistry, Properties and Green Packaging Technology: A Review. *Int. J. Food Prop.* **2011**, *14*, 37–58. [CrossRef]
7. Mulla, M.Z.; Rahman, M.R.T.; Marcos, B.; Tiwari, B.; Pathania, S. Poly Lactic Acid (PLA) Nanocomposites: Effect of Inorganic Nanoparticles Reinforcement on Its Performance and Food Packaging Applications. *Molecules* **2021**, *26*, 1967. [CrossRef]
8. Leonés, A.; Lieblich, M.; Benavente, R.; Gonzalez, J.L.; Peponi, L. Potential Applications of Magnesium-Based Polymeric Nanocomposites Obtained by Electrospinning Technique. *Nanomaterials* **2020**, *10*, 1524. [CrossRef]
9. Nofar, M.; Sacligil, D.; Carreau, P.J.; Kamal, M.R.; Heuzey, M.-C. Poly (Lactic Acid) Blends: Processing, Properties and Applications. *Int. J. Biol. Macromol.* **2019**, *125*, 307–360. [CrossRef]
10. Bergström, J.S.; Hayman, D. An Overview of Mechanical Properties and Material Modeling of Polylactide (PLA) for Medical Applications. *Ann. Biomed. Eng.* **2016**, *44*, 330–340. [CrossRef]
11. Wu, Y.; Gao, X.; Wu, J.; Zhou, T.; Nguyen, T.T.; Wang, Y. Biodegradable Polylactic Acid and Its Composites: Characteristics, Processing, and Sustainable Applications in Sports. *Polymers* **2023**, *15*, 3096. [CrossRef] [PubMed]
12. Murariu, M.; Dubois, P. PLA Composites: From Production to Properties. *Adv. Drug Deliv. Rev.* **2016**, *107*, 17–46. [CrossRef] [PubMed]
13. Song, L.; Li, Y.; Meng, X.; Wang, T.; Shi, Y.; Wang, Y.; Shi, S.; Liu, L. Crystallization, Structure and Significantly Improved Mechanical Properties of PLA/PPC Blends Compatibilized with PLA-PPC Copolymers Produced by Reactions Initiated with TBT or TDI. *Polymers* **2021**, *13*, 3245. [CrossRef] [PubMed]
14. Stefaniak, K.; Masek, A. Green Copolymers Based on Poly(Lactic Acid)—Short Review. *Materials* **2021**, *14*, 5254. [CrossRef]
15. Sanusi, O.M.; Benelfellah, A.; Bikiaris, D.N.; Aït Hocine, N. Effect of Rigid Nanoparticles and Preparation Techniques on the Performances of Poly(Lactic Acid) Nanocomposites: A Review. *Polym. Adv. Technol.* **2021**, *32*, 444–460. [CrossRef]
16. Bikiaris, N.D.; Koumentakou, I.; Samiotaki, C.; Meimaroglou, D.; Varytimidou, D.; Karatza, A.; Kalantzis, Z.; Roussou, M.; Bikiaris, R.D.; Papageorgiou, G.Z. Recent Advances in the Investigation of Poly(Lactic Acid) (PLA) Nanocomposites: Incorporation of Various Nanofillers and Their Properties and Applications. *Polymers* **2023**, *15*, 1196. [CrossRef]
17. Leonés, A.; Sonseca, A.; López, D.; Fiori, S.; Peponi, L. Shape Memory Effect on Electrospun PLA-Based Fibers Tailoring Their Thermal Response. *Eur. Polym. J.* **2019**, *117*, 217–226. [CrossRef]
18. Leonés, A.; Peponi, L.; García-Martínez, J.-M.; Collar, E.P. Compositional Influence on the Morphology and Thermal Properties of Woven Non-Woven Mats of PLA/OLA/MgO Electrospun Fibers. *Polymers* **2022**, *14*, 2092. [CrossRef]
19. Jia, P.; Xia, H.; Tang, K.; Zhou, Y. Plasticizers Derived from Biomass Resources: A Short Review. *Polymers* **2018**, *10*, 1303. [CrossRef]
20. Leonés, A.; Peponi, L.; Lieblich, M.; Benavente, R.; Fiori, S. In Vitro Degradation of Plasticized PLA Electrospun Fiber Mats: Morphological, Thermal and Crystalline Evolution. *Polymers* **2020**, *12*, 2975. [CrossRef]
21. Raquez, J.M.; Habibi, Y.; Murariu, M.; Dubois, P. Polylactide (PLA)-Based Nanocomposites. *Prog. Polym. Sci.* **2013**, *38*, 1504–1542. [CrossRef]
22. Reddy, V.S.; Tian, Y.; Zhang, C.; Ye, Z.; Roy, K.; Chinnappan, A.; Ramakrishna, S.; Liu, W.; Ghosh, R. A Review on Electrospun Nanofibers Based Advanced Applications: From Health Care to Energy Devices. *Polymers* **2021**, *13*, 3746. [CrossRef] [PubMed]
23. Leonés, A.; Salaris, V.; Mujica-Garcia, A.; Arrieta, M.P.; Lopez, D.; Lieblich, M.; Kenny, J.M.; Peponi, L. PLA Electrospun Fibers Reinforced with Organic and Inorganic Nanoparticles: A Comparative Study. *Molecules* **2021**, *26*, 4925. [CrossRef]
24. Xu, X.; Yang, Q.; Wang, Y.; Yu, H.; Chen, X.; Jing, X. Biodegradable Electrospun Poly(l-Lactide) Fibers Containing Antibacterial Silver Nanoparticles. *Eur. Polym. J.* **2006**, *42*, 2081–2087. [CrossRef]
25. Aničić, N.; Kurtjak, M.; Jeverica, S.; Suvorov, D.; Vukomanović, M. Antimicrobial Polymeric Composites with Embedded Nanotextured Magnesium Oxide. *Polymers* **2021**, *13*, 2183. [CrossRef] [PubMed]
26. Gan, Y.X. Effect of Interface Structure on Mechanical Properties of Advanced Composite Materials. *Int. J. Mol. Sci.* **2009**, *10*, 5115–5134. [CrossRef]
27. Martin, T.B.; Jayaraman, A. Identifying the Ideal Characteristics of the Grafted Polymer Chain Length Distribution for Maximizing Dispersion of Polymer Grafted Nanoparticles in a Polymer Matrix. *Macromolecules* **2013**, *46*, 9144–9150. [CrossRef]
28. Huang, J.; Zhou, J.; Liu, M. Interphase in Polymer Nanocomposites. *JACS Au* **2022**, *2*, 280–291. [CrossRef]
29. Hore, M.J.A. Polymers on Nanoparticles: Structure & Dynamics. *Soft Matter* **2019**, *15*, 1120–1134. [CrossRef]

30. Pitsa, D.; Danikas, M.G. Interfaces Features in Polymer Nanocomposites: A Review of Proposed Models. *Nano* **2011**, *6*, 497–508. [CrossRef]
31. Smith, J.S.; Bedrov, D.; Smith, G.D. A Molecular Dynamics Simulation Study of Nanoparticle Interactions in a Model Polymer-Nanoparticle Composite. *Compos. Sci. Technol.* **2003**, *63*, 1599–1605. [CrossRef]
32. Avolio, R.; Castaldo, R.; Avella, M.; Cocca, M.; Gentile, G.; Fiori, S.; Errico, M.E. PLA-Based Plasticized Nanocomposites: Effect of Polymer/Plasticizer/Filler Interactions on the Time Evolution of Properties. *Compos. B Eng.* **2018**, *152*, 267–274. [CrossRef]
33. Cicogna, F.; Coiai, S.; De Monte, C.; Spiniello, R.; Fiori, S.; Franceschi, M.; Braca, F.; Cinelli, P.; Fehri, S.M.K.; Lazzeri, A.; et al. Poly(Lactic Acid) Plasticized with Low-Molecular-Weight Polyesters: Structural, Thermal and Biodegradability Features. *Polym. Int.* **2017**, *66*, 761–769. [CrossRef]
34. Avolio, R.; Castaldo, R.; Gentile, G.; Ambrogi, V.; Fiori, S.; Avella, M.; Cocca, M.; Errico, M.E. Plasticization of Poly(Lactic Acid) through Blending with Oligomers of Lactic Acid: Effect of the Physical Aging on Properties. *Eur. Polym. J.* **2015**, *66*, 533–542. [CrossRef]
35. Burgos, N.; Tolaguera, D.; Fiori, S.; Jiménez, A. Synthesis and Characterization of Lactic Acid Oligomers: Evaluation of Performance as Poly(Lactic Acid) Plasticizers. *J. Polym. Environ.* **2014**, *22*, 227–235. [CrossRef]
36. Arrieta, M.; Díez García, A.; López, D.; Fiori, S.; Peponi, L. Antioxidant Bilayers Based on PHBV and Plasticized Electrospun PLA-PHB Fibers Encapsulating Catechin. *Nanomaterials* **2019**, *9*, 346. [CrossRef]
37. Greco, A.; Ferrari, F. Thermal Behavior of PLA Plasticized by Commercial and Cardanol-Derived Plasticizers and the Effect on the Mechanical Properties. *J. Therm. Anal. Calorim.* **2021**, *146*, 131–141. [CrossRef]
38. Li, H.; Huneault, M.A. Effect of Nucleation and Plasticization on the Crystallization of Poly(Lactic Acid). *Polymer* **2007**, *48*, 6855–6866. [CrossRef]
39. Box, G.E.P.; Hunter, W.G.; Hunter, J.S. Response Surface Methods. In *Statistics for Experimenters*, 1st ed.; Wiley & Sons: New York, NY, USA, 1978.
40. Fisher, R.A. *The Desing of Experiments*, 1st ed.; Hafner: New York, NY, USA, 1960.
41. Baji, A.; Mai, Y.W.; Wong, S.C.; Abtahi, M.; Chen, P. Electrospinning of Polymer Nanofibers: Effects on Oriented Morphology, Structures and Tensile Properties. *Compos. Sci. Technol.* **2010**, *70*, 703–718. [CrossRef]
42. Curgul, S.; Van Vliet, K.J.; Rutledge, G.C. Molecular Dynamics Simulation of Size-Dependent Structural and Thermal Properties of Polymer Nanofibers. *Macromolecules* **2007**, *40*, 8483–8489. [CrossRef]
43. Arinstein, A.; Burman, M.; Gendelman, O.; Zussman, E. Effect of Supramolecular Structure on Polymer Nanofibre Elasticity. *Nat. Nanotechnol.* **2007**, *2*, 59–62. [CrossRef] [PubMed]
44. Kim, G.M.; Lach, R.; Michler, G.H.; Pötschke, P.; Albrecht, K. Relationships between Phase Morphology and Deformation Mechanisms in Polymer Nanocomposite Nanofibres Prepared by an Electrospinning Process. *Nanotechnology* **2006**, *17*, 963–972. [CrossRef] [PubMed]

polymers

MDPI

Article

Thermomechanical Properties of Virgin and Recycled Polypropylene—High-Density Polyethylene Blends

Hannah Jones [1], Jake McClements [2], Dipa Ray [1], Colin S. Hindle [3], Michail Kalloudis [4] and Vasileios Koutsos [1,*]

1 School of Engineering, Institute for Materials and Processes, The University of Edinburgh, Sanderson Building, King's Buildings, Edinburgh EH9 3FB, UK
2 School of Engineering, Newcastle University, Merz Court, Claremont Road, Newcastle upon Tyne NE1 7RU, UK
3 School of Engineering and The Built Environment, Edinburgh Napier University, Merchiston Campus, 10 Colinton Road, Edinburgh EH10 5DT, UK
4 Impact Laboratories Ltd. (Impact Solutions), Impact Technology Centre, Fraser Road, Kirkton Campus, Livingston EH54 7BU, UK
* Correspondence: vasileios.koutsos@ed.ac.uk; Tel.: +44-(0)131-650-8704

Abstract: This paper provides evidence and discusses the variability in the thermomechanical behaviour of virgin and recycled polypropylene/high-density polyethylene blends without the addition of other components, which is sparse in the literature. Understanding the performance variability in recycled polymer blends is of critical importance in order to facilitate the re-entering of recycled materials to the consumer market and, thus, contribute towards a circular economy. This is an area that requires further research due to the inhomogeneity of recycled materials. Therefore, the thermal and mechanical properties of virgin and recycled polypropylene/high-density polyethylene blends were investigated systematically. Differential scanning calorimetry concludes that both the recycled and virgin blends are immiscible. Generally, recycled blends have lower overall crystallinity and melting temperatures compared with virgin blends while, remarkably, their crystallisation temperatures are compared favourably. Dynamical mechanical analysis showed little variation in the storage modulus of recycled and virgin blends. However, the alpha and beta relaxation temperatures are lower in recycled blends due to structural deterioration. Deterioration in the thermal and mechanical properties of recycled blends is thought to be caused by the presence of contaminants and structural degradation during reprocessing, resulting in shorter polymeric chains and the formation of imperfect crystallites. The tensile properties of recycled blends are also affected by the recycling process. The Young's modulus and yield strength of the recycled blends are inferior to those of virgin blends due to the deterioration during the recycling process. However, the elongation at break of the recycled blends is higher compared with the virgin blends, possibly due to the plasticity effect of the low-molecular-weight chain fragments.

Keywords: polypropylene; polyethylene; blends; recycled; thermomechanical properties

check for updates

Citation: Jones, H.; McClements, J.; Ray, D.; Hindle, C.S.; Kalloudis, M.; Koutsos, V. Thermomechanical Properties of Virgin and Recycled Polypropylene—High-Density Polyethylene Blends. *Polymers* **2023**, *15*, 4200. https://doi.org/10.3390/polym15214200

Academic Editors: Jesús-María García-Martínez and Emilia P. Collar

Received: 31 August 2023
Revised: 9 October 2023
Accepted: 17 October 2023
Published: 24 October 2023

1. Introduction

Plastic waste is a major environmental issue, with only 9% of the world's plastics being recycled [1]. Polyolefins, encompassing polypropylene (PP) and polyethylene (PE), are the main components in municipal waste due to their abundant use in commodity applications as they possess good mechanical properties and processability, in addition to having high availability and low manufacturing costs [2–5]. The complete separation of PP and PE during mechanical waste recovery is uneconomical due to their close densities and structural similarity; therefore, PP and PE usually remain mixed [6–8]. During the mechanical recycling process, PP and PE undergo irreversible thermomechanical degradation processes such as chain scission, which lowers the mechanical properties of the recycled PP (rPP) and recycled PE (rPE) compared with the virgin polymers [9].

Blends of PP and PE are of great commercial interest as they have the potential to reduce the deficient characteristics of PP and PE, such as low impact resistance at low temperatures and poor environmental stress cracking resistance, respectively [6,10,11]. However, blends of PP: High-Density PE (HDPE) and PP: Low-Density PE (LDPE) are thermodynamically immiscible, resulting in poor material performance due to their phase-separated morphology and low interfacial adhesion between the phases [12,13]. The mechanical performance of immiscible blends are dependent upon the blend components' crystallisation behaviour and final blend morphology [14–16]. Several factors are important for morphology development during polymer processing such as composition, viscosity ratio of the components, interfacial properties, crystallinity and processing conditions [6,14,17–22]. Several studies have reported the mechanical properties, crystallisation behaviour and morphology of PP:PE blends [6,7,11,22–28].

Although it is important to determine and understand the mechanical and thermal behaviour of the virgin PP (vPP): virgin polyethylene (vPE) blends in order to optimise the properties of recycled blends, comparative studies are also important. A comparison of the thermomechanical properties of vPP:vHDPE and rPP:rHDPE blends over a wide range of compositions is lacking as the literature mainly focuses on ternary systems (e.g., PP:HDPE and compatibilisers/co-polymers/other polymers/fillers [5,29–32]). It is important to understand the variability in performance of rPP:rHDPE blends before the addition of further components.

Studies into the crystallinities of rPP, rHDPE and rPP:rHDPE blends have found that crystallinity is affected by degradation mechanisms during the recycling process [33–35]. Interestingly, rPP crystallinity has been found to be higher than that of vPP by several authors [33,35,36]. da Costa et al. [35] suggested that the higher value of crystallinity of rPP compared with vPP was caused by a decrease in the molecular weight, which resulted in an increase in chain mobility. Increased chain mobility improved the ability of chains to fold into thicker lamella and, hence, an increased crystallisation rate and crystallinity. Therefore, it is important to understand the variation in crystallinity in rPP:rHDPE blends in order to optimise the mechanical performance of the recyclates.

Research has been undertaken in the literature to understand the effect of recycling cycles on the mechanical properties of PP, PE and their blends. For example, Aurrekoetxea et al. [33] subjected PP to 10 successive injection moulding cycles at 200 °C and found that the degree of crystallinity increased with each cycle. This caused an increase in Young's modulus and yield stress. On the other hand, Oliveira et al. [37], who subjected PP to seven successive cycles at 175–190 °C, observed a decrease in Young's modulus and yield stress after the third cycle that was attributed to a reduction in tie molecules between the crystalline and amorphous phases. Conflicting observations by Aurrekoetxea et al. [33] and Oliveira et al. [37] for the Young's modulus and yield stress of rPP could be due to differences in the processing methodology. Aurrekoetxea et al. [33] used injection moulding, whereas Oliveira et al. [37] opted for a single screw extruder followed by compression moulding. This highlights the importance of the reprocessing methodology but also demonstrates the difficulty of comparing the performance of recycled materials in the literature. Furthermore, chemical analysis studies of recycled PE and PP have revealed their variability, degradation and the presence of impurities and contaminants [38–40].

PE can be subjected to a higher number of extrusion cycles before any deterioration in the mechanical properties is observed. Jin et al. [41] found no significant change in crystallinity and, hence, in the mechanical properties of LDPE up to the 40th extrusion cycle. However, a decrease in crystallinity was observed between the 40 and 50th cycles, either caused by short side branches in the backbone chain or side groups, or by crosslinking. Oblak et al. [42] subjected HDPE to 100 consecutive extrusion cycles at 220–270 °C. They found that chain branching and chain scission, which occurred up to the 60th cycle, resulted in a decrease in crystallinity and Young's modulus. However, crystallinity and Young's modulus remained stable after the 60th extrusion cycle due to crosslinking. After the 100th cycle, the Young's modulus of the rHDPE had only reduced by 20% compared with that of the vHDPE.

Studies have been carried out to understand the mechanical properties of PP:PE blends subjected to recycling cycles [9,26,43–47]. Saikrishnan et al. [43] reported that recycling affected the melt flow behaviour of PP:LDPE blends but found that the tensile properties were not substantially affected (subjected to up to five recycling cycles). Interestingly, PP underwent chain scission on each recycling cycle but the overall properties of the blend were maintained. However, they only investigated the PP:LDPE up to 10 wt% of LDPE. The literature is typically limited in the blend composition range investigated. However, due to the variability in the waste streams, it is important to understand the mechanical properties for all blend compositions without the addition of a third component initially. This would enable the recycling industry to be reactive to changes in waste stream composition in different locations and batches and enable more recyclate to re-enter the market. Therefore, this paper aims to understand the variability in thermomechanical properties for virgin and recycled PP:HDPE blends.

This study reports the thermal and mechanical properties of vPP:vHDPE and rPP:rHDPE blends through differential scanning calorimetry (DSC), dynamical mechanical analysis (DMA) and tensile testing. This comparative study enables the mapping of not only the challenges but also the potentially unique opportunities of the recycled systems. The recycling industry is looking to improve the plastic circular economy by obtaining recycled commingled waste blends with desirable end-use properties acceptable for commercial applications but at low cost.

2. Experimental Section

2.1. Materials

PP (Moplen EP440G), supplied by LyondellBasell (London, UK), had a melt flow index (MFI) of 1.3 g 10 min^{-1} and density of 900 kg m^{-3}. HDPE (HDPE, K46-06-185), supplied by Ineos (Grangemouth, UK), had an MFI of 4.2 g 10 min^{-1} and a density of 946 kg m^{-3}. Post-consumer rPP and rPE were supplied by Impact Solutions Recycled (Livingston, UK). The rPE was mainly composed of HDPE, but small quantities of LDPE were present. rPP and rHDPE had MFIs of 15 and 1.5 g 10 min^{-1}, respectively. As shown by the MFI values, the grades of virgin and recycled PP and HDPE used are quite different; therefore, the properties are not directly comparable. The comparisons made through the study are more general between virgin and recycled grades.

Virgin and recycled blends of different compositions of PP and HDPE (P10, P20, P25, P40, P50, P60, P75, P80 and P90) were prepared, where P denotes PP and the number corresponds to the percentage composition by weight of PP in the blend PP:HDPE. The pure 100 weight percentage (wt%) PP and HDPE will be denoted as PP and HDPE, respectively. To denote virgin or recycled, the symbols of v and r will be used, respectively, before the blend composition, e.g., virgin P10 would be represented as vP10.

2.2. Preparation

Extrusion and Injection Moulding

vPP and vHDPE were in the form of pellets, whereas rPP and rPE were in the form of flakes. Blends were prepared using a lab scale Haake MiniCTW twin screw extruder (Karlsruhe, Germany) for 5 min with feeder and mixing speeds of 50 rpm and 100 rpm, respectively. The conical screws were 4–15 mm in diameter, 109.4 mm in length and co-rotate. The barrel temperature was 180–185 °C. Molten blends were transferred to the Haake MiniJet injection moulder (Karlsruhe, Germany), where the cylinder temperature was 210 °C, mould temperature was 35 °C, injection pressure was 50 MPa and hold-on pressure time was 10 s. The ISO 527-2-1BA [48] and 557–2296 moulds were used for the dog-bone-shaped and DMA rectangular samples, respectively.

2.3. Characterisation

2.3.1. DSC

The melting and crystallisation behaviour of the vPP:vHDPE and rPP:rHDPE blends were evaluated using a Perkin Elmer DSC 8000 (Waltham, MA, USA). The instrument was calibrated using an indium sample. Approximately 5–6 mg of the sample was scanned under a nitrogen atmosphere. Samples were exposed to the following thermal cycle: heated from 25 to 200 °C at 10 °C min^{-1}, isothermal at 200 °C for 5 min, cooled from 200 to 25 °C at 10 °C min^{-1}, isothermal at 25 °C for 2 min and heated from 25 to 200 °C at 10 °C min^{-1}. The melting temperature T_m and enthalpy of fusion ΔH_f were obtained from the first heating ramp. The crystallisation temperature T_c was taken from the cooling ramp.

The degree of crystallinity was calculated by Equation (1),

$$\%\ Crystallinity = \frac{\Delta H_f^{obs}}{\Delta H_f^0} \times 100 \quad (1)$$

where ΔH_f^{obs} is the observed enthalpy of fusions for the individual PP and HDPE peaks, and ΔH_f^0 is the 100% crystalline HDPE or PP, which are 287 and 207 J g^{-1}, respectively [6]. ΔH_f^{PP} and ΔH_f^{HDPE} were taken from the first heating ramp to calculate the crystallinity. The thermal history of the sample was erased after the first heating ramp [49]. However, very little difference was found when comparing the crystallinity obtained from the first and second heating ramps for the virgin and recycled blends (Tables S1 and S2).

2.3.2. DMA

DMA was used to determine the viscoelastic properties of the virgin and recycled blends. A Triton DMA (Leicestershire, UK) in dual cantilever mode at a frequency of 1 Hz was used. A temperature sweep from −50 to 150 °C at a heating rate of 5 °C min^{-1} was implemented. Sample dimensions were approximately 45 mm (l) × 10 mm (w) × 2.7 mm (d). A minimum of three samples were tested, and the average and standard deviation were calculated for each blend ratio.

The dynamic response was given as the elastic (storage modulus, E'), viscous (loss modulus, E'') and damping (tan delta, tanδ) components. The glass transition (T_g) and transition relaxation processes can be seen as changes in the E'' or tanδ traces [50]. The tanδ trace was used to quote the T_g and other relaxation peaks present [51].

2.3.3. Tensile Testing

Tensile properties were determined using an Instron Tensile Machine (Buckinghamshire, UK) with a crosshead speed of 5 mm min^{-1} and a 10 kN load cell. Tensile properties were carried out at ambient temperature in accordance with the ISO 527-2 standard. Young's modulus was determined using a Zwick Roell Tensile Machine with a video-extensometer. A crosshead speed of 1 mm min^{-1}, gauge length of 25 mm and a 10 kN load cell were used. A minimum of five samples were tested, and the average and standard deviation were calculated.

The "rule of mixtures" was used to predict the Young's modulus of the virgin and recycled PP:HDPE blend samples compared to the experimental data. The rule of mixtures was calculated by Equation (2),

$$E_{Blend} = W_{PP} E_{PP} + W_{HDPE} E_{HDPE} \quad (2)$$

where E_{Blend} is the Young's modulus of the polymer blend; W_{PP} and E_{PP} are the weight fraction and Young's modulus of PP, respectively; and W_{HDPE} and E_{HDPE} are the weight fraction and Young's modulus of component HDPE, respectively [52]. The experimental Young's modulus of the virgin and recycled homogenous PP and HDPE systems was used for the virgin and recycled E_{PP} and E_{HDPE}, respectively.

3. Results and Discussion

3.1. Thermal Properties of Virgin and Recycled PP:HDPE Blends

The melting behaviour of both rPP:rHDPE and vPP:vHDPE blends presented two separate peaks assigned to PP and HDPE, suggesting that the blends were immiscible (Table 1 and Figure 1). The two melting peaks present in the rPP and rPE indicated that contaminants were present due to the challenge of complete separation of PP and PE during the recycling process [26,29,53]. Little variation in the T_m of virgin and recycled PP, HDPE and their blends suggested that the blending of PP and HDPE did not significantly alter the T_m of PP and HDPE (Table 1, Figure S1) [6,29]. In some cases, the rPP:rHDPE blends had lower T_m than the respective vPP:vHDPE blends, indicating structural deteriorations of the polymeric components during the mechanical recycling process [54], which could imply that less-perfect crystallites formed [55].

Table 1. Melting and crystallisation behaviour of vPP:vHDPE and rPP:rHDPE blends obtained from the first heating and cooling cycles using DSC. Bold values are for vPP:vHDPE blends and un-bolded values are for the rPP:rHDPE blends. Related graphs are presented in the Supporting Information (Figures S1–S5).

PP wt% Content in PP:HDPE	Peak Melting Temperature (°C)		Enthalpy of Fusion (J g^{-1})		Peak Crystallisation Temperature (°C)		Enthalpy of Crystallisation (J g^{-1})	Percentage Crystallinity (%)	
	PP	HDPE	PP	HDPE	PP	HDPE	(PP + HDPE)	PP	HDPE
0	**-**	**134.6**	**0.0**	**161.2**	**-**	**118.4**	**161.4**	**0.0**	**56.2**
	159.0	132.9	11.8	115.4	125.0	116.6	147.4	7.3	36.2
10	**173.9**	**132.6**	**3.8**	**161.0**	**-**	**115.6**	**161.6**	**1.8**	**56.1**
	166.4	131.7	27.3	87.8	122.5	118.0	131.4	12.5	30.8
20	**166.1**	**130.5**	**11.4**	**105.8**	**-**	**115.9**	**159.7**	**5.5**	**36.9**
	166.1	131.5	18.9	87.0	123.1	118.4	137.3	14.0	30.4
25	**165.8**	**130.9**	**12.9**	**114.2**	**-**	**116.4**	**148.3**	**6.2**	**36.8**
	165.2	131.5	27.2	72.4	123.4	117.7	116.5	13.9	23.6
40	**167.6**	**130.5**	**31.1**	**78.5**	**124.5**	**117.1**	**131.4**	**15.0**	**27.4**
	163.2	131.3	27.6	79.5	122.0	118.4	129.3	12.0	27.2
50	**172.8**	**134.6**	**41.5**	**79.7**	**126.5**	**117.5**	**110.4**	**20.1**	**27.8**
	162.2	130.7	30.3	68.6	122.4	117.9	125.8	13.6	21.9
60	**165.9**	**127.9**	**38.2**	**55.2**	**124.1**	**117.7**	**117.4**	**18.4**	**19.2**
	162.7	130.8	21.2	60.6	123.1	118.0	116.3	15.7	20.2
75	**168.5**	**128.9**	**50.0**	**36.6**	**123.6**	**116.9**	**108.6**	**24.2**	**12.7**
	164.6	130.5	37.5	50.0	122.0	118.5	110.5	16.9	16.4
80	**168.0**	**128.7**	**55.2**	**24.0**	**124.4**	**116.9**	**94.5**	**26.7**	**8.4**
	163.2	130.2	31.4	53.7	122.2	118.1	105.3	15.6	17.7
90	**168.2**	**128.7**	**66.9**	**5.4**	**121.8**	**116.1**	**88.6**	**32.3**	**1.9**
	161.8	130.3	32.7	41.0	121.8	117.8	107.1	17.0	13.9
100	**167.4**	**-**	**64.4**	**0.0**	**122.3**	**-**	**83.0**	**31.1**	**0.0**
	164.9	129.5	42.8	31.1	122.7	118.2	100.2	19.6	10.6

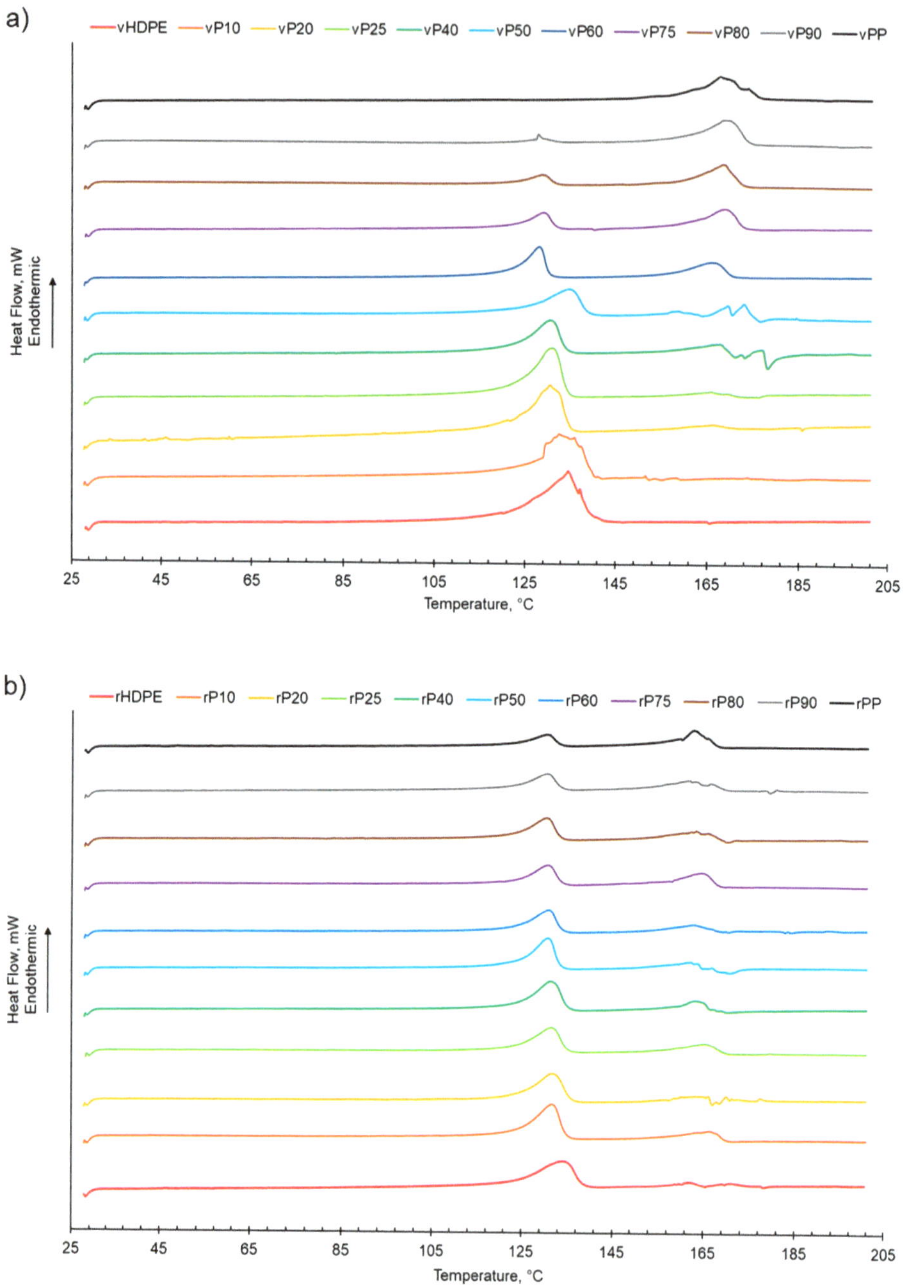

Figure 1. The melting behaviour of PP:HDPE blends obtained from DSC: (**a**) vPP:vHDPE blends and (**b**) rPP:rHDPE blends.

As shown in Table 1 and Figure S2, increasing the PP wt% in the virgin or recycled PP:HDPE blends results a decrease in the enthalpy of crystallisation of HDPE while PP generally presents an increasing trend. This is in agreement with Jose et al. [6], who suggested that the decrease in enthalpy of crystallisation could be attributed to the differing rates of crystallisation for PP and HDPE and the resulting size of crystallites. PP crystallises at a slower rate compared with HDPE, which enables the formation of large spherulites. The large spherulites in PP liberate less energy during crystallisation compared with the smaller crystallites in HDPE [6]. The crystallisation behaviour of semi-crystalline polymers is more complex compared with their melting behaviour due to the numerous factors that can affect the phase structure, such as polymer composition and distribution, intra- and inter-molecular interactions, and processing conditions [56]. The presence of a second semi-crystalline material also complicates the crystallisation behaviour [8,57]. Typically, PP and HDPE crystallise separately and at different rates. HDPE has a quicker nucleation and growth rate compared with PP due to the HDPE's flexible chain and limited intermolecular interactions [6,26,56]. In PP, crystallisation is hindered by the bulky methyl groups on the polymer chain backbone [56]. There was little variation observed in the crystallisation temperature (T_c) for the recycled and virgin blends (Figure 2, Table 1 and Figure S3). Two crystallisation peaks were observed for rPP:rHDPE blends at the approximate individual PP and HDPE crystallisation temperatures. Phase separation is caused by the PP and HDPE crystals growing at different rates. Crystallisation peaks were observed in the rPP and rPE due to the presence of PE and PP contaminants, respectively. One crystallisation peak was observed for vPP:vHDPE blends up to vP25, suggesting co-crystallisation and/or partial miscibility. However, upon increasing the PP wt% further, two peaks were observed at the approximate individual PP and HDPE crystallisation temperatures, suggesting an onset of independent crystallisation and incompatibility. There is literature reporting a single crystallisation peak for vPP:vHDPE blends over a wide composition range. Lin et al. [27] and Sutar et al. [30] suggested that the addition of HDPE affected the PP crystallisation rate, resulting in one crystallisation peak. Jose et al. [6], who studied a range of PP:HDPE blends, reported only one crystallisation temperature, which possessed an intermediary T_c value between the T_c values of PP and HDPE. Aumunate et al. [26] found a single crystallisation peak for vPP:vHDPE blends caused by the merging of the vPP and vHDPE peaks due to their close T_c. However, they suggested that bimodal behaviour was present at higher vHDPE contents due to the presence of a slight shoulder peak.

As the PP wt% increased for both the rPP:rHDPE and vPP:vHDPE blends, the PP crystallinity in the blend increased and the HDPE crystallinity decreased (Table 1 and Figure 3). rPP crystallinity was higher than vPP crystallinity in HDPE up to the P25 blend. rHDPE crystallinity was higher than vHDPE crystallinity from P60 to PP. The crystallinity is affected by the presence of PP and HDPE contaminants in the recycled materials. The quantity of other plastic contaminants will be dependent on the waste stream composition and the quality of sorting at the material recycling facility [58,59]. Thermomechanical degradation, which occurs during recycling, results in an increase in polydispersity caused by the presence of shorter polymeric chains [60]. Shorter polymeric chains form crystals more easily compared with long chains due to their low degree of entanglement, which may lead to an increase in crystallinity [61]. On the other hand, the crystallinities of rPP and rHDPE were found to be lower than the vPP and vHDPE crystallinities for blends between P40 and PP, and HDPE and P50, respectively. The presence of other plastics, varying chains lengths and branching, and impurities such as oxidative moieties and additives can lead to the formation of imperfect crystallites and a heterogeneous crystalline morphology, hence reducing crystallinity [58,62]. Therefore, determining the exact cause of (enhanced or reduced) crystallinity in recycled blends is a challenge. It has to be noted that the T_g values of PP and HDPE were not observed in the thermograms as they are located below the starting temperature of the DSC thermograms.

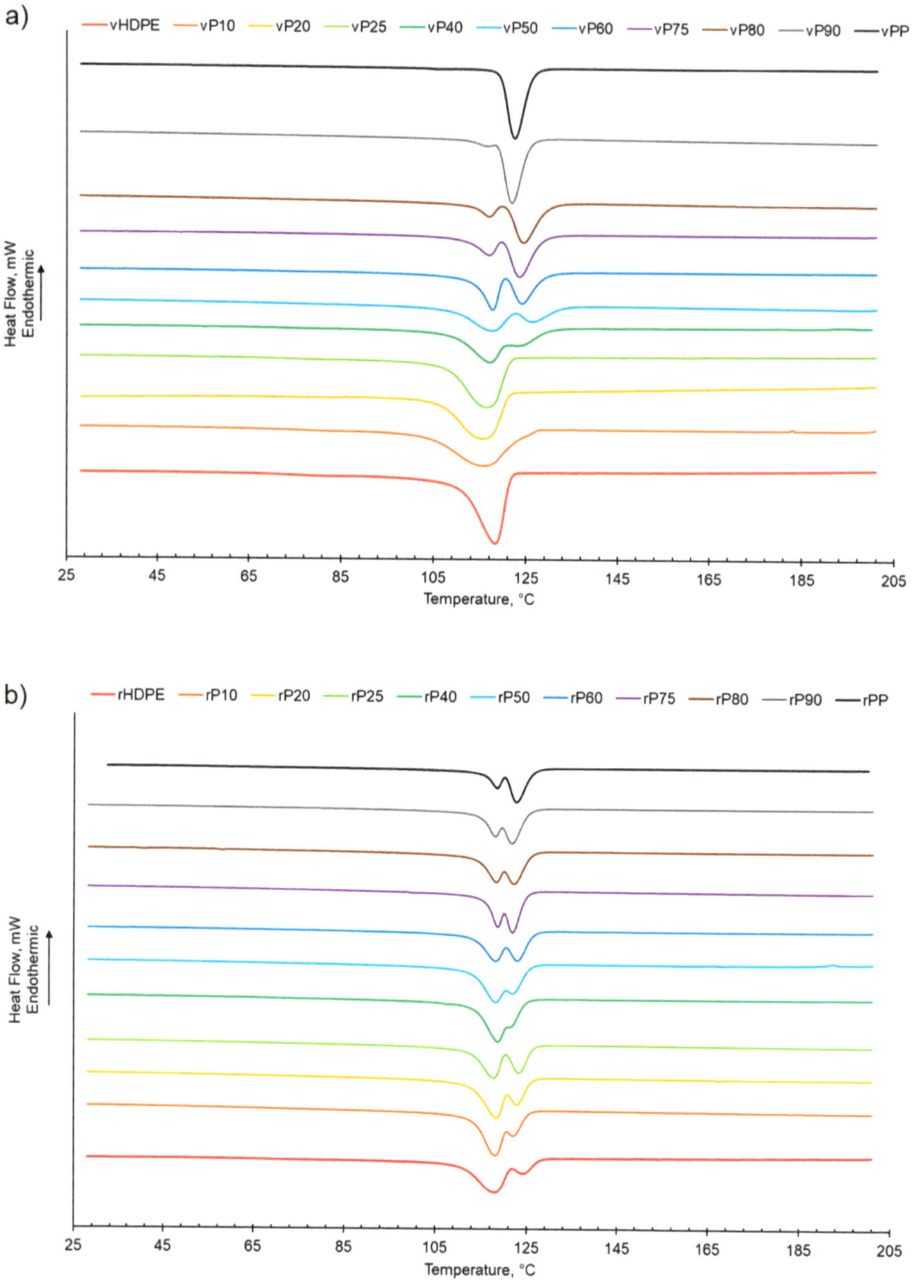

Figure 2. The crystallisation behaviour of PP:HDPE blends obtained from DSC: (**a**) vPP:vHDPE blends and (**b**) rPP:rHDPE blends.

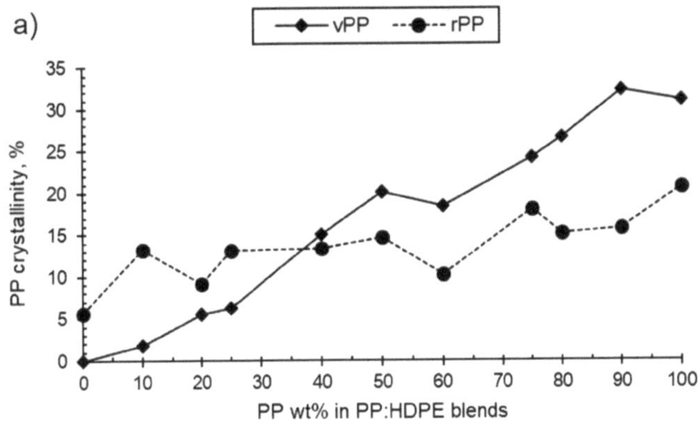

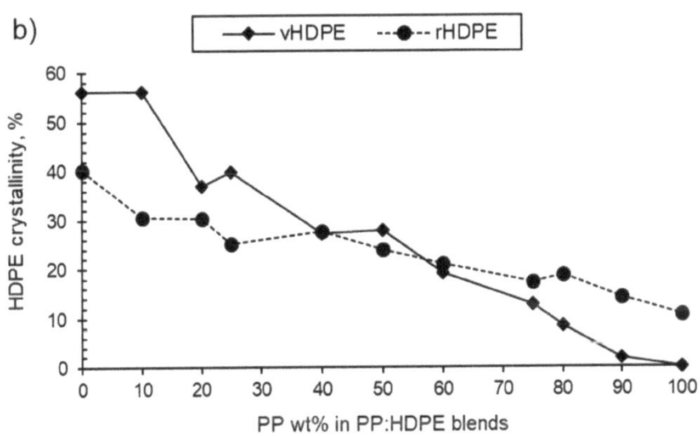

Figure 3. Variation of (**a**) virgin and recycled PP crystallinity in PP:HDPE blends and (**b**) virgin and recycled HDPE crystallinity in PP:HDPE blends.

3.2. Mechanical Properties of Virgin and Recycled PP:HDPE Blends

3.2.1. DMA Measurements of Virgin and Recycled PP:HDPE Blends

DMA was used to determine the viscoelastic response of the blends as a function of temperature. E' indicates the relative dynamic stiffness of the material and E'' indicates the ability to dissipate energy (Figure 4). No large variation was observed in E' at different blend compositions for both the virgin and recycled blends. However, there was a decreasing trend for the loss modulus for the virgin blends. As the PP wt% increased, the vPP:vHDPE blend's E'' decreased, whereas the rPP:rHDPE blends did not show such an obvious decrease in E'' with variation in composition. Interestingly, Fang et al. [47], who investigated the storage and loss moduli of rPP:rPE blends without the addition of a compatibiliser or filler, found an increase in moduli with rPP content. For example, the rP60 blend presented an E' at 40 °C, which was twice that of the rP45 blend. They concluded that an increase in stiffness occurs with an increase in PP wt%. The difference in the temperature at which the moduli were taken, different manufacturing processes and MFI of the rPE and rPP could account for the differences observed. Structural deteriorations caused by the recycling process can introduce flexibility and mobility due to the shorter chains. However, impurities can act as fillers in the recycled materials, imposing a mechanical restraint that increases the stiffness [63].

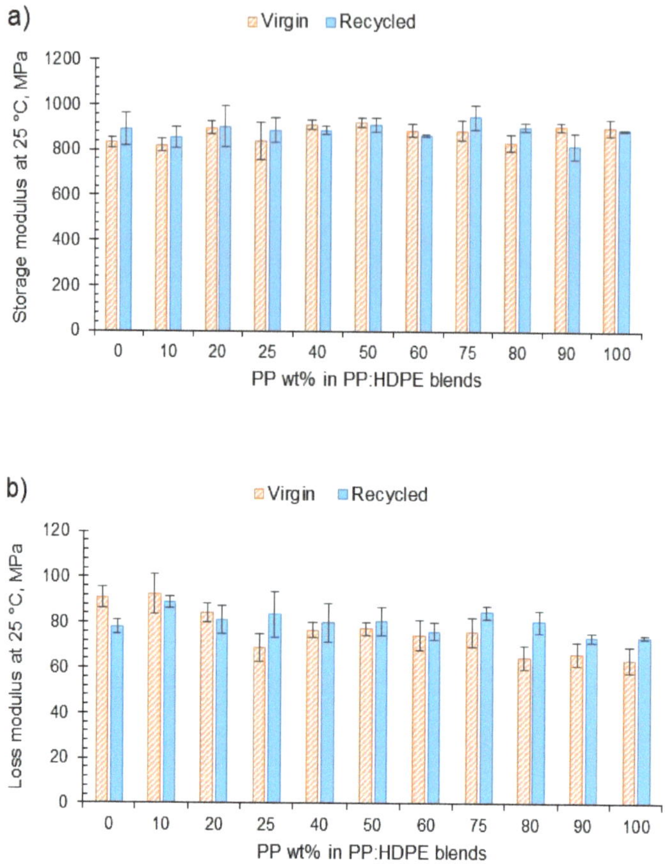

Figure 4. (**a**) The variation in storage modulus at 25 °C for vPP:vHDPE and rPP:rHDPE blends and (**b**) the variation in loss modulus at 25 °C for vPP:vHDPE and rPP:rHDPE blends.

For both virgin and recycled PP, HDPE and their blends, as temperature increased, E' decreased and E'' increased, shown in the tanδ vs. temperature graphs presented in Figure 5. This is due to material softening and the beginning of relaxation processes with increasing temperature [64]. HDPE and PP exhibit three relaxation processes: alpha (α), beta (β) and gamma (γ) [51,65]. Both the virgin and recycled PP:HDPE blends exhibited α and β relaxation processes in the tanδ vs. T graphs, as shown in Figure 5. γ relaxation was not observed as it typically occurs below −100 °C, which is outside the experimental temperature range. γ relaxation is associated with the motions of the side chain groups attached to the main chain in the amorphous region [65]. α relaxation is associated with the crystalline region, where the CH_2 groups within the crystallites have vibrational and re-orientation motion. The chains are flexible and freely rotating [65–67]. The rHDPE and rPP alpha relaxation temperatures (T_α) were lower compared with the vHDPE and vPP, respectively, possibly caused by imperfect crystallite formation due to recycling (Figures 5 and 6, see also Table S1) [55]. The higher T_α of HDPE compared with PP could be due to HDPE's higher crystallinity and amount of crystalline domains compared with PP [68]. The T_α of the virgin blends decrease as the PP wt% increases, with a similar trend observed in the recycled blends. The T_α are intermediary between the T_α of PP and HDPE. As suggested by Karaagac et al. [32], the observed relaxation temperatures of the blends are likely following the rule mixtures, and caution must be taken before suggesting partial miscibility at the interface due to the observation of a single peak. As the T_α of PP and

HDPE are close in value, it is possible that there is an overlap in the peaks causing the blend to have a broad T_α peak.

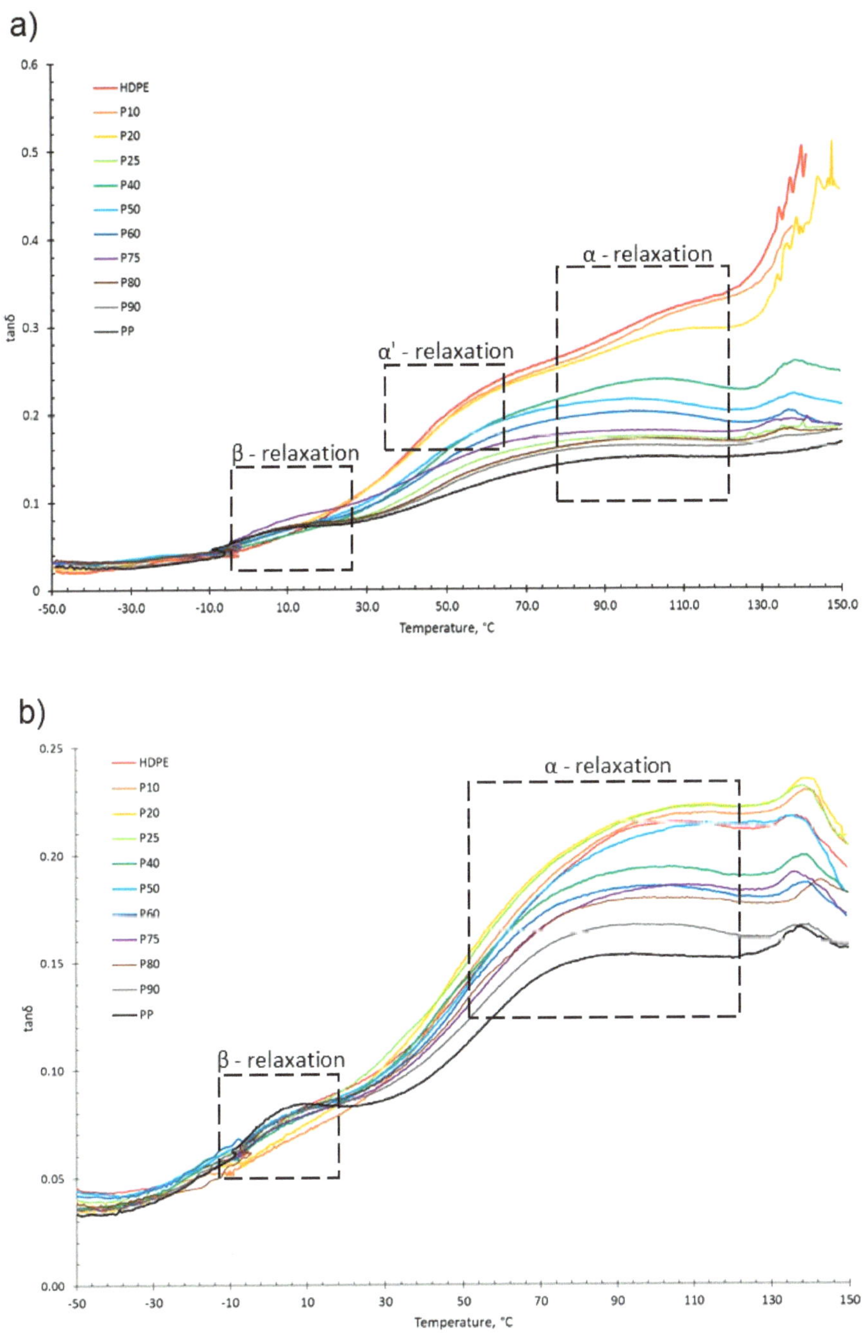

Figure 5. tanδ graphs as a function of temperature obtained from DMA of (**a**) vPP:vHDPE blends and (**b**) rPP:rHDPE blends with the relaxation regions highlighted.

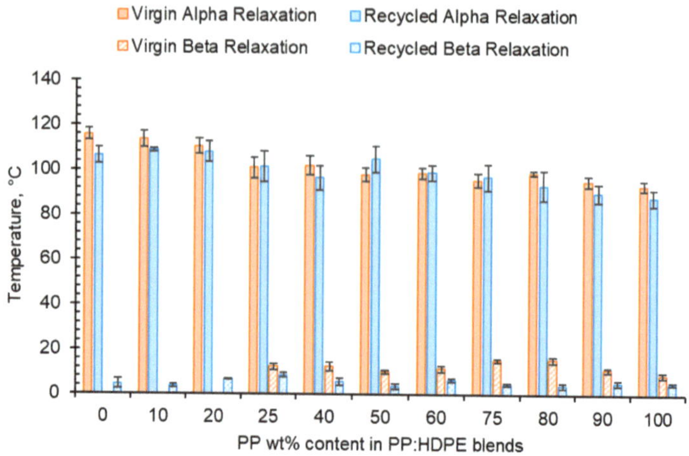

Figure 6. Comparison of T_α and T_β taken from tanδ traces for virgin and recycled PP:HDPE blends.

HDPE exhibits an additional relaxation process, α', which is associated with the crystalline region and partially overlaps into the α region [51]. α' is observed in vHDPE at approximately 40–50 °C (Figure 5a). As the vHDPE content in the virgin blends decreases, α' decreases in prominence. α' is not observed in the rHDPE, possibly due to the recycling process, which causes the formation of imperfect crystallites and the presence of contaminants, thus decreasing the peak prominence.

β relaxation is associated with the motion of the branches in the amorphous region and is connected to the T_g [63,66,69]. The PP β relaxation temperature (T_β) is the T_g. There are many opposing viewpoints surrounding where the T_g of PE is: (a) in the β region just below 0 °C, (b) in the region of −81 °C and (c) in the γ region below −100 °C [65,70]. The magnitude of the T_β is dependent on the amount of amorphous domains, as the relaxation occurs in the amorphous domain. The T_β in HDPE may not always be observed due to the low proportion of amorphous domains compared with crystalline domains. Additionally, tie molecules between the crystalline and amorphous domains restrict the complete relaxation of amorphous chains [63,67]. The T_β observed will be that of vPP as the vHDPE T_β is not seen (Figure 5). The vPP T_β is not visible in the vP10 and vP20 blends due to the small magnitude of the relaxation. The T_β of PP becomes visible at 12.4 °C for vP25. The T_β is present in the rHDPE due to the presence of PP impurities, which cannot be completely removed in the recycling process [53]. The T_β of the recycled blends were lower than the virgin blends and had little variation. The recycling process results in a decrease in molecular weight. The presence of the low-molecular-weight chains causes an increase in free volume and reduced chain packing [71]. An increase in free volume lowers the T_β as less thermal energy is required for chain mobility.

3.2.2. Tensile Measurements of Virgin and Recycled PP:HDPE Blends

PP, HDPE and their blends undergo macroscopic deformation during a tensile test and typically exhibit strain hardening and a ductile fracture, as shown in Figure 7. Initially, the polymers undergo elastic deformation; however, as the force applied continues to increase, the polymer sample reaches the yield point and enters the region of plastic deformation. At the yield point, a small neck forms within the gauge section and the polymer chains align in the direction of elongation. Continuing beyond the yield point, the virgin and recycled PP, HDPE and their blends exhibit the strain-hardening phenomenon. Strain hardening occurs when there is resistance to deformation and the neck region propagates and extends, which is termed necking. The polymer chains continue to orientate and align in the direction of elongation, which results in an increase in the strength of the plastic. Necking continues until fracture.

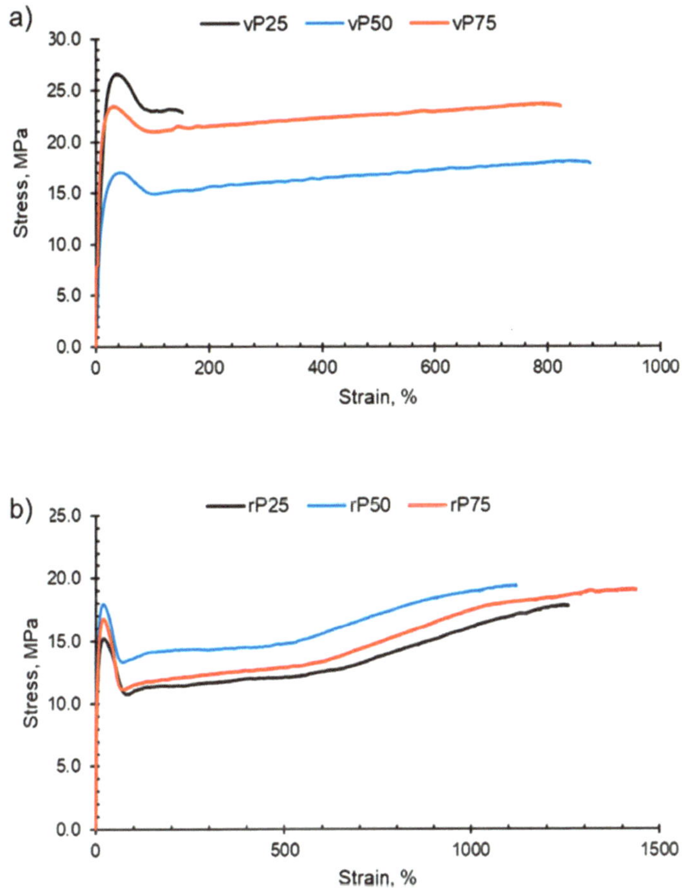

Figure 7. Examples of typical stress–strain curves obtained for PP:HDPE blends. The blends P25, P50 and P75 have been shown for (**a**) vPP:vHDPE blends and (**b**) rPP:rHDPE blends.

The recycled blends exhibited deteriorated tensile properties compared with the virgin blends in terms of Young's modulus and yield strength (Figure 8). The Young's modulus values of rPP and rHDPE were lower than those of the virgin polymers due to structural deterioration caused by the recycling process [60,72]. However, above rP75, the rPP:rHDPE blend's yield strength values approach the virgin blend's values. Studies have found that yield strength increases with crystallinity and lamellar thickness, with little or no effect of molecular weight [73]. The crystallinity of the recycled blends is lower compared with the virgin blends, and the recycling process results in the formation of imperfect crystals (which has been discussed in the thermal properties section), thus causing a reduction in the yield strength. Generally, there was little variation in the yield strength for the rPP:rHDPE blends.

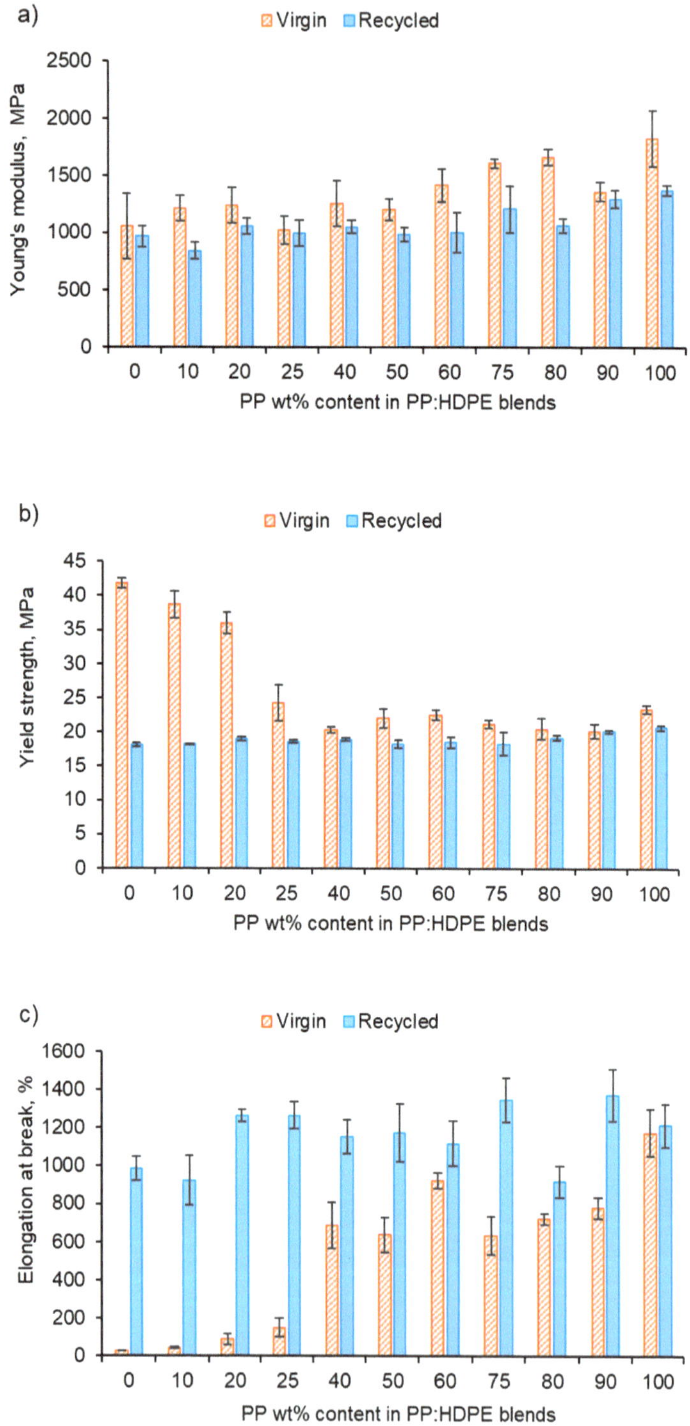

Figure 8. Tensile properties of vPP:vHDPE and rPP:rHDPE blends: (**a**) Young's modulus, (**b**) yield strength and (**c**) elongation at break.

The vHDPE demonstrated unexpected behaviour in yield strength (42 MPa) and elongation at break (29%) (Figure 8). vHDPE did not show typical necking behaviour and a brittle fracture was observed. No change in crystallinity was found by DSC when comparing the crystallinity of the HDPE before and after extrusion and injection moulding. During the injection moulding process, the polymer melt is exposed to a strong shear and elongational flow in which the chains are stretched and become highly orientated [74]. Flow-induced crystallisation increases HDPE's crystallisation rate and forms a highly orientated shish-kebab structure, which improves the strength of HDPE [75]. Lei et al. [76] found no necking behaviour when vHDPE was blended with 4% ultra-high molecular weight PE prepared by twin screw extruder and dynamic injection moulding. An increase in the tensile strength in the flow direction was observed from 23 to 76 MPa, which was caused by the formation of a web-like shish-kebab morphology and chain orientation. Therefore, the high chain orientation of the vHDPE could result in an interlocking of the shish-kebab to form a rigid structure, which affected the yield strength and elongation at break up to the vP25 blend (Figure 8) [76–78]. The rHDPE did not exhibit the same unexpected behaviour as vHDPE in its yield strength and elongation at break. This is most likely due to the presence of lower-molecular-weight chains caused by the recycling processing, which have a reduced packing ability and degree of orientation. Additionally, the presence of micro-voids can result in a decrease in compatibility between polymeric components [79].

The comparison between the elongation at break for the virgin and recycled blends presents interesting results (Figure 8c). It was expected that recycled blends would have a lower elongation at break compared with the virgin blends due to the structural deterioration during recycling causing a reduction in molecular weight [80]. For example, Fang et al. [47] found that with the addition of rPP up to 15 wt% in a PP:PE blend, the elongation at break decreased, and with over 30 wt%, the elongation at break reached a minimum. However, the longer elongation at break observed for the recycled blend could be due to the presence of lower-molecular-weight polymer chains caused by the recycling process [81]. It is possible that the low-molecular-weight polymer chains locate at the interface between PP and HDPE phases and lower the interfacial tension [81]. Additionally, the lower-molecular-weight chains increase the capability of molecules sliding over each other, resulting in an increase in deformability [80]. The vHDPE up to vP25 presented extremely low elongation at break and samples exhibited brittle fractures. The data sheet provided by Ineos suggests an elongation at break value of 800% for vHDPE at 2 in min^{-1}. As discussed, this behaviour could be due to the formation of a very rigid crystalline structure for vHDPE (and up to vP25), which would explain the brittle fracture observed. Due to this behaviour, a comparison between the virgin and recycled blends is more complex.

The virgin and recycled PP:HDPE blends gave intermediary Young's moduli values between PP and HDPE. Comparing the predicted rules of mixtures to the experimental Young's modulus shows a negative deviation for most recycled blends (Figure 9b). A negative deviation suggests poor compatibility and weak adhesion between the phases [6]. It is important to note that the rule of mixtures does not take into account interactions between components. The Young's modulus values of the virgin blends showed positive and negative deviations from the rule of mixtures with composition (Figure 9a): positive deviation for blends vP10 and vP20, negative deviation between vP25 and vP60, a minor positive deviation at vP75 and vP80, and a negative deviation at vP90. Lovinger and Williams [82] observed a maximum deviation at P80 and suggested that PE can play the role of stiffener to the PP matrix; in sufficient quantities, it enhances the intercrystalline links. The alternating changes of positive and negative deviations suggest a complex interplay of morphological factors and crystallinity as the composition varied.

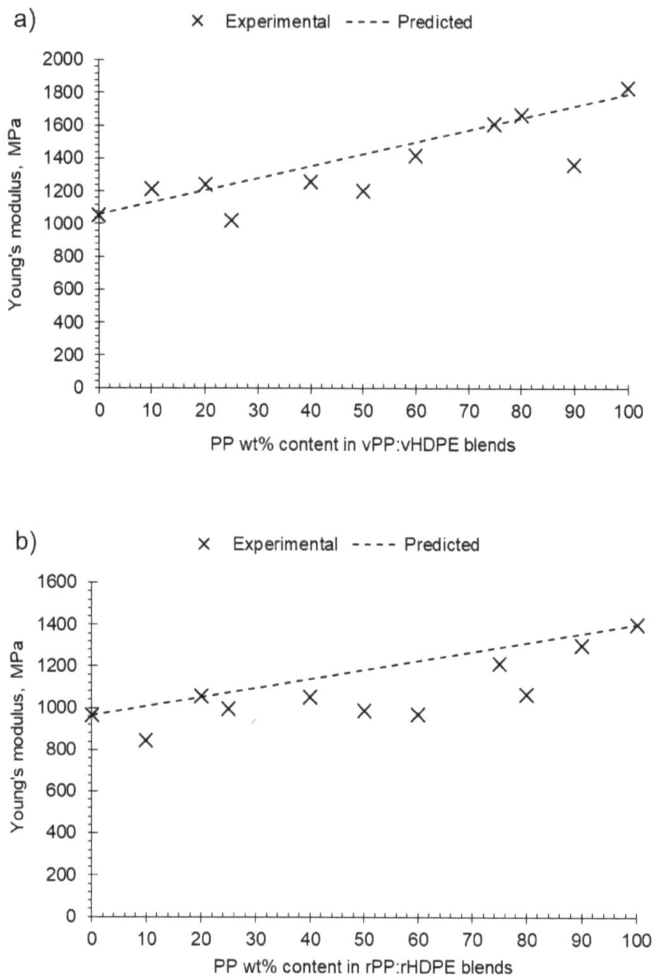

Figure 9. Plot of experimental and predicted Young's modulus against PP wt% content in PP:HDPE blends: (**a**) vPP:vHDPE blends and (**b**) rPP:rHDPE blends.

4. Conclusions

This study investigated the thermal and mechanical properties of virgin and recycled PP:HDPE blends. Thermal studies carried out by DSC confirmed that both the virgin and recycled PP:HDPE blends were immiscible. The recycling process was found to lower the T_m values of the rPP:rHDPE blends due to structural deterioration and the formation of imperfect crystallites. Interestingly, there was little difference in the T_c values when comparing the virgin and recycled blends. The PP and HDPE crystallinities were dependent upon the blend composition. As the ratio of PP increased, the crystallinity of PP increased and that of HDPE decreased in the PP:HDPE blends. Generally, the rPP:rHDPE blends had a lower overall crystallinity compared with the vPP:vHDPE blends, suggesting the formation of imperfect crystallites and a heterogeneous crystalline morphology. However, the crystallisation of the individual polymers was more complex. vPP crystallinity was enhanced (compared with rPP) at higher PP content; conversely, rHDPE crystallinity was enhanced (compared with vHDPE) at higher PP content. rPP and rHDPE could contain contaminants due to the difficulty of separating PP and HDPE during the recycling process, thus affecting the crystallinity behaviour.

DMA analysis found little variation in the E' of the virgin and recycled blends with composition. However, a decreasing trend was observed for the virgin blends E'' as the PP wt% increased, while the recycled blends' E'' was relatively constant. The interplay between the structure and dissipation mechanisms can be complex. Chain scission caused by the recycling process can introduce plasticising shorter chains. However, impurities can act as fillers in the recycled materials that impose a mechanical restraint. Recycled blends were found to have lower T_α and T_β due to structural deterioration caused by the recycling process. Recycled blends gave a reduced Young's modulus and yield strength in comparison with virgin blends due to deterioration during the recycling process. Generally, the recycled blends gave a higher elongation at break compared with the virgin blends, possibly due to the plasticity effect of the low-molecular-weight chain fragments. However, a comparison between the virgin and recycled blends' elongation at break was not straightforward in all cases due to the highly orientated vHDPE induced by the injection moulding.

This work explored the variability in the thermomechanical behaviour of vPP:vHDPE and rPP:rHDPE blends without the addition of other components. Understanding the performance variability in recycled blends is key to increasing the quantity of recycled material re-entering the consumer market to contribute towards a circular plastic economy.

Supplementary Materials: The following supporting information can be downloaded at: https://www.mdpi.com/article/10.3390/polym15214200/s1, Table S1: Summary of the crystallinities obtained from the first and second heating ramps for vPP:vHDPE blends; Table S2: Summary of the crystallinities obtained from the first and second heating ramps for rPP:rHDPE blends; Table S3: Summary of T_α and T_β taken from tanδ traces for virgin and recycled PP:HDPE blends; Figure S1: Melting temperatures of vPP, rPP, vHDPE and rHDPE of vPP:vHDPE and rPP:rHDPE blends obtained from DSC; Figure S2: Enthalpy of fusion of vPP, rPP, vHDPE and rHDPE of vPP:vHDPE and rPP:rHDPE blends obtained from DSC; Figure S3: Crystallisation temperatures of vPP, rPP, vHDPE and rHDPE of vPP:vHDPE and rPP:rHDPE blends obtained from DSC; Figure S4: Enthalpy of crystallisation of vPP:vHDPE and rPP:rHDPE blends obtained from DSC; Figure S5: Percentage crystallinity of vPP, rPP, vHDPE and rHDPE, and total crystallinity of vPP:vHDPE and rPP:rHDPE blends obtained from DSC.

Author Contributions: Conceptualization, V.K., D.R., C.S.H. and M.K.; methodology, V.K., D.R. and J.M.; software, H.J.; validation, H.J. and D.R.; formal analysis, H.J.; investigation, H.J.; resources, V.K., D.R. and C.S.H.; data curation, H.J.; writing—original draft preparation, H.J.; writing—review and editing, V.K., D.R, H.J., J.M., C.S.H. and M.K.; visualization, H.J.; supervision, V.K. and D.R.; project administration, V.K., D.R. and M.K.; funding acquisition, V.K. and H.J. All authors have read and agreed to the published version of the manuscript.

Funding: This research was funded by EPSRC SOFI CDT (Grant ref No. EP/L015536/1) and partially by Impact Solutions.

Institutional Review Board Statement: Not applicable.

Data Availability Statement: Data available upon request.

Acknowledgments: We thank Impact Solutions for their partial financial support of the project as well as providing the materials used in the experiments. We acknowledge the financial support of the EPSRC and the SOFI CDT (Grant ref No. EP/L015536/1). For the purpose of open access, the author has applied a Creative Commons Attribution (CC BY) licence to any Author Accepted Manuscript version arising from this submission.

Conflicts of Interest: Author M.K. was employed by the company Impact Solutions. The remaining authors declare that the research was conducted in the absence of any commercial or financial relationships that could be construed as a potential conflict of interest.

References

1. Geyer, R.; Jambeck, J.R.; Law, K.L. Production, use, and fate of all plastics ever made. *Sci. Adv.* **2017**, *3*, e1700782. [CrossRef] [PubMed]
2. Jones, H.; Saffar, F.; Koutsos, V.; Ray, D. Polyolefins and Polyethylene Terephthalate Package Wastes: Recycling and Use in Composites. *Energies* **2021**, *14*, 7306. [CrossRef]
3. Maddah, H.A. Polypropylene as a Promising Plastic: A Review. *Am. J. Polym. Sci.* **2016**, *6*, 1–11.
4. Kumar, S.; Panda, A.K.; Singh, R.K. A review on tertiary recycling of high-density polyethylene to fuel. *Resour. Conserv. Recycl.* **2011**, *55*, 893–910. [CrossRef]
5. Parameswaranpillai, J.; Pulikkalparambil, H.; Sanjay, M.R.; Siengchin, S. Polypropylene/high-density polyethylene based blends and nanocomposites with improved toughness. *Mater. Res. Express* **2019**, *6*, 075334. [CrossRef]
6. Jose, S.; Aprem, A.S.; Francis, B.; Chandy, M.C.; Werner, P.; Alstaedt, V.; Thomas, S. Phase morphology, crystallisation behaviour and mechanical properties of isotactic polypropylene/high density polyethylene blends. *Eur. Polym. J.* **2004**, *40*, 2105–2115. [CrossRef]
7. Graziano, A.; Jaffer, S.; Sain, M. Review on modification strategies of polyethylene/polypropylene immiscible thermoplastic polymer blends for enhancing their mechanical behavior. *J. Elastomers Plast.* **2019**, *51*, 291–336. [CrossRef]
8. Strapasson, R.; Amico, S.C.; Pereira, M.F.R.; Sydenstricker, T.H.D. Tensile and impact behavior of polypropylene/low density polyethylene blends. *Polym. Test.* **2005**, *24*, 468–473. [CrossRef]
9. Madi, N.K. Thermal and mechanical properties of injection molded recycled high density polyethylene blends with virgin isotactic polypropylene. *Mater. Des.* **2013**, *46*, 435–441. [CrossRef]
10. Feldman, D. Polyblend Nanocomposites. *J. Macromol. Sci. Part A* **2015**, *52*, 648–658. [CrossRef]
11. Salih, S.E.; Hamood, A.F.; Alsabih, A.H. Comparison of the Characteristics of LDPE:PP and HDPE:PP Polymer Blends. *Mod. Appl. Sci.* **2013**, *7*, 33–42. [CrossRef]
12. Ajji, A.; Utracki, L.A. Interphase and compatibilization of polymer blends. *Polym. Eng. Sci.* **1996**, *36*, 1574–1585. [CrossRef]
13. John, B.; Varughese, K.T.; Oommen, Z.; Pötschke, P.; Thomas, S. Dynamic mechanical behavior of high-density polyethylene/ethylene vinyl acetate copolymer blends: The effects of the blend ratio, reactive compatibilization, and dynamic vulcanization. *J. Appl. Polym. Sci.* **2003**, *87*, 2083–2099. [CrossRef]
14. Scott, C.E.; Macosko, C.W. Morphology development during the initial stages of polymer-polymer blending. *Polymer* **1995**, *36*, 461–470. [CrossRef]
15. Souza, A.M.C.; Demarquette, N.R. Influence of composition on the linear viscoelastic behavior and morphology of PP/HDPE blends. *Polymer* **2002**, *43*, 1313–1321. [CrossRef]
16. Souza, A.M.C.; Demarquette, N.R. Influence of coalescence and interfacial tension on the morphology of PP/HDPE compatibilized blends. *Polymer* **2002**, *43*, 3959–3967. [CrossRef]
17. Favis, B.D. The effect of processing parameters on the morphology of an immiscible binary blend. *J. Appl. Polym. Sci.* **1990**, *39*, 285–300. [CrossRef]
18. Jordan, A.M.; Kim, K.; Soetrisno, D.; Hannah, J.; Bates, F.S.; Jaffer, S.A.; Lhost, O.; Macosko, C.W. Role of Crystallization on Polyolefin Interfaces: An Improved Outlook for Polyolefin Blends. *Macromolecules* **2018**, *51*, 2506–2516. [CrossRef]
19. Albano, C.; González, J.; Ichazo, M.; Rosales, C.; Urbina de Navarro, C.; Parra, C. Mechanical and morphological behavior of polyolefin blends in the presence of CaCO3. *Compos. Struct.* **2000**, *48*, 49–58. [CrossRef]
20. Kallel, T.; Massardier-Nageotte, V.; Jaziri, M.; Gérard, J.-F.; Elleuch, B. Compatibilization of PE/PS and PE/PP blends. I. Effect of processing conditions and formulation. *J. Appl. Polym. Sci.* **2003**, *90*, 2475–2484. [CrossRef]
21. Cao, W.; Wang, K.; Zhang, Q.; Du, R.; Fu, Q. The hierarchy structure and orientation of high density polyethylene obtained via dynamic packing injection molding. *Polymer* **2006**, *47*, 6857–6867. [CrossRef]
22. Huang, D.E.; Kotula, A.P.; Snyder, C.R.; Migler, K.B. Crystallization Kinetics in an Immiscible Polyolefin Blend. *Macromolecules* **2022**, *55*, 10921–10932. [CrossRef] [PubMed]
23. Li, J.; Shanks, R.A.; Long, Y. Mechanical properties and morphology of polyethylene–polypropylene blends with controlled thermal history. *J. Appl. Polym. Sci.* **2000**, *76*, 1151–1164. [CrossRef]
24. Şirin, K.; Doğan, F.; Çanlı, M.; Yavuz, M. Mechanical properties of polypropylene (PP) + high-density polyethylene (HDPE) binary blends: Non-isothermal degradation kinetics of PP + HDPE (80/20) Blends. *Polym. Adv. Technol.* **2013**, *24*, 715–722. [CrossRef]
25. Xie, M.; Chen, J.; Li, H. Morphology and mechanical properties of injection-molded ultrahigh molecular weight polyethylene/polypropylene blends and comparison with compression molding. *J. Appl. Polym. Sci.* **2009**, *111*, 890–898. [CrossRef]
26. Aumnate, C.; Rudolph, N.; Sarmadi, M. Recycling of Polypropylene/Polyethylene Blends: Effect of Chain Structure on the Crystallization Behaviors. *Polymers* **2019**, *11*, 1456. [CrossRef]
27. Lin, J.-H.; Pan, Y.-J.; Liu, C.-F.; Huang, C.-L.; Hsieh, C.-T.; Chen, C.-K.; Lin, Z.-I.; Lou, C.-W. Preparation and Compatibility Evaluation of Polypropylene/High Density Polyethylene Polyblends. *Materials* **2015**, *8*, 8850–8859. [CrossRef]
28. Chiu, F.-C.; Yen, H.-Z.; Lee, C.-E. Characterization of PP/HDPE blend-based nanocomposites using different maleated polyolefins as compatibilizers. *Polym. Test.* **2010**, *29*, 397–406. [CrossRef]
29. Kazemi, Y.; Ramezani Kakroodi, A.; Rodrigue, D. Compatibilization efficiency in post-consumer recycled polyethylene/polypropylene blends: Effect of contamination. *Polym. Eng. Sci.* **2015**, *55*, 2368–2376. [CrossRef]

30. Sutar, H.; Sahoo, P.C.; Sahu, P.S.; Sahoo, S.; Murmu, R.; Swain, S.; Mishra, S.C. Mechanical, Thermal and Crystallization Properties of Polypropylene (PP) Reinforced Composites with High Density Polyethylene (HDPE) as Matrix. *Mater. Sci. Appl.* **2018**, *9*, 502–515. [CrossRef]
31. Hammache, Y.; Serier, A.; Chaoui, S. The effect of thermoplastic starch on the properties of polypropylene/high density polyethylene blend reinforced by nano-clay. *Mater. Res. Express* **2020**, *7*, 025308. [CrossRef]
32. Karaagac, E.; Koch, T.; Archodoulaki, V.-M. The effect of PP contamination in recycled high-density polyethylene (rPE-HD) from post-consumer bottle waste and their compatibilization with olefin block copolymer (OBC). *Waste Manag.* **2021**, *119*, 285–294. [CrossRef] [PubMed]
33. Aurrekoetxea, J.; Sarrionandia, M.A.; Urrutibeascoa, I.; Maspoch, M.L. Effects of recycling on the microstructure and the mechanical properties of isotactic polypropylene. *J. Mater. Sci.* **2001**, *36*, 2607–2613. [CrossRef]
34. Yin, S.; Tuladhar, R.; Shi, F.; Shanks, R.A.; Combe, M.; Collister, T. Mechanical reprocessing of polyolefin waste: A review. *Polym. Eng. Sci.* **2015**, *55*, 2899–2909. [CrossRef]
35. da Costa, H.M.; Ramos, V.D.; de Oliveira, M.G. Degradation of polypropylene (PP) during multiple extrusions: Thermal analysis, mechanical properties and analysis of variance. *Polym. Test.* **2007**, *26*, 676–684. [CrossRef]
36. Ha, K.H.; Kim, M.S. Application to refrigerator plastics by mechanical recycling from polypropylene in waste-appliances. *Mater. Des.* **2012**, *34*, 252–257. [CrossRef]
37. Oliveira, T.A.; Oliveira, R.R.; Barbosa, R.; Azevedo, J.B.; Alves, T.S. Effect of reprocessing cycles on the degradation of PP/PBAT-thermoplastic starch blends. *Carbohydr. Polym.* **2017**, *168*, 52–60. [CrossRef]
38. Bunjes, A.; Arndt, J.-H.; Geertz, G.; Barton, B. Characterization and chemometric modelling of mechanically recycled polypropylene for automotive manufacturing. *Polymer* **2022**, *249*, 124823. [CrossRef]
39. Chen, Z.-F.; Lin, Q.-B.; Su, Q.-Z.; Zhong, H.-N.; Nerin, C. Identification of recycled polyethylene and virgin polyethylene based on untargeted migrants. *Food Packag. Shelf Life* **2021**, *30*, 100762. [CrossRef]
40. Jmal, H.; Bahlouli, N.; Wagner-Kocher, C.; Leray, D.; Ruch, F.; Munsch, J.-N.; Nardin, M. Influence of the grade on the variability of the mechanical properties of polypropylene waste. *Waste Manag.* **2018**, *75*, 160–173. [CrossRef]
41. Jin, H.; Gonzalez-Gutierrez, J.; Oblak, P.; Zupančič, B.; Emri, I. The effect of extensive mechanical recycling on the properties of low density polyethylene. *Polym. Degrad. Stab.* **2012**, *97*, 2262–2272. [CrossRef]
42. Oblak, P.; Gonzalez-Gutierrez, J.; Zupančič, B.; Aulova, A.; Emri, I. Processability and mechanical properties of extensively recycled high density polyethylene. *Polym. Degrad. Stab.* **2015**, *114*, 133–145. [CrossRef]
43. Saikrishnan, S.; Jubinville, D.; Tzoganakis, C.; Mekonnen, T.H. Thermo-mechanical degradation of polypropylene (PP) and low-density polyethylene (LDPE) blends exposed to simulated recycling. *Polym. Degrad. Stab.* **2020**, *182*, 109390. [CrossRef]
44. Cecon, V.S.; Da Silva, P.F.; Vorst, K.L.; Curtzwiler, G.W. The effect of post-consumer recycled polyethylene (PCRPE) on the properties of polyethylene blends of different densities. *Polym. Degrad. Stab.* **2021**, *190*, 109627. [CrossRef]
45. Noor Hasanah, T.I.T.; Wijeyesekera, D.C.; Lim, A.J.M.S.; Ismail, B. Recycled PP/HDPE Blends: A Thermal Degradation and Mechanical Properties Study. *Appl. Mech. Mater.* **2014**, *465–466*, 932–936. [CrossRef]
46. Atiqah, A.A.S.M.; Salmah, H.; Firuz, Z.; Uy Lan, D.N. Properties of Recycled High Density Polyethylene/Recycled Polypropylene Blends: Effect of Maleic Anhydride Polypropylene. *Key Eng. Mater.* **2014**, *594–595*, 837–841. [CrossRef]
47. Fang, C.; Nie, L.; Liu, S.; Yu, R.; An, N.; Li, S. Characterization of polypropylene–polyethylene blends made of waste materials with compatibilizer and nano-filler. *Compos. Part B Eng.* **2013**, *55*, 498–505. [CrossRef]
48. *ISO 527-2*; Tensile Testing for Plastics. International Organization for Standardization: Geneva, Switzerland, 2012.
49. Watt, E.; Abdelwahab, M.A.; Snowdon, M.R.; Mohanty, A.K.; Khalil, H.; Misra, M. Hybrid biocomposites from polypropylene, sustainable biocarbon and graphene nanoplatelets. *Sci. Rep.* **2020**, *10*, 10714. [CrossRef]
50. Shrivastava, A. 3-Plastic Properties and Testing. In *Introduction to Plastics Engineering*; Shrivastava, A., Ed.; William Andrew Publishing: Oxford, UK, 2018; pp. 49–110.
51. Wetton, R.E. Thermal analysis. In *Polymer Characterisation*; Hunt, B.J., James, M.I., Eds.; Springer Netherlands: Dordrecht, The Netherlands, 1993; pp. 178–221.
52. Nielsen, L.E. *Predicting the Properties of Mixtures: Mixture Rules in Science and Engineering*; Marcel Dekker: New York, NY, USA, 1978.
53. Hopewell, J.; Dvorak, R.; Kosior, E. Plastics recycling: Challenges and opportunities. *Philos. Trans. R. Soc. B Biol. Sci.* **2009**, *364*, 2115–2126. [CrossRef]
54. Momanyi, J.; Herzog, M.; Muchiri, P. Analysis of Thermomechanical Properties of Selected Class of Recycled Thermoplastic Materials Based on Their Applications. *Recycling* **2019**, *4*, 33. [CrossRef]
55. Gu, J.; Xu, H.; Wu, C. Thermal and Crystallization Properties of HDPE and HDPE/PP Blends Modified with DCP. *Adv. Polym. Technol.* **2014**, *33*, 21384. [CrossRef]
56. Mileva, D.; Tranchida, D.; Gahleitner, M. Designing polymer crystallinity: An industrial perspective. *Polym. Cryst.* **2018**, *1*, e10009. [CrossRef]
57. Samanta, P.; Srivastava, R.; Nandan, B.; Chen, H.-L. Crystallization behavior of crystalline/crystalline polymer blends under confinement in electrospun nanofibers of polystyrene/poly(ethylene oxide)/poly(ε-caprolactone) ternary mixtures. *Soft Matter* **2017**, *13*, 1569–1582. [CrossRef] [PubMed]

58. Stangenberg, F.; Ågren, S.; Karlsson, S. Quality Assessments of Recycled Plastics by Spectroscopy and Chromatography. *Chromatographia* **2004**, *59*, 101–106. [CrossRef]
59. Ruj, B.; Pandey, V.; Jash, P.; Srivastava, V. Sorting of plastic waste for effective recycling. *Int. J. Appl. Sci. Eng. Res.* **2015**, *4*, 564–571.
60. Pinheiro, L.A.; Chinelatto, M.A.; Canevarolo, S.V. The role of chain scission and chain branching in high density polyethylene during thermo-mechanical degradation. *Polym. Degrad. Stab.* **2004**, *86*, 445–453. [CrossRef]
61. Shirkavand, M.J.; Azizi, H.; Ghasemi, I.; Karabi, M. Effect of Molecular Structure Parameters on Crystallinity and Environmental Stress Cracking Resistance of High-Density Polyethylene/TiO2 Nanocomposites. *Adv. Polym. Technol.* **2018**, *37*, 770–777. [CrossRef]
62. Vilaplana, F.; Karlsson, S. Quality Concepts for the Improved Use of Recycled Polymeric Materials: A Review. *Macromol. Mater. Eng.* **2008**, *293*, 274–297. [CrossRef]
63. Sewda, K.; Maiti, S.N. Dynamic mechanical properties of high density polyethylene and teak wood flour composites. *Polym. Bull.* **2013**, *70*, 2657–2674. [CrossRef]
64. Hidalgo-Salazar, M.A.; Correa-Aguirre, J.P.; García-Navarro, S.; Roca-Blay, L. Injection Molding of Coir Coconut Fiber Reinforced Polyolefin Blends: Mechanical, Viscoelastic, Thermal Behavior and Three-Dimensional Microscopy Study. *Polymers* **2020**, *12*, 1507. [CrossRef]
65. McCrum, N.G.; Read, B.E.; Williams, G. *Anelastic and Dielectric Effects in Polymeric Solids*; John Wiley: London, UK; New York, NY, USA, 1967.
66. Sethi, M.; Gupta, N.K.; Srivastava, A.K. Dynamic mechanical analysis of polyethylene and ethylene vinylacetate copolymer blends irradiated by electron beam. *J. Appl. Polym. Sci.* **2002**, *86*, 2429–2434. [CrossRef]
67. Suljovrujic, E.; Micic, M.; Milicevic, D. Structural Changes and Dielectric Relaxation Behavior of Uniaxially Oriented High Density Polyethylene. *J. Eng. Fibers Fabr.* **2013**, *8*, 131–143. [CrossRef]
68. Popli, R.; Glotin, M.; Mandelkern, L.; Benson, R.S. Dynamic mechanical studies of α and β relaxations of polyethylenes. *J. Polym. Sci. Polym. Phys. Ed.* **1984**, *22*, 407–448. [CrossRef]
69. Khanna, Y.P.; Turi, E.A.; Taylor, T.J.; Vickroy, V.V.; Abbott, R.F. Dynamic mechanical relaxations in polyethylene. *Macromolecules* **1985**, *18*, 1302–1309. [CrossRef]
70. Fakirov, S.; Krasteva, B. On the Glass Transition Temperature of Polyethylene as Revealed by Microhardness Measurements. *J. Macromol. Sci. Part B* **2000**, *39*, 297–301. [CrossRef]
71. Morris, B.A. 4-Commonly Used Resins and Substrates in Flexible Packaging. In *The Science and Technology of Flexible Packaging*; Morris, B.A., Ed.; William Andrew Publishing: Oxford, UK, 2017; pp. 69–119.
72. Ragaert, K.; Delva, L.; Van Geem, K. Mechanical and chemical recycling of solid plastic waste. *Waste Manag.* **2017**, *69*, 24–58. [CrossRef]
73. Chivers, R.A.; Moore, D.R. The effect of molecular weight and crystallinity on the mechanical properties of injection moulded poly(aryl-ether-ether-ketone) resin. *Polymer* **1994**, *35*, 110–116. [CrossRef]
74. Yang, H.-R.; Lei, J.; Li, L.; Fu, Q.; Li, Z.-M. Formation of Interlinked Shish-Kebabs in Injection-Molded Polyethylene under the Coexistence of Lightly Cross-Linked Chain Network and Oscillation Shear Flow. *Macromolecules* **2012**, *45*, 6600–6610. [CrossRef]
75. Zhao, B.; Li, X.; Huang, Y.; Cong, Y.; Ma, Z.; Shao, C.; An, H.; Yan, T.; Li, L. Inducing Crystallization of Polymer through Stretched Network. *Macromolecules* **2009**, *42*, 1428–1432. [CrossRef]
76. Lei, J.; Zhang, Z.; Jiang, C.; Shen, K. Bi-axial self-reinforcement of high-density polyethylene induced by high-molecular weight polyethylene through dynamic packing injection molding. *Polym. Int.* **2006**, *55*, 1021–1026. [CrossRef]
77. Deng, C.; Lei, J.; Gao, X.; Chen, Z.; Shen, K. Study on the Improvement of Crystallization in HDPE Induced by High–Molecular-Weight Polyethylene Through Dynamic Packing Injection Molding. *Polym. Plast. Technol. Eng.* **2008**, *47*, 716–721. [CrossRef]
78. Bayer, R.K.; Baltá Calleja, F.J.; López Cabarcos, E.; Zachiviann, H.G.; Paulsen, A.; Brüning, F.; Meins, W. Properties of elongational flow injection moulded polyethylene. *J. Mater. Sci.* **1989**, *24*, 2643–2652. [CrossRef]
79. Murugan, D.; Varughese, S.; Swaminathan, T. Recycled Polyolefin-Based Plastic Wastes for Sound Absorption. *Polym. Plast. Technol. Eng.* **2006**, *45*, 885–888. [CrossRef]
80. La Mantia, F.P.; Botta, L.; Mistretta, M.C.; Di Fiore, A.; Titone, V. Recycling of a Biodegradable Polymer Blend. *Polymers* **2020**, *12*, 2297. [CrossRef] [PubMed]
81. Dorigato, A. Recycling of polymer blends. *Adv. Ind. Eng. Polym. Res.* **2021**, *4*, 53–69. [CrossRef]
82. Lovinger, A.J.; Williams, M.L. Tensile properties and morphology of blends of polyethylene and polypropylene. *J. Appl. Polym. Sci.* **1980**, *25*, 1703–1713. [CrossRef]

polymers

MDPI

Article

A Comparison of Three-Layer and Single-Layer Small Vascular Grafts Manufactured via the Roto-Evaporation Method

Gualberto Antonio Zumbardo-Bacelis [1,2], Laura Peponi [3,*], Rossana Faride Vargas-Coronado [1], Eustolia Rodríguez-Velázquez [4,5], Manuel Alatorre-Meda [6], Pascale Chevallier [7], Francesco Copes [7], Diego Mantovani [7], Gustavo A. Abraham [8] and Juan Valerio Cauich-Rodríguez [1,*]

1 Unidad de Materiales, Centro de Investigación Científica de Yucatán, Calle 43 #130 x 32 y 34, Colonia Chuburná de Hidalgo, Mérida C.P. 97205, Mexico; gualberto-antonio.zumbardo-bacelis.1@ulaval.ca (G.A.Z.-B.); ross@cicy.mx (R.F.V.-C.)
2 Department of Chemical Engineering, Laval University, Quebec, QC G1V 0A6, Canada
3 Instituto de Ciencia y Tecnología de Polímeros (ICTP-CSIC), C/Juan de la Cierva 3, 28006 Madrid, Spain
4 Facultad de Odontología, Universidad Autónoma de Baja California, Tijuana 22390, Mexico; eustolia.rodriguez@uabc.edu.mx
5 Centro de Graduados e Investigación en Química-Grupo de Biomateriales y Nanomedicina, Tecnológico Nacional de México, Instituto Tecnológico de Tijuana, Tijuana 22510, Mexico
6 Centro de Graduados e Investigación en Química-Grupo de Biomateriales y Nanomedicina, CONAHCYT-Tecnológico Nacional de México, Instituto Tecnológico de Tijuana, Tijuana 22510, Mexico; manuel.alatorre@tectijuana.edu.mx
7 Laboratory for Biomaterials and Bioengineering (CRC-I), Department of Min-Met-Materials Engineering & CHU de Quebec Research Center, Laval University, Quebec, QC G1V0A6, Canada; pascale.chevallier@crchudequebec.ulaval.ca (P.C.)
8 Research Institute for Materials Science and Technology, INTEMA (UNMdP-CONICET). Av. Colón 10850, Mar del Plata B7606BWV, Argentina
* Correspondence: lpeponi@ictp.csic.es (L.P.); jvcr@cicy.mx (J.V.C.-R.)

Citation: Zumbardo-Bacelis, G.A.; Peponi, L.; Vargas-Coronado, R.F.; Rodríguez-Velázquez, E.; Alatorre-Meda, M.; Chevallier, P.; Copes, F.; Mantovani, D.; Abraham, G.A.; Cauich-Rodríguez, J.V. A Comparison of Three-Layer and Single-Layer Small Vascular Grafts Manufactured via the Roto-Evaporation Method. *Polymers* **2024**, *16*, 1314. https://doi.org/10.3390/polym16101314

Academic Editors: Young-Sam Cho and Shanfeng Wang

Received: 26 March 2024
Revised: 18 April 2024
Accepted: 2 May 2024
Published: 8 May 2024

Abstract: This study used the roto-evaporation technique to engineer a 6 mm three-layer polyurethane vascular graft (TVG) that mimics the architecture of human coronary artery native vessels. Two segmented polyurethanes were synthesized using lysine (SPUUK) and ascorbic acid (SPUAA), and the resulting materials were used to create the intima and adventitia layers, respectively. In contrast, the media layer of the TVG was composed of a commercially available polyurethane, Pearlbond 703 EXP. For comparison purposes, single-layer vascular grafts (SVGs) from individual polyurethanes and a polyurethane blend (MVG) were made and tested similarly and evaluated according to the ISO 7198 standard. The TVG exhibited the highest circumferential tensile strength and longitudinal forces compared to single-layer vascular grafts of lower thicknesses made from the same polyurethanes. The TVG also showed higher suture and burst strength values than native vessels. The TVG withstood up to 2087 ± 139 mmHg and exhibited a compliance of 0.15 ± 0.1%/100 mmHg, while SPUUK SVGs showed a compliance of 5.21 ± 1.29%/100 mmHg, akin to coronary arteries but superior to the saphenous vein. An indirect cytocompatibility test using the MDA-MB-231 cell line showed 90 to 100% viability for all polyurethanes, surpassing the minimum 70% threshold needed for biomaterials deemed cytocompatibility. Despite the non-cytotoxic nature of the polyurethane extracts when grown directly on the surface, they displayed poor fibroblast adhesion, except for SPUUK. All vascular grafts showed hemolysis values under the permissible limit of 5% and longer coagulation times.

Keywords: roto-evaporation; vascular graft; burst pressure; compliance; mechanical properties; biomedical applications

1. Introduction

Cardiovascular diseases (CVDs) remain the leading cause of death worldwide, even during the COVID-19 era. In fact, COVID-19 impacts the cardiovascular system in multiple ways, resulting in increased morbidity in patients with pre-existing cardiovascular conditions, as well as causing myocardial injury and dysfunction [1]. According to the World Health Organization [2], 33% of deaths worldwide can be attributed to CVDs. These diseases are typically characterized by the obstruction of blood flow through vessels and damage to the endothelial tissue, leading to a lack of nutrient supply [3]. The prevailing complications are coronary heart disease (CHD), peripheral artery disease (PAD), thrombosis, intimal hyperplasia, atherosclerosis and myocardial ischemia [4]. Common treatments for CVDs include the drastic modification of risk factors such as smoking cessation, blood pressure management, lipid control, weight loss, physical activity and dietary modifications [5]. However, in advanced cases of cardiovascular pathologies, a vascular surgery intervention (minimal or invasive) is required involving open surgical and endovascular procedures such as angioplasty, stenting, atherectomy and, in the worst scenario, the replacement or bypass of a damaged or occluded vessel [6]. Coronary artery bypass graft (CABG) surgery is the gold-standard procedure in patients with complex multivessel coronary artery disease [7]. The preferred conduits for vascular grafting are native veins and arteries, with saphenous vein (SV) being the most used autograft vessel [8]. Despite substantial advances in CABG surgery in the past decade, graft patency continues to be the 'chink in the armor' of this procedure. Moreover, autologous vessels are not always the best option due to the limited availability and poor quality of tissue from donors [4,9]. Failures related to saphenous vein grafts include mechanical mismatch, thrombosis, neointimal hyperplasia, stenosis and complete occlusion [10,11]. Synthetic vascular grafts are often considered the best alternative when autologous vessels are not available. The first successful vascular prosthesis replacement was performed by Voorhees in 1952 [12] using a porous fabric made from Vinyon "N" (PVC, polyvinylchloride). Ever since, the development of synthetic grafts has evolved significantly, and prostheses made from Dacron® (PET, polyethylene terephthalate) [13] and expanded Teflon® (ePTFE, polytetrafluoroethylene) [14] have been used as middle- and large-caliber grafts when autologous grafts are not available. These materials have been widely adopted due to their biocompatibility, durability and availability [15]. Not showing the same clinical progress as medium and large-diameter vascular grafts, small-diameter vascular grafts ($\leq$6 mm) are seldom used due to their poor patency rates, which limit their application in coronary artery bypass graft surgery [3,16,17]. The optimal vascular substitute must be capable of mimicking native vessels; in other words, it should be mechanically strong, compliant, kink-resistant with good suture retention, non-toxic, non-immunogenic, non-thrombogenic and biocompatible [18].

Polyurethanes are versatile biomaterials that have found applications in various fields, including maxillofacial implants, non-adhesive barriers, controlled-release devices, and blood contact applications like catheters, aortic valves, coatings for pacemakers, and vascular grafts [19,20]. Among the different types of polyurethanes, segmented polyurethanes (SPU) have attracted significant attention due to their ability to adjust mechanical and biological properties by manipulating the chemical architecture using different hard and soft segments [21,22]. Polyurethanes can be used in their prepared/synthesized form [23], in a surface-functionalized form [24] and as scaffolds for soft [25] and hard tissue engineering [26], making them a versatile class of biomaterials. Taking advantage of the versatility of polyurethanes to match the properties required in the cardiovascular field, it is possible to design a small-caliber vascular graft, composed of a three-layer structure with different thicknesses and compositions comparable to the vascular wall of vessels. Artery and vein walls are organized into three layers, where the tunica intima is composed of endothelial cells covering the lumen; the tunica media is composed of elastin, collagen fibers and smooth muscle cells (SMCs) aligned circumferentially; and the tunica adventitia is formed of connective tissue and fibroblasts. The proportion of each tunica and the wall thickness is variable among small-, medium- and large-caliber veins and arteries [27]. With this structure in mind, several techniques have been proposed to develop vascular grafts, including electrospinning, bioprinting, 3D-printing [28,29], salt-leaching/the solvent casting method, films wrapped in a rod, molding [30–32] and even a combination of various techniques in a sequential manner with the same polymer composition [32,33].

Considering the low compliances exhibited by high-caliber vascular grafts made from expanded Teflon and knitted Dacron to date, there is no equivalent of these materials for their use in small vascular grafts. Therefore, the need for alternatives to PTFE and PET vascular grafts remains. In this regard, polyurethanes are an excellent alternative, as they have been used in several cardiovascular applications since 1960. Although polyurethane vascular grafts' long-term stability and safety have been a significant concern due to uncontrolled biodegradation problems and biocompatibility issues without eliciting a chronic inflammatory response, they can be manufactured by various simple techniques and can be modified for improved performance.

In this work, we explored a technique based on the rotational solvent casting method (referred to as roto-evaporation) to form a three-layer vascular graft in a cylindrical mold, using different polyurethanes for each layer deposited sequentially. The inner layer (intima) was formed with a lysine-based segmented polyurethane (SPUUK), as it has been proven that alkaline amino acid-based polyurethanes are biocompatible with endothelial cells [30]. The media layer was deposited by using commercial polyurethane (Pearlbond 703 EXP). The outer layer (adventitia) was formed with an ascorbic acid-based segmented polyurethane (SPUAA), which has been proven to be non-cytotoxic to fibroblasts and osteoblasts [34]. We hypothesize that a semi-crystalline polyurethane can exhibit high compliances (similar or superior to common PTFE/PET grafts) despite having a high Young's modulus by tailoring polyurethane composition and thickness. Therefore, the objectives of this paper are as follows: (i) to obtain three-layer small-caliber vascular grafts where each layer has a different composition; (ii) to compare the three-layer construct with small-caliber vascular grafts made from a single layer (only one segmented polyurethane) or a single layer obtained from physical blends of the three segmented polyurethanes; (iii) to assess their mechanical performance in simple tension and according to the ISO 7198 standard for cardiovascular grafts (longitudinal and circumferential strength, suture strength, burst strength and compliance); and (iv) to determine their cytocompatibility and their hemocompatibility (coagulation time and hemolysis).

2. Materials and Methods

2.1. Materials

Poly (ε-caprolactone) diol (PCL diol, Mn = 2000), 4,4′ (metylene-bis-cyclohexyl) diisocyanate ($H_{12}MDI$), tin (II) 2-ethylhexanoate, L-lysine dihydrochloride (K) and L-ascorbic acid (AA) were purchased from Sigma-Aldrich (Milwaukee, USA). Dimethylformamide (DMF) from Sigma-Aldrich (Steinheim, Germany) and tetrahydrofuran (THF) from JT Baker (Phillipsburg, USA) were used as received. Thermoplastic Polyurethane Pearlbond 703 EXP was obtained from Lubrizol® and used as a model polyurethane for the core of the vascular graft. This commercial thermoplastic polyurethane is a hydrolytic-resistant aromatic-based polymer (toluene diisocyanate) with polycaprolactone diol as the soft segment, as demonstrated by FTIR, Raman and ^{1}H NMR. It has a T_g of −36 °C (measured by DMA), a melting range between 40 °C and 84 °C (measured by DSC) and a decomposition temperature at 389 °C (measured by TGA), and it is considered as EU-Food-compliant (10/2011 regulation).

2.2. Synthesis of Segmented Polyurethanes

Segmented polyurethanes and urea were synthesized with a molar ratio of 1:2:1 (PCL: $H_{12}MDI$: K or AA), as previously reported [34]. During the pre-polymer formation, PCL diol was dissolved in DMF and then reacted with a molar excess of $H_{12}MDI$ added dropwise in the presence of tin (II) 2-ethylhexanoate as a catalyst. The reaction was conducted under a dry nitrogen atmosphere and stirred for 4 h at 80 °C. For the chain extension reaction, either K or AA, previously dissolved in DMF, was added to the pre-polymer and left to react for another 2 h under constant stirring. To stop the reaction, the DMF dissolved polymer was precipitated in distilled water and stirred overnight. The polymer was thoroughly washed to eliminate the residual monomers and dried at 65 °C in vacuum. The obtained polymers were named as SPUUK for lysine-based polyurea urethane and SPUAA for ascorbic acid-based polyurethane. The chemical structure proposed for the SPUUK and SPUAA is shown in Scheme 1. Polyurethane synthesis was confirmed by FT-IR and Raman spectroscopy.

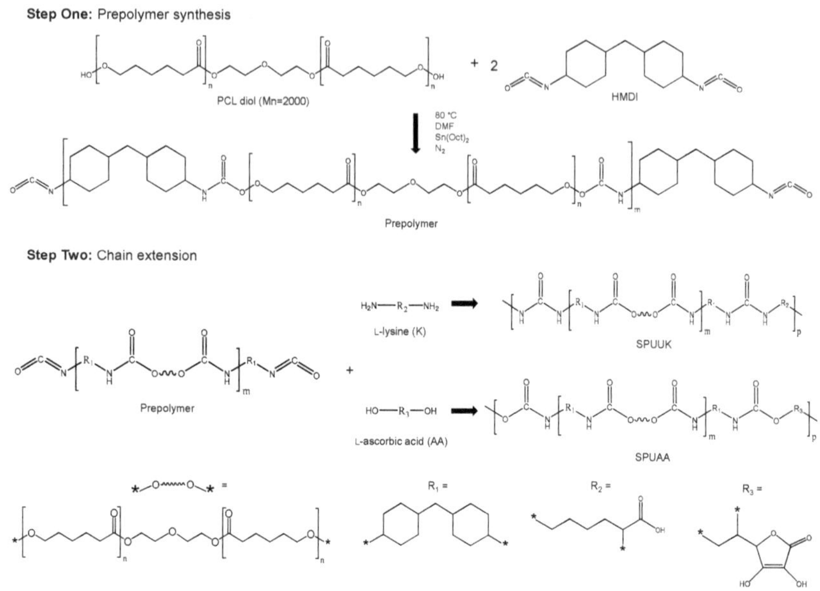

Scheme 1. Synthesis route for SPUUK and SPUAA.

2.3. Physicochemical Characterization

Fourier transform infrared (FT-IR) spectra were recorded with a Nicolet 8700 spectrometer (Thermo Scientific, Madison, WI, USA) using attenuated total reflectance (ATR). A ZnSe crystal was pressed against the samples and each spectrum was collected between 4000 and 650 cm^{-1}, with 100 scans and a resolution of 4 cm^{-1}. Raman spectra were acquired using a Renishaw inVia Reflex Raman spectrometer (Wotton-under-Edge, Gloucestershire, UK) in the spectral range of 3200–200 cm^{-1}. A 633 nm laser was used as the excitation radiation source using 50% of the total power, a 50× objective lens and 10 ms of exposure.

2.4. Rationale/Justification of the Roto-Evaporation Technique for Manufacturing Vascular Grafts

Most standard techniques for manufacturing experimental vascular grafts use electrospinning or wrapping onto a mandrel. The first provides suitable wall thickness with long processing times, while in the second, delamination occurs. Roto-evaporation offers a good alternative incorporating a different composition and thickness layer without blending the other biodegradable polyurethanes. The addition of the second layer did not dissolve the first layer as they have various polymer concentrations and, therefore, different densities and viscosities. In addition, the fast evaporation of THF did not allow the penetration and dissolution of a larger polyurethane block (typically, dissolution time for film forming was 24 h at 25 °C). Because of this, the different layers do not mix, demonstrating the adequacy of the technique for manufacturing vascular grafts where the intima layer can be thin, whereas the media layer can be thicker. However, this method is limited as it does not provide polymer alignment in other directions other than circumferential directions.

To obtain dense tubular structures, a 6 mm diameter glass rod was immersed for 5–10 s in a low viscosity polyurethane solution (1 g polymer/30 mL THF). Then, the glass rod was placed horizontally on a Caframo® overhead stirrer rotating at 60 rpm within a glass test tube to slow the rate of evaporation, as depicted in Figure 1a,b. The diameter of the glass rod plus the coating of the polymer was measured with a micrometer Mitutoyo model No.

547-512. This procedure was repeated until coronary artery thickness was achieved (see Figure 1c); for the single-layer grafts, the wall thickness was 0.132 ± 0.01 mm, according to the mean thickness of the intimal tunica in the coronary artery [35].

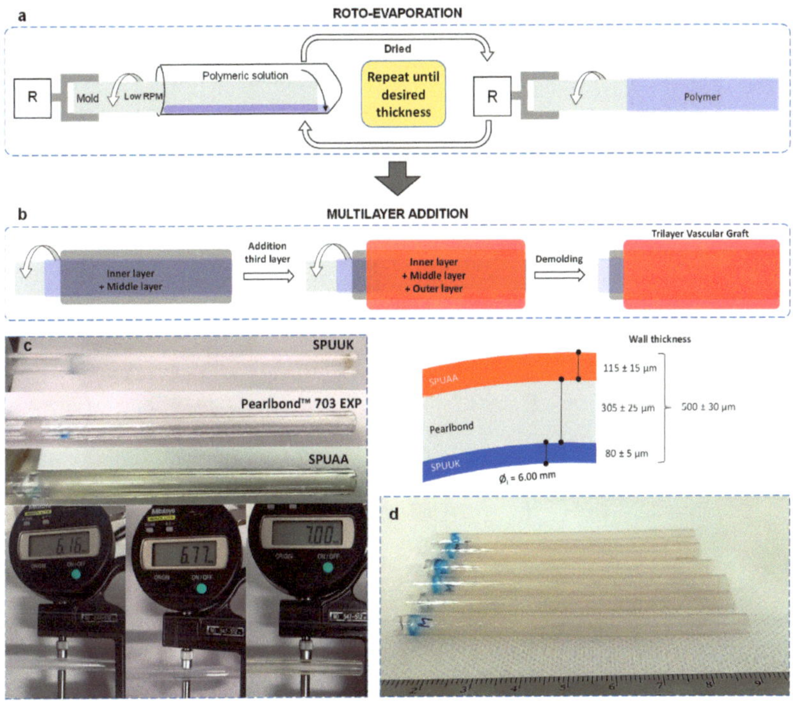

Figure 1. (**a**) Process flowchart of "Roto-evaporation" to fabricate vascular grafts, (**b**) multilayer addition to obtain multilayer vascular grafts, (**c**) digital images measuring the diameter of different layers added and (**d**) three-layer vascular grafts (TVGs) demolded.

2.5. Fabrication of Polyurethane Three-Layer Vascular Graft

Based on high-frequency ultrasound imaging, the coronary artery has a wall thickness of 0.5 ± 0.1 mm, of which 16 ± 2% corresponds to the tunica intima, 62 ± 10% corresponds to the tunica media and 21 ± 5% corresponds to the tunica adventitia [35–38]. Therefore, to obtain the three different polymer solutions of TVG based on SPUUK, Pearlbond 703 EXP and SPUAA were prepared and deposited in a sequential manner. The inner layer of the vascular graft was made using SPUUK until reaching a diameter of 6.16 ± 0.01 mm (w_{SPUUK} = 0.08 mm for a 6 mm glass rod diameter), while the middle layer was made using Pearlbond 703 EXP, and the diameter of the SPUUK + Pearlbond was $Ø_{\text{SPUUK+Pearlbond}}$ = 6.77 ± 0.05 mm ($w_{\text{SPUUK+Pearlbond}}$ = 0.385 mm). Finally, the outer layer was made from SPUAA, and the final diameter was $Ø_{\text{SPUUK+Pearlbond+SPUAA}}$ = 7.00 ± 0.03 mm ($w_{(\text{SPUUK+Pearlbond+SPUAA})}$ = 0.500 mm). All roto-evaporated tubular grafts in this study had an internal diameter of 6.00 mm. A summary of the process is showed in Figure 1. To assess the contribution of each individual polyurethane, tubular structures were fabricated with only SPUUK, Pearlbond 703 EXP and SPUAA, referred to as SVG. In addition, a physical blend of the three types of polyurethanes (MVG) was prepared by dissolving each polyurethane in THF (1:1:1 weight ratio of SPUUK/Pearlbond 703 EXP/SPUAA).

2.6. Mechanical Properties of Polyurethane Films and Vascular Grafts

2.6.1. Tensile Properties of Polyurethane Films

The films used for the tensile tests were prepared by means of solvent evaporation. For this, each polyurethane was dissolved in tetrahydrofuran assisted by magnetic stirring for 24 h. Then, the polymeric solution was poured onto a 120×120 mm^2 non-stick mold and left for 24 h to evaporate the solvent. Mechanical tensile tests were carried out in a universal testing machine MiniShimadzu AGS-X (Kyoto, Japan) equipped with a 100 N load cell with a resolution of 0.01 N and operated at a crosshead speed of 50 mm/min according to ASTM-D882 [39] at room temperature. The samples were dumbbell specimens 25 mm in length, 5 mm in width in the narrow section and 100 to 150 μm in thickness. The Young's modulus (E) was calculated from the initial slope of the strain–stress curves in the strain range of 10%, the ultimate tensile strength (σ_{max}) and strain to failure (ε_{max}) are reported for at least 10 specimens.

2.6.2. Tensile Properties of Vascular Grafts

The circumferential and longitudinal tensile properties of tubular structures (vascular grafts) were determined according to Sections 8.3.1 and 8.3.2, respectively, of the ISO 7198 standard [40] for cardiovascular implants and tubular vascular prostheses (ISO 7198:1998). Figure 2a shows the fixture used for measuring the circumferential tensile strength of 7 mm long specimens (L_C = 7 mm) loaded along the tube radial direction and using two metallic hooks (diameter (d_{hook}) = 2.4 mm) until the failure of at least 10 samples. For the circumferential tensile test, the ISO 7198 standard recommends reporting the maximum force reached (F_C^{max}) divided by 2 times the length of the specimen ($F_C^{max}/2L_C$). This was referred to as the "circumferential tensile strength", while the maximum displacement during the circumferential test was labeled as δ_C^{max}. For longitudinal tensile strength (see Figure 2b), 8 tubular specimens with a nominal length of 50 mm (L_L = 50 mm) were used and loaded along the tube axial direction (lengthwise) until failure. Both tests were carried out in a universal machine, MiniShimadzu AGS-X (Kyoto, Japan), equipped with a 1 kN load cell and operated at a crosshead speed of 50 mm/min according to ISO 7198 standard. The tensile force (F) as a function of crosshead speed displacement (δ) was recorded and the parameters reported were the maximum force or "longitudinal tensile strength", labeled as F_L^{max}, and the "maximum displacement" of the crosshead machine upon failure during the test, labeled as δ_L^{max}.

2.6.3. Burst Strength

The burst strength test was conducted using a custom-made piece of equipment designed and manufactured for this type of test [41], as shown in Figure 2c. An 80 mm long vascular graft was pressurized (0.3 psi/s, ~15.5 mmHg/s, for all tests herein) using compressed air. By means of an optoCONTROL 1200-30 micrometer from Micro-Epsilon, Ortenburg, Germany, the external diameter (D_0) displacement was measured. The burst strength was calculated from the maximum pressure average of 8 specimens when these ruptured. Furthermore, the theoretical burst pressure was also calculated with the simplified Lame's equation (Equation (1)):

$$P_i = \sigma_\theta \frac{\left(D_{ex}^2 - D_{in}^2\right)}{\left(D_{ex}^2 + D_{in}^2\right)} [\text{MPa}] \quad (1)$$

where σ_θ, P_i, D_{in} and D_{ex} are the hoop stress (≈Tensile stress, MPa), internal pressure (MPa), failure inner diameter (mm) and failure outer diameter (mm) of the theoretical tube, respectively [42]. The hoop stress at failure was calculated using Equation (2), where F_C^{max} is the maximal load at failure (N), L_C is the initial length (mm) and t is the thickness (mm) of the ring specimen.

$$\sigma_\theta^{max} = \frac{F_C^{max}}{2L_C t} [\text{MPa}] \quad (2)$$

To calculate the failure inner diameter (D_{in}) in the specimens during the circumferential test, Equation (3) was used where C_{fail} is the real internal circumference of the ring specimen at failure, d_{hook} is the diameter of the hooks and δ_c^{max} is the distance of displacement between the hooks [43].

$$D_{in}^{fail} = \frac{C_{in}^{fail}}{\pi} = \frac{d_{hook}(\pi + 2) + 2\delta_c^{max}}{\pi} [\text{mm}] \quad (3)$$

The failure external diameter was calculated with Equation (4) assuming the incompressibility of the wall thickness of the tube, where t is the wall thickness.

$$D_{ex} = D_{in} + 2t \; [\text{mm}] \quad (4)$$

Circumferential strain ($\Delta D/D_0$) was calculated as the change in the external diameter divided by the initial external diameter.

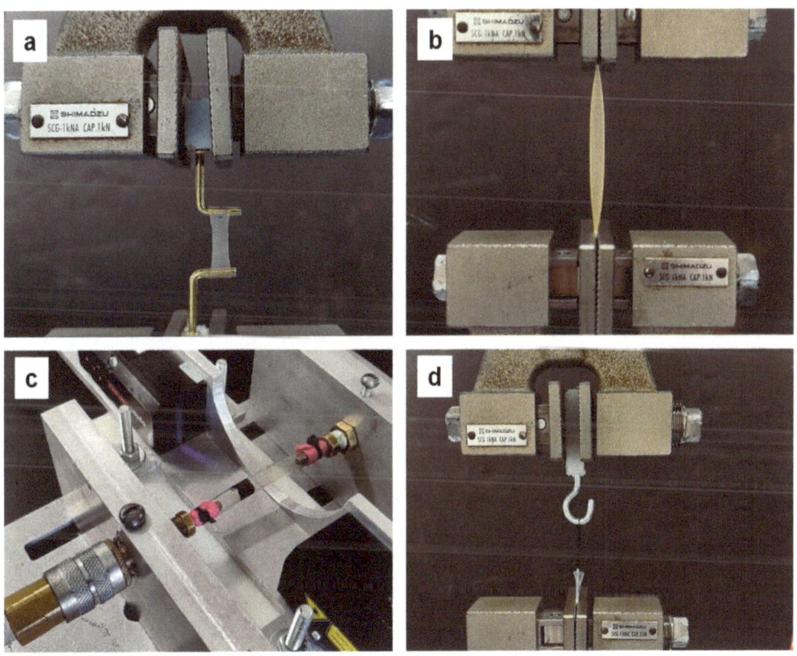

Figure 2. The fixtures for different mechanical tests of the tubular vascular grafts: (**a**) circumferential tensile strength; (**b**) longitudinal tensile strength; (**c**) burst strength in the custom-made equipment; and (**d**) suture retention strength.

2.6.4. Compliance

Circumferential compliance (C) was calculated from the experimental data at three different pressure ranges—low pressure (50 to 90 mmHg), medium pressure (80 to 120 mmHg) and high pressure (110 to 150 mmHg). Compliance was computed from recordings of inner pressure, P_i, and a computed inner diameter, D_P^{in}, through Equation (5) [42]:

$$C = \frac{\frac{D_H^{in} - D_L^{in}}{D_L^{in}}}{P_H - P_L} \times 10^4 \; [\%/100 \text{ mmHg}] \quad (5)$$

where D_H^{in} is the internal diameter corresponding to the higher-pressure value and D_L^{in} is the internal diameter corresponding to the lower-pressure value. The internal diameter was calculated using Equation (3). Pressure and diameter values were used to calculate a dimensional stiffness index, β, by means of Equation (6) [42,44]:

$$\beta = \frac{\ln\left(\frac{P_H}{P_L}\right)}{\left[\frac{D_H^{ex} - D_L^{ex}}{D_L^{ex}}\right]} \quad (6)$$

2.6.5. Suture Retention Strength

According to the ISO 7198 standard (Section 8.8), the suture retention strength test determines the force necessary to pull a suture from the prosthesis or cause the wall of the prosthesis to fail (ISO 7198:1998). The suture retention strength was measured using a 3-0 Nylon suture (Johnson & Johnson), as depicted in Figure 2d. For this, a tubular specimen was cut normal to the long axis and the suture was stitched at 2 mm from the end of the vascular graft forming a loop. Then, sutured specimens were placed on the universal testing machine MiniShimadzu AGS-X (Kyoto, Japan) equipped with a 1 kN load cell operated at a crosshead speed of 50 mm/min.

2.7. Scanning Electronic Microscopy (SEM)

The morphology and imaging of the grafts before and after testing were obtained using a scanning electron microscope, JEOL, JMS 6360LV (Akishima, Tokyo, Japan), operating in low-vacuum mode. The microscope was operated at 20 kV. All samples were coated with gold in a Desk II Sputter Coater (Moorestown, NJ, USA) at 60 s and 40 mA.

2.8. Biological Characterization of Pristine Polyurethanes

2.8.1. Screening Test for Cytocompatibility

Cytocompatibility screening tests were carried out by means of elution (indirect) assays employing MDA-MB-231 as testing cells (breast cancer cells) and CCK-8 and crystal violet as testing probes. MDA-MB-231 (purchased from the American Type Culture Collection) cells were chosen for this experiment because of their highly proliferative behavior. Working with robust and highly proliferative cells in screening cytocompatibility/cytotoxicity tests compensates for parameters/conditions that might lead to false results (either positive or negative), which might be observed when employing cells with regular (moderate) growth. Examples of such parameters/conditions include the presence of reductive chemicals and metal ions in the extract or culture media, even at trace concentrations (which alter the reduction and color conversion of colorimetric testing probes, CCK-8 in our case) [45,46] and minute gradients in topography and water contact angle of cell substrates (e.g., tissue culture plates, which are known to impact the adhesion and proliferation of viable cells) [47]. Parameters like these or their combinations have been found to cause interferences when working with colorimetric probes, leading to either negative or positive false results from cytocompatibility tests when caution is not observed [48]. These tests were carried out according to the ISO 10993−5:2009 elution method with subtle variations [45]. Briefly, 20 mg of each polyurethane was cut into small pieces and immersed in 2 mL of culture medium and incubated under standard culture conditions for 48 h to produce extracts. MDA-MB-231 cells were seeded in 96-well plates at a density of 1000 cells/well with 100 µL of the cell culture medium and incubated for 24 h. Then, the culture medium was withdrawn and 100 µL of the extracts were added to the wells and incubated for 24 and 48 h; the cell culture medium was used as a control. Afterwards, the culture medium was replaced with 100 µL of fresh culture medium administered 10 µL of CCK-8 cell proliferation reagent and incubated for 4 h. Optical absorbance was measured afterwards at 450 nm using an ELISA microplate reader (BIO-RAD model iMark, USA). Cytocompatibility after the exposure of cells to sample extracts, represented as the percentage of cell viability, was calculated using Equation (7):

$$Cell\ viability(\%) = \frac{A_{Ext}}{A_{Ctrl}} \times 100 \quad (7)$$

where A_{Ext} is the absorbance of cells after exposure to extracts and A_{Ctrl} is the absorbance of the cell control, non-exposed cells (100% viability). The results were the average of six independent experiments for each time. The cytocompatibility of the prepared samples was also determined via the well-known crystal violet staining method [49]. To this end, the MDA-MB-231 cells were seeded in 96-well plates (100 µL, 15,000 cells/well) and grown for 24 h under standard culture conditions. Afterwards, the samples were internalized (100 µL) and incubated for the chosen period (24 h and 48 h). Then, the culture medium was discarded and the cells were attached to the plate in the presence of 10 µL of a glutaraldehyde solution (11 *w/v*% in water). The solution was discarded and cells were washed two times with Milli-Q water. The cells were then shaken at room temperature (300 rpm, 15 min) in the presence of 100 µL of a crystal violet solution (0.1 *w/v*% in 200 mM orthophosphoric acid, 200 mM formic acid and 200 mM MES, pH 6). The solution was discarded and the cells were again washed twice with Milli-Q water. Once washed, the cells were incubated at room temperature overnight for drying. Once dried, the cells were shaken at room temperature (300 rpm, 15 min) in the presence of 100 µL of acetic acid (10 *w/w*% in water). Immediately after this, the absorbance of the resulting solution was measured at 595 nm. The percentage of cell viability was quantified using Equation (7). The results were the average of six independent experiments. Sterile conditions were preserved up until cell fixation with glutaraldehyde.

2.8.2. Cell Adhesion and Proliferation

A live/dead assay was performed to investigate the viability of cells in direct contact with the samples. For this, flat samples of 5 × 5 mm^2 of each polyurethane were cut and placed in a 96-well plate, where fibroblast L–929 cells (1.5 × 10^4 cells/well, 100 µL) were seeded and incubated under culture conditions for 24 h. Then, the culture medium was withdrawn, 3 µL of a solution 98:1:1 of PBS/calcein-AM/propidium iodide aqueous solution was pipetted on top of each sample and the samples were incubated for 10 min at room temperature and dark. Experiments were carried out in duplicates for each kind of sample. The samples were mounted on coverslips and observed using a LEICA SP8 confocal microscope (LEICA Microsystems Heidelberg GmbH) operating at wavelengths of 488/495–545 and 552/590–700 nm for the detection of calcein AM (green) and propidium iodide (red), respectively.

Both cell lines were cultured in a glucose-rich DMEM culture medium containing phenol red, supplemented with 10% FBS, 1% pen-strep, 1% non-essential amino acids (NEAA) and 1% sodium pyruvate. The materials were acquired from Sigma-Aldrich (DMEM, pen-strep, NEAA and sodium pyruvate) and Biotecfron (FBS). Throughout the experiments, sterile Milli-Q water was utilized. All samples were sterilized to ensure sterility by immersing them in 70% ethanol for 10 min before characterization. The humidified atmosphere was kept at 5% CO_2 and the temperature was maintained at 37 °C.

2.8.3. Clot Formation

To evaluate clotting times, the free hemoglobin method was used on 15 mm-diameter samples of the different polyurethanes [50]. First, samples were cleaned in an ultrasonic bath with 70% alcohol (v/v) for 10 min and then sterilized with UV light for 30 min on each side. For each sample, 100 µL of citrated blood was placed onto the surfaces and 20 µL of $CaCl_2$ was added immediately. Finally, samples were incubated at 37 °C for 5, 10, 20, 30 and 40 min to study the kinetics of clotting. After these incubation times, 2 mL of deionized water was added and blood cells not entrapped in a thrombus were hemolyzed. One minute later, 100 µL aliquots of the solution containing the hemoglobin were transferred to a 96-well plate to perform absorbance reading. Free hemoglobin molecules in water were measured by monitoring the absorbance at 545 nm using a spectrophotometer ELISA reader (BioRad mod. 450 Mississauga, ON, Canada). The higher detected quantity of hemoglobin is correlated with the low level of clot formation in the presence of the samples. The test was performed in triplicate with 6 samples per condition each time. Blood from a different donor was used for each experiment.

2.8.4. Hemolysis

For an early assessment of the possible hemotoxic effects of the different polymers used, a hemolysis test was conducted based on the ASTM Standard F756-17 (Standard Practice for Assessment of Hemolytic Properties of Materials) [51]. We collected whole human blood from healthy donors using citrate-containing blood collection tubes for this experiment. We placed three samples for each condition in a 15 mL tube and added 10 mL of sterile PBS 1X to each tube. PBS 1X was used as a negative control and deionized H_2O was used as a positive control. Samples and controls were incubated at 37 °C for 30 min. In the meantime, the collected blood was diluted in PBS 1X to a final ratio of 4:5 (4 parts of citrated blood and 5 parts of PBS 1X). After the incubation, 200 µL of diluted blood were added in each tube and carefully mixed by inverting each tube. After that, samples and controls were incubated at 37 °C for 1 h. All tubes were carefully mixed via inversion after 30 min of incubation. At the end of the incubation, the tubes containing the samples and the controls underwent a centrifugation step at 800 g for 5 min. The supernatant was collected and 100 µL aliquots were placed in a 96-well plate. The absorbance (OD) at a wavelength of 545 nm was recorded and the hemolysis percentage was calculated according to Equation (8).

$$Hemolysis(\%) = \frac{A_S - A_N}{A_P - A_N} \times 100 \quad (8)$$

where A_S represents the sample absorption value, A_N is the negative control absorption value and A_P is the positive control absorption value.

2.9. Statistical Analysis

Data were expressed as means ± the standard deviations of means. Significant differences were determined by running a one-way ANOVA followed by Tukey's post hoc method to test all possible pairwise comparisons and determine where the differences lay. A value of p-value < 0.05 was considered statistically significant.

3. Results and Discussion

Segmented polyurethane-ureas were synthesized with a molar ratio of 1:2:1 (PCL: H_{12}MDI: K or AA), leading to a rigid content of approximately 38 wt%. During prepolymer formation and by using an FDA-approved catalyst (a commonly used catalyst for the ring opening polymerization of L-lactide and glycolide used in the synthesis of PLA or PGA), the urethane group is formed, but due to the H_{12}MDI excess used in the first step, prepolymers and free isocyanates are present. Then, the isocyanate-terminated soft segment was extended into a polymer by adding either lysine or ascorbic acid. In the case of lysine, the reaction of isocyanates with amines introduces a urea group. In contrast, an amide is also formed in the second reaction between isocyanates and carboxylic acids. Furthermore, chemical reactions between either the urethane or urea groups and isocyanates might form allophanates and Biuret products.

3.1. Physicochemical Characterization

FT-IR spectra of SPUUK, Pearlbond 703 EXP and SPUAA are shown in Figure 3a, demonstrating the suggested structures proposed in Scheme 1. FT-IR spectra of the SPUUK and SPUAA show the 'amide bands', free –NH of urethane bond absorbance at 3300–3400 and amide II band (C–N stretching + N–H bending) at 1552 cm^{-1} [52]. The peaks observed at ~2930 and ~2862 cm^{-1} are associated with the asymmetric and symmetric $-CH_2$ groups. The peak at ~1725 cm^{-1} is assigned to the stretching vibration of C=O ester groups. The peaks located at ~3365, 1725, 1462 and 1095 cm^{-1} correspond to the –NH, –C=O, $-CH_2$

and –C–O absorptions and confirm the presence of NHCOO groups in the synthesized polyurethanes [53]. In FT-IR spectrum of Pearlbond 703 EXP, the band at 3338 cm^{-1} is caused by the stretching vibration of the N–H bond of the urethane. The bands at 2944 and 2855 cm^{-1} can be ascribed to asymmetric and symmetric –CH_2 groups. The intense band at 1713 cm^{-1} represents stretching vibrations of the carbonyl groups and may be associated with the ester and to urethane carbonyls. The band at 1599 cm^{-1} can be associated at –C=C– stretching vibration of the aromatic ring in the diisocyanate [54]. The Raman spectra of the polyurethanes is shown in Figure 3b. The Raman bands at ~2930 and ~2862 cm^{-1} is associated with asymmetric and symmetric CH stretching vibrations of the –CH_2 groups [55]. C=O stretching vibrations of urethane and ester groups can be observed at 1731 cm^{-1} in the SPUUK and SPUAA and 1724 cm^{-1} in the Pearlbond 703 EXP. The intense band at 1617 cm^{-1} in the Pearlbond 703 EXP spectrum is associated with –C=C– stretching vibrations and the non-intense band at 3061 cm^{-1} is associated with the =C– H of aromatic ring [56].

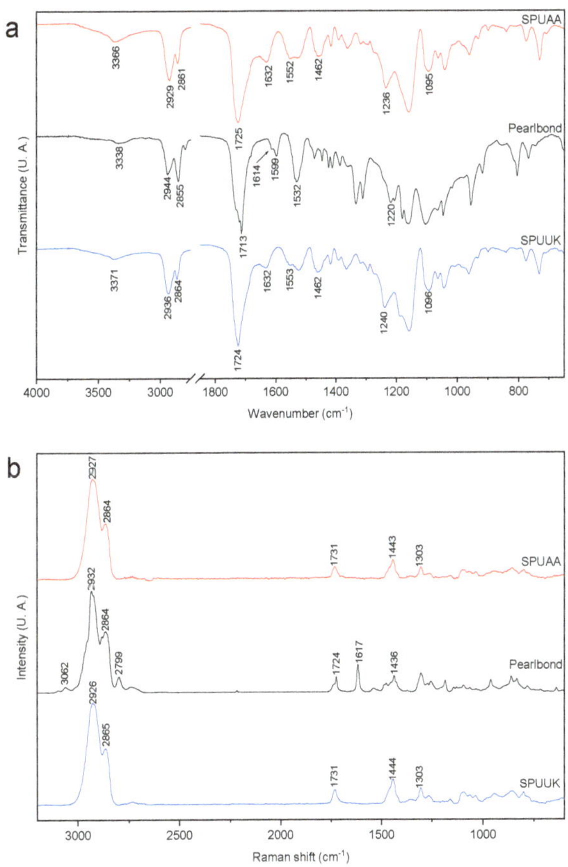

Figure 3. FT-IR spectra (**a**) and Raman spectra (**b**) of SPUUK, Pearlbond 703 EXP and SPUAA.

The absence of absorption peaks at 2260 cm^{-1} proved the absence of residual isocyanate. Furthermore, DMF and tin (II) 2-ethylhexanoate were not detected, either by FTIR/Raman or during biological studies, possibly because of their low concentration and thorough washing during polymer synthesis. However, it is recommended that ^{1}H NMR be used for better quantification. Therefore, the safety of these polyurethanes is guaranteed.

From a chemical point of view, few differences were detected between SPUUK and SPUAA, considering that they share the same soft segment content. In addition, as demonstrated later, they exhibited different mechanical behavior and biological properties.

3.2. Tensile Properties of Films and Tubes

Uniaxial tensile testing representative curves of pristine SPUUK, Pearlbond 703 EXP, SPUAA and films made of a polyurethane blend are shown in Figure 4a. It was observed that under tension, SPUUK and SPUAA had a similar mechanical behavior, i.e., a deformation higher than 1000%, because they share 95% of their chemical structure. In contrast, Pearlbond 703 EXP showed lower deformation and ultimate tensile strength but a higher Young's modulus than SPUUK and SPUAA. Therefore, the physical mixture or polyurethane blend showed intermediate values of the ultimate tensile strength (UTS) and deformation between SPUUK/SPUAA and Pearlbond 703 EXP. A rectangular strip cut from the TVG was also loaded under tension; only samples that broke in the middle section were considered. Assuming that there is no significant difference between a dumbbell and a rectangular sample in terms of the Young's modulus but that failure properties may be affected, TVG tensile properties were lower in terms of tensile strength and deformation than SPUUK and SPUAA. Figure 4b displays the single-layer grafts assessed under longitudinal deformation, revealing a consistent trend, SPUAA exhibited the highest longitudinal load (49.8 ± 8.3 N), while the lowest was observed for Pearlbond 703 EXP (18.1 ± 4.3 N), being lower compared to three-layer vascular grafts, TVGs (101.3 ± 4.0 N).

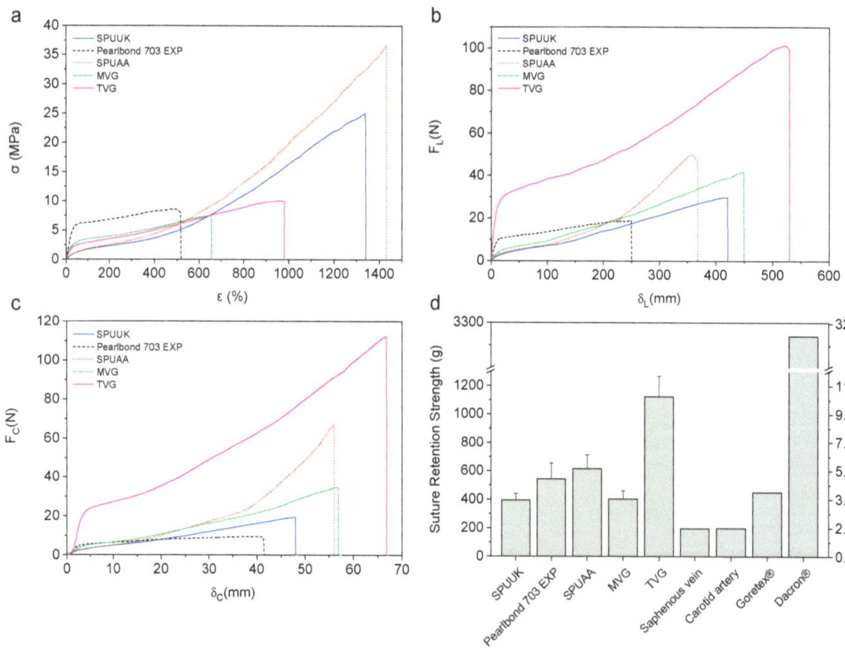

Figure 4. Representative curves of the mechanical behavior of films, tubes, and rings under the tensile loading of SPUUK, Pearlbond 703 EXP, SPUAA, polymer blends (MVGs) and three-layer vascular graft (TVGs) for (**a**) films under uniaxial tensile loading, (**b**) tubes under longitudinal load, (**c**) rings under circumferential load and (**d**) the suture retention strength test.

In agreement with previous results (see Figure 4c), SPUUA showed the maximum circumferential strength (5.00 ± 0.61 N/mm), followed by SPUUK (1.49 ± 0.39 N/mm), while the lowest circumferential force was achieved by the polyurethane Pearlbond 703 EXP

(0.87 ± 0.11 N/mm), which seems to be composed of MDI as the isocyanate used for its synthesis. Polymer blending led to an intermediate value between SPUAA and Pearlbond 703 EXP, even when they were at the same weight percentage. However, the TVG exhibited the highest circumferential tensile strength (7.17 ± 0.37 N/mm), suggesting the effect of the thicker layer but also the contributing effect of each layer with a dominance of SPUAA properties.

Figure 4d depicts the suture retention strength of the single-layer and three-layer polyurethane vascular grafts. From here it is seen that the highest suture strength was achieved by the TVG (1123 ± 144 g), followed by single-layer SPUAA (616 ± 97 g), with the lowest being those exhibited by single-layer SPUUK (396 ± 46 g) and the single-layer polyurethane blend (404 ± 57 g). These values are lower than those reported for Dacron but superior in the case of TVG to those reported for Goretex® using a 2-0 catgut suture [57]. A summary of the tensile mechanical properties is reported in Table 1.

Table 1. Tensile mechanical properties of films and tubular grafts made from single-layer SPUUK, Pearlbond 703 EXP, SPUAA, the mixture vascular graft and the three-layer vascular graft [a].

	SPUUK	Pearlbond 703 EXP	SPUAA	MVG	TVG [b]
			Film		
$E_{10\%}$ [MPa]	3.4 ± 0.4	17.9 ± 3.3	3.3 ± 0.2	10.6 ± 0.3	9.3 ± 2.4
σ_{max} [MPa]	12.76 ± 1.42	7.04 ± 0.73	30.77 ± 4.11	6.8 ± 0.4	9.4 ± 1.6
ε_{max} [%]	1330 ± 125	521 ± 101	1430 ± 88	650 ± 83	930 ± 242
F_{max} [N]	14.5 ± 2.5	3.5 ± 0.5	18.8 ± 2.3	4.0 ± 1.4	13.0 ± 3.1
δ_{max} [mm]	158 ± 14	59 ± 14	182 ± 17	65 ± 14	120 ± 31
			Graft (longitudinal)		
F_L^{max} [N]	26.1 ± 6.5	18.1 ± 4.3	49.8 ± 8.3	40.8 ± 5.2	101.3 ± 4.0
δ_L^{max} [mm]	404 ± 40	307 ± 86	373 ± 18	443 ± 26	484 ± 29
			Graft (circumferential)		
wall thickness [mm]	0.130 ± 0.034	0.130 ± 0.019	0.130 ± 0.027	0.140 ± 0.01	0.499 ± 0.021
$F_C^{max}/2L_C$ [N/mm]	1.49 ± 0.39	0.87 ± 0.11	5.00 ± 0.61	2.70 ± 0.20	7.17 ± 0.37
F_C^{max} [N]	19.2 ± 4.9	11.3 ± 1.4	68.7 ± 8.5	37.3 ± 3.4	107.5 ± 6.1
δ_C^{max} [mm]	48.7 ± 4.5	42.2 ± 7.6	56.2 ± 2.8	56.9 ± 2.2	65.6 ± 1.8
† σ_θ^{max} [MPa]	11.55 ± 1.48	6.76 ± 0.88	40.10 ± 6.24	18.97 ± 1.49	14.49 ± 0.71
			Suture retention strength		
F [N]	3.9 ± 0.5	5.3 ± 1.1	6.0 ± 1.0	4.0 ± 0.6	11.0 ± 1.4
F [g]	396 ± 46	544 ± 111	616 ± 97	404 ± 57	1123 ± 144

[a] Obtained according to ISO 7198 standard. [b] Coronary artery-like wall thickness ~0.50 mm. ‡ Calculated using Equation (2).

The above results show that the TVG exhibited better mechanical properties (longitudinal and circumferential strength) than MVG and single-layer vascular grafts due to its higher thickness. The TVG's thickness was chosen because of the reported coronary artery wall thickness (0.500 mm) and because some PTFE grafts reach up to 0.650 mm, while Dacron grafts are reported to have a wall thickness of 0.33 mm. Considering the complex stress state of hyperelastic rings (tension, bending and shear forces), it was not possible to establish equivalent stresses for rings of different thicknesses. Nevertheless, analytical solutions exist to address the difference in thickness for circumferential tension, such as treating the rings as curved beams. However, this approach was not pursued for the current paper. It is important to note that standard 7198 only considers the structural and geometrical properties, such as force and diameter, rather than intrinsic properties tied to the vascular graft wall thickness, i.e., only those properties with commercial or real-world relevance are considered. However, in the case of longitudinal tension, it can be normalized by its corresponding area, assuming that there is no compression at the grips, that the testing area is in the middle or far from the grips where the shape is retained and that there is no misalignment. Hence, the outcomes can be standardized based on their thickness or area. Due to this standardization, during longitudinal testing, MVG and

SPUAA experienced higher stress as compared to TVG, possibly due to the creation of a single bulk system (miscible or immiscible) instead of an anisotropic layered structure, indicating a combined contribution of various polyurethanes.

A drawback of our research pertains to the absence of assessment concerning the viscoelastic characteristics of the polyurethanes utilized in creating the vascular grafts. These properties are known to be time-dependent and were not measured due to the lack of a suggested method by the ISO 7198 standard. However, a review of the literature indicated that stress relaxation and viscoelastic recovery are meaningful, as demonstrated by Amabili et al. [58], who underscored the significance of cyclic axisymmetric diameter alterations acquired under physiological pulsatile conditions. Studies have shown that Dacron grafts are noticeably stiffer around their circumference than the aortic vessels they replace, resulting in a significant mechanical mismatch between the grafts and the native tissue. While we did not perform dynamic experiments, our circumferential and longitudinal quasi-static testing observations aligned with the ISO 7198 standard, indicating that longitudinal deformation is more significant than circumferential deformation, with no distinct J-shape observed. Additionally, the slopes (N/mm) at any deformation were higher in the circumferential direction than in the longitudinal direction, demonstrating the anisotropic stiffness of our polyurethanes and supporting the results of prior dynamic experiments.

3.3. Burst Strenght

The changes in the circumferential strain with respect to the internal pressure of the vascular grafts can be seen in Figure 5a. From this figure, it is evident that single-layer SPUUK grafts experienced a high change in diameter (circumferential strain) under pressure while exhibiting the lowest burst pressure (373 $\pm$ 120 mmHg) compared to the other grafts with the same wall thickness. In contrast, single-layer SPUAA grafts showed the maximum circumferential strain ($\varepsilon_\theta = 151.5 \pm 49.4\%$) of all grafts but needed more pressure to reach the same strain as SPUUK. The Pearlbond 703 EXP graft showed the lowest change in diameter under pressure, i.e., it was the least compliant and most stiff. The measured burst pressure of Pearlbond 703 EXP exhibited the highest value (884 $\pm$ 120 mmHg), which is about twice as high as that measured for the SPUUK (323 $\pm$ 120 mmHg) and SPUAA (517 $\pm$ 43 mmHg). The vascular grafts prepared via polymer blending (MVG) showed a burst pressure located between SPUUK/SPUAA and Pearlbond 703 EXP, with a value of 523 $\pm$ 78 mm Hg, which is more than four times the upper physiological pressure of 120 mm Hg and more than twice the hypertensive crisis pressure (180–220 mm Hg) [27]. In summary, the burst strength of single-layer grafts had the following order: $P_{BS}^{SPUUK} < P_{BS}^{SPUAA} < P_{BS}^{M.V.G.} < P_{BS}^{Pearlbond}$. Finally, the burst strength of the three-layer vascular graft (2087 $\pm$ 139 mmHg) with a wall thickness mimicking a coronary artery (~0.50 mm) was higher than the other types of vascular grafts. This value is much higher than any possible physiological blood pressure but consistent with saphenous veins and carotid arteries, which have burst pressures between 1680 and 2273 mmHg and between 2031 and 4225 mmHg [57,59], respectively.

The burst pressures measured during this study and those estimated according to the failure diameter (Equation (1)) are shown in Figure 5b. In some cases, burst strength is overestimated, while in other cases it is underestimated, suggesting a poor correlation with the experimental results. Figure 6a depicts the compliance values of the five types of vascular prostheses at different pressure levels: low (50–90 mmHg), medium (80–120 mmHg) and high (110–150 mmHg). The SPUUK, SPUAA, MVG and TVG samples showed higher compliance at high pressure levels and lower compliance at low pressure levels, indicating that compliance increases with increasing pressure. In contrast, the Pearlbond 703 EXP graft exhibited its highest compliance at a low pressure and its lowest compliance at a high pressure, indicating a decrement in compliance with increasing pressure.

The SPUUK and SPUAA grafts exhibited the highest compliance, measured at 80–120 mmHg of 5.2 $\pm$ 1.3 and 4.1 $\pm$ 0.5%/100 mmHg, respectively. This finding is noteworthy because it is closer to the physiological value for human saphenous veins, with a compliance of 4.4 $\pm$ 0.4%/100 mmHg [60]. The MVG graft showed lower compliance than SPUUK and SPUAA but a higher compliance than Pearlbond 703 EXP. Figure 6b illustrates a comparison of the stiffness measured at different pressure ranges. Consistent with the compliance values, the stiffness index was higher for TVG, indicating that the lower the compliance, the higher the stiffness index. Table 2 summarizes the burst strength and compliance values of the various polyurethane vascular grafts.

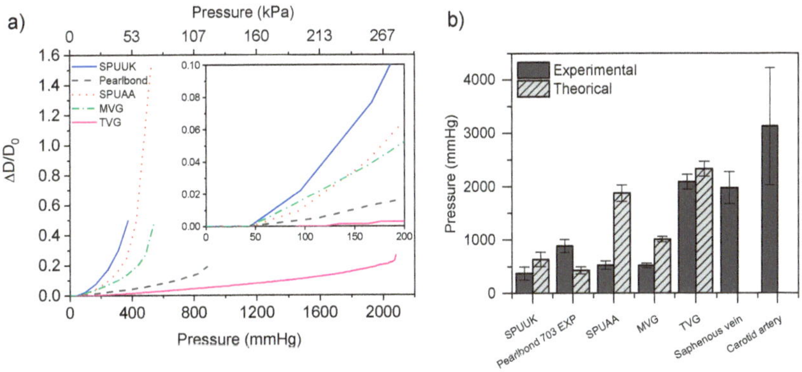

Figure 5. (**a**) Change in external diameter (circumferential strain) as a function of internal pressure for SVG, MVG and TVG and (**b**) experimental and theoretical burst pressure estimated via a ring tensile test using the failure diameter and those reported for human carotid arteries and human saphenous veins.

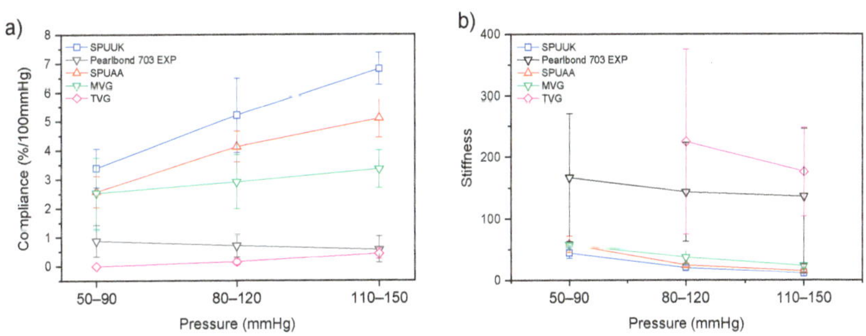

Figure 6. (**a**) Compliance variation with pressure (50 to 150 mmHg) for single (SPUUK, Pearlbond 703 EXP, SPUAA, MVG) and three-layer vascular grafts (TVG). (**b**) Stiffness values obtained with Equation (6).

Table 2. Burst strength and compliances of vascular grafts based on single-layer SPUUK, Pearlbond 703 EXP, SPUAA, the mixture blend vascular graft and the three-layer vascular graft [a].

	SPUUK	Pearlbond 703 EXP	SPUAA	MVG	TVG [b]
Wall thickness [mm]	0.132 ± 0.008	0.130 ± 0.012	0.142 ± 0.010	0.140 ± 0.004	0.500 ± 0.010
External diameter [mm]	6.27 ± 0.02	6.26 ± 0.02	6.29 ± 0.03	6.28 ± 0.01	7.01 ± 0.01
Burst pressure [mmHg]	373 ± 120	884 ± 120	517 ± 43	523 ± 78	2087 ± 139
P_i^{BS} **[mmHg] Df** [c]	634 ± 134	428 ± 66	1877 ± 155	1005 ± 51	2330 ± 136
P_i^{BS} **[mmHg] Du** [c]	3642 ± 936	2126 ± 267	12,230 ± 1517	6350 ± 628	15,834 ± 1372
Deformation [%]	49.3 ± 8.2	18.8 ± 2.6	151.5 ± 49.4	48.2 ± 7.6	40 ± 15.3
***C* 50–90 [%100/mmHg]**	3.03 ± 1.06	0.88 ± 0.55	2.58 ± 0.54	2.53 ± 1.22	0.0 ± 0.0
***C* 80–120 [%100/mmHg]**	5.21 ± 1.29	0.71 ± 0.4	4.13 ± 0.53	2.92 ± 0.93	0.15 ± 0.1
***C* 110–150 [%/100 mmHg]**	6.81 ± 0.55	0.57 ± 0.46	5.1 ± 0.66	3.35 ± 0.65	0.44 ± 0.18
Stiffness index 80–120 mmHg	20.3 ± 3.9	143 ± 80	24.8 ± 3.2	37.6 ± 11.1	225 ± 150

[a] Obtained according to ISO 7198 standard. [b] This graft had a wall thickness mimicking coronary artery ~0.5 mm. [c] Burst pressure computed with Lame's equation for unload diameter (Du) and failure diameter (Df).

3.4. Fracture Surfaces of Vascular Grafts after Mechanical Testing

Figure 7a–e present representative SEM cross-sections of the liquid-nitrogen-fractured surfaces of SVG (SPUUK, Pearlbond 703 EXP, SPUAA and MVG) and TVG obtained via roto-evaporation. For TVG, good adhesion between the three different polyurethanes that compose the graft was observed, with a clear demarcation of the inner, middle and outer layers, (SPUUK, Pearlbond 703 EXP and SPUAA). The inner surface (Figure 7i-Inner) of the TVG based on SPUUK showed a smooth surface where the polymer was highly compacted in contact with the glass rod, while the outer surface (Figure 7i-Outer) of the TVG based on SPUAA showed a rougher surface caused by the evaporation of the solvent during the rotation. In addition, Figure 7f–h show the three-layer vascular graft after the longitudinal, circumferential and burst test, respectively. After longitudinal or circumferential failure, no delamination of the polyurethanes composing the graft was observed. However, the different compositions can be identified according to the different types of fracture. For example, after longitudinal tensile tests, the TVG showed a ductile failure in the middle thicker layer composed of Pearlbond 703 EXP, while brittle failure was observed after circumferential tensile tests. Figure 7h depicts the surface of the vascular graft after the burst strength test. The surface exhibited various fracture patterns depending on the layer composition. Specifically, the outer and inner layers, composed of SPUAA and SPUUK, respectively, displayed rougher surfaces (indicative of a more brittle fracture). In contrast, the middle layer, consisting of a commercial polyurethane called Pearlbond 703 EXP, exhibited a smoother surface (suggesting a less brittle fracture).

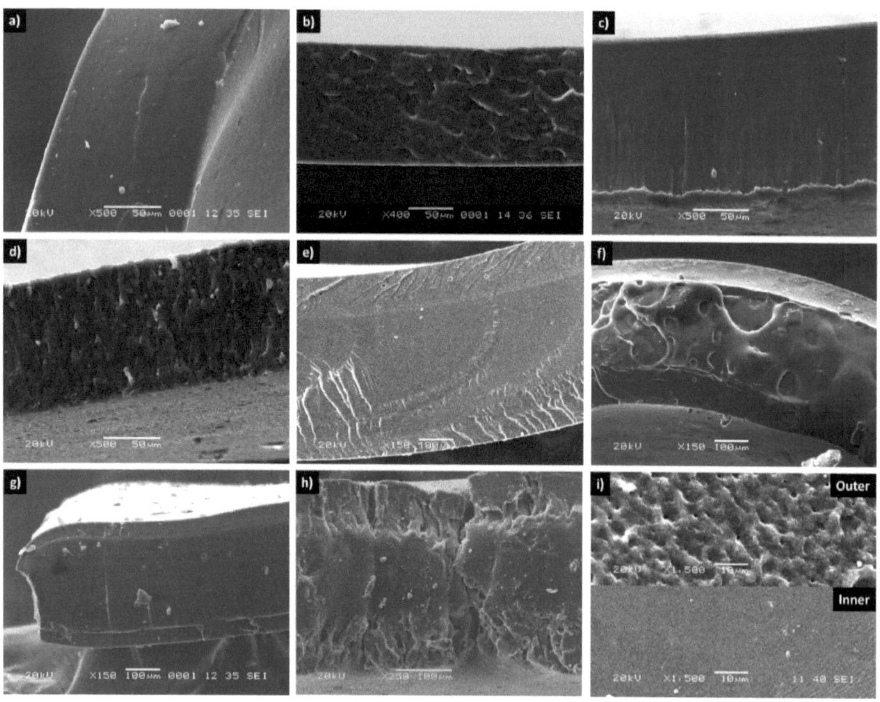

Figure 7. SEM micrographs of the cross-section of single-layer vascular grafts made from (**a**) SPUUK, (**b**) Pearlbond 703 EXP, (**c**) SPUAA and (**d**) MVG. (**e**) Three-layer vascular graft prepared with SPUUK, Pearlbond 703 EXP and SPUAA. (**f**) Cross-section of TVG after longitudinal testing, (**g**) circumferential testing and (**h**) the burst strength test. (**i-Inner**) The inner surface of the TVG based on SPUUK and (**i-Outer**) the outer surface based on SPUAA.

3.5. Biological Performance

3.5.1. MDA-MB-231 Cytocompatibility

Figure 8a shows the viability of MDA-MB-231 cells determined by means of the CCK-8 assay after 24 and 48 h of exposure to extracts of the different polyurethanes. Metabolically active cells can reduce the WST-8 tetrazolium salt present in CCK-8 into a water-soluble formazan product. Nonviable cells rapidly lose their ability to reduce the WST-8. Therefore, the production of the colored formazan products is proportional to the number of viable cells [61]. The three layers comprising the TVG exhibited values ranging from 90% to 100%, surpassing the minimum threshold of 70% required for biomaterials to be considered cytocompatible [49]. Figure 8b shows the viability of MDA-MB-231 cells obtained using the crystal violet assay to assess cell membrane integrity, upon exposure to extracts of the different polyurethanes, normalized with respect to the control (100%) cells at 24 and 48 h. All samples exhibited cell viability above the 70% required, with values around 100%.

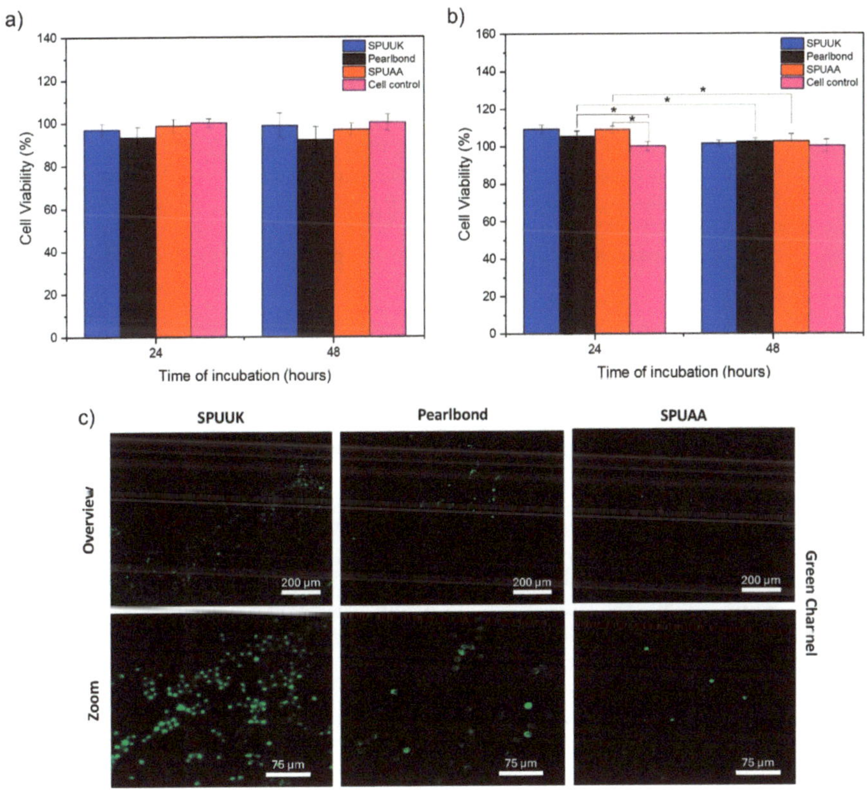

Figure 8. Viability of MDA-MB-231 cells upon exposure to extracts from the different polyurethanes. (**a**) CCK-8 and (**b**) crystal violet assay. (*) The values shown are the mean ± SD. A value of p-value < 0.05 was considered significant. (**c**) Confocal fluorescence micrographs of fibroblast L-929 cells 24 h after seeding onto selected films. Live cells (stained with calcein-AM) appear in green.

3.5.2. Cell Adhesion and Proliferation

The cytocompatibility of the polyurethane films was also evaluated via the well-known calcein AM/propidium iodide assay (Live/Dead). Calcein-AM is a small molecule dye that fluoresces when cleaved by esterases in viable cells, staining them in green [62], whereas propidium iodide is a fluorescent molecule that binds preferentially to DNA, staining membrane-disrupted cells in red. As such, the staining by these two dyes proves useful to discriminate between populations of live and dead cells [61]. The results from the live/dead assays are shown in Figure 8c, where few fibroblast cells appeared on the surface with little proliferation of Pearlbond 703 EXP and SPUAA.

Even when phalloidin staining was not performed to assess proper cell morphology, SPUUK exhibited good fibroblast adhesion. In terms of the water contact angle, there was no difference between the three polyurethanes (87.9 ± 3.3°, 84 ± 2.8° and 87.3 ± 1.4° for SPUUK, Pearlbond 703 EXP and SPUAA, respectively). Still, in terms of surface roughness, there was a slight difference (Ra = 1.02 ± 0.3 μm, 0.52 ± 0.09 μm and 0.31 ± 0.11 μm for SPUUK, Pearlbond 703 EXP and SPUAA, respectively) that can explain the observed outcome. As demonstrated by the screening tests of cytocompatibility presented in the previous section, the moderate adhesion and proliferation of fibroblast cells onto Pearlbond 703 EXP and SPUAA are not expected to be related to any toxic effect of these materials. Instead, they could be related to their surface's structural parameters, such as the topography, elasticity and water contact angle.

Taking all of these results together, it was observed that both cell metabolic activity and membrane integrity were uncompromised in the presence of all polyurethane extracts, even when solvents (DMF), monomers (isocyanates) or catalysts (tin (II) 2-ethylhexanoate) were used during their synthesis, probably due to their low concentration and exposure time. In contrast, the direct cytotoxicity test showed that only SPUUK could support fibroblast adhesion due to a slightly rougher surface. In this regard, many polymers are not cytotoxic but show poor or little cell adhesion. However, it must be noted that cell adhesion is mainly mediated by protein adsorption on the surface and not only by chemical composition, hydrophilic/hydrophobic balance, surface roughness, etc. Therefore, the extra lysine amino groups in SPUUK promoted the interaction with a negatively charged cell membrane.

3.5.3. Hemocompatibility

Figure 9a illustrates the progression of thrombus formation over time, with observations made at 5, 10, 20, 30 and 40 min of contact between whole blood and the various polyurethane surfaces. At 5 min, no significant difference between surfaces was noted, and the thrombus barely started to form. However, after 10 min, the amount of free hemoglobin was significantly higher in the SPUUK polyurethane than in the other polyurethanes for all incubation times evaluated. This is advantageous for the TVG because the inner layer, which will be in direct contact with blood, comprises SPUUK. Once free hemoglobin values reach 25%, clot formation becomes evident. According to studies by Boccafoschi et al. [63], Teflon® shows values of 25% at 20 min. In our research, SPUUK exhibited values of approximately 25% at around 15 min, whereas Pearlbond 703 EXP and SPUAA exhibited equivalent values at around 10 min. The blood compatibility of polyurethanes was also evaluated by measuring the cell hemolysis percentage during contact with red blood cells in vitro [64]. The percentage of hemolysis depicts the degree of destruction of erythrocytes when they meet the polyurethane films. Therefore, the higher the number of broken erythrocytes, the higher the value of the hemolysis percentage is. A higher degree of hemolysis indicates poor hemocompatibility of the biomaterial. Figure 9b illustrates the hemolysis rate values of different polyurethane films. The degree of hemolysis in the presence of SPUAA was 0.12 ± 0.38%, for Pearlbond 703 EXP it was 1.07 ± 0.96% and for SPUUK it was 1.82 ± 0.96%. However, the hemolysis values of all three kinds of materials were under the permissible limit of 5% [65].

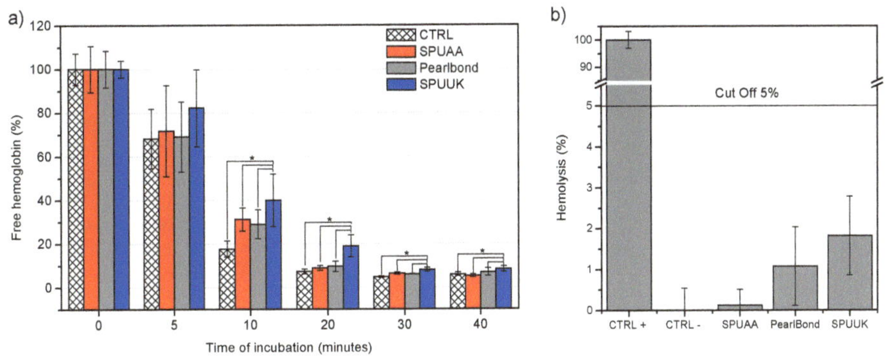

Figure 9. (**a**) Clotting time study on polyurethane films. A p-value < 0.05 was considered significant (*) and was found at 10, 20, 30 and 40 min between SPUUK and the other conditions. The results are normalized to the plastic control. The values shown are the mean ± SD. Experiments were repeated three times for three different donors, with n = 3 each time. (**b**) Hemolysis percentage of different polyurethanes.

4. Conclusions

A three-layered polyurethane vascular graft that mimics the thickness of each tunica in an artery was developed through the roto-evaporation method. The properties of these vascular grafts were assessed using the ISO 7198 standard. It was found that the 500 μm-wall-thickness TVG made from SPUUK (tunica intima), Pearlbond 703 EXP (tunica media) and SPUAA (tunica adventitia) exhibited the highest circumferential tensile strength (7.17 ± 0.37 N/mm) and longitudinal forces (101.3 ± 4.0 N) when compared to single-layer vascular grafts (100 μm wall thickness) made from the same polyurethanes. In addition, the three-layer vascular graft showed the highest suture strength (11.0 ± 1.4 N) and exhibited higher burst strength values (2087 ± 139 mmHg) than native vessels. Lower compliances (0.15 ± 0.1 and 0.44 ± 0.18 at 80–120 mmHg and 110–150 mmHg, respectively) and higher stiffness indexes (4054.7 ± 0.01) were achieved for TVG. Although the three polyurethane extracts exhibited non-cytotoxic behavior when cultured directly on the surface, they showed poor fibroblast adhesion, except for the lysine-based polyurethane (SPUUK). In agreement with this, the inner layer of the TVG, a lysine-based polyurethane, exhibited 80% free hemoglobin after 5 min (longer than the plastic control), which was reduced to 10% after 40 min, indicating a long clotting time. Additionally, the inner layer showed low hemolysis (1.82 ± 0.96%), which is below the acceptable 5% threshold for this parameter.

Determining mechanical properties is a suitable way of assessing the performance of a vascular graft, with higher only sometimes being better. The tensile mechanical test of SPUUK and SPUAA films showed that they have higher strength and deformation than Pearlbond 703 EXP but a lower modulus. Therefore, the three-layer structure improves the grafts' tensile properties and allows them to perform better than the corresponding immiscible polymer blend. However, this led to compliances that were lower than most natural tissues.

Based on these results, it is concluded that for achieving higher compliances or compliances akin to those of native tissues, the thickness and composition of the media layer (the thicker one) are the most important factors in the development of synthetic grafts. Therefore, a three-layer synthetic vascular graft must be at least 50–100 μm thick for high compliance when the segmented polyurethanes used as a media layer (Pearlbond 703 EXP in this study) exhibit a high Young's modulus. Alternatively, it is suggested that compliances can be increased in a tubular form by increasing porosity. Finally, the polyurethane used for the media layer can be replaced with a low-modulus polyurethane (with no phase separation) for higher compliance.

In conclusion, the roto-evaporation method is a good alternative for preparing layered tubular structures as it allows for different material compositions and thicknesses in a shorter period. However, further studies are needed to evaluate these synthetic grafts' long-term performance and biocompatibility before they can be used in clinical applications. To develop a final vascular substitute, it is recommended to include future studies related to their performance in animal models, their viscoelastic properties and more hemocompatibility tests that include platelet or protein adhesion and to observe it via SEM.

Author Contributions: Conceptualization, J.V.C.-R. and L.P.; methodology, G.A.Z.-B., R.F.V.-C., P.C., F.C., E.R.-V. and M.A.-M.; software, G.A.Z.-B. and R.F.V.-C.; validation, G.A.Z.-B.; formal analysis, G.A.Z.-B., E.R.-V. and M.A.-M.; investigation, G.A.Z.-B.; resources, J.V.C.-R., L.P. and D.M.; data curation, G.A.Z.-B.; writing—original draft preparation, G.A.Z.-B. and J.V.C.-R.; writing—review and editing, G.A.Z.-B., G.A.A., M.A.-M. and J.V.C.-R.; visualization, J.V.C.-R. and G.A.Z.-B.; supervision, J.V.C.-R. and L.P.; project administration, J.V.C.-R.; funding acquisition, J.V.C.-R., L.P. and D.M. All authors have read and agreed to the published version of the manuscript.

Funding: This work was supported by the Consejo Nacional de Ciencia y Tecnología (CONACYT) Mexico (Projects 1360 and 248378). We also acknowledge the funding provided by MCIN/AEI/10.13039/501100011033 with project PID2021-123753NB-C31.

Institutional Review Board Statement: This study was conducted according to the guidelines of the Declaration of Helsinki and approved by the Institutional Review Board (or Ethics Committee) of Centre Hospitalier Universitaire Québec—Université Laval, Quebec, Canada (FWA00000329, IRB00001242).

Data Availability Statement: The data presented in this study are available upon request from the corresponding author.

Acknowledgments: Gualberto A. Zumbardo-Bacelis gratefully acknowledges the financial support from the Consejo Nacional de Ciencia y Tecnología (CONACYT) and the Natural Sciences and Engineering Research Council of Canada (Discovery Program). The support for Project PID2021-123753NB-C31 funded by MCIN/AEI/10.13039/501100011033 and "ERDF A way of making Europe" is also gratefully acknowledged.

Conflicts of Interest: The authors declare no conflicts of interests.

References

1. Clerkin, K.J.; Fried, J.A.; Raikhelkar, J.; Sayer, G.; Griffin, J.M.; Masoumi, A.; Jain, S.S.; Burkhoff, D.; Kumaraiah, D.; Rabbani, L.R.; et al. COVID-19 and Cardiovascular Disease. *Circulation* **2020**, *141*, 1648–1655. [CrossRef] [PubMed]
2. WHO. World Health Statistics 2022: Monitoring Health for the SDGs, Sustainable Development Goals. Available online: https://www.who.int/publications/i/item/9789240051157 (accessed on 19 January 2023).
3. Pashneh-Tala, S.; MacNeil, S.; Claeyssens, F. The Tissue-Engineered Vascular Graft—Past, Present, and Future. *Tissue Eng. Part. B Rev.* **2016**, *22*, 68–100. [CrossRef] [PubMed]
4. Mallis, P.; Kostakis, A.; Stavropoulos-Giokas, C.; Michalopoulos, E. Future Perspectives in Small-Diameter Vascular Graft Engineering. *Bioengineering* **2020**, *7*, 160. [CrossRef] [PubMed]
5. Conte, S.M.; Vale, P.R. Peripheral Arterial Disease. *Heart Lung Circ.* **2018**, *27*, 427–432. [CrossRef] [PubMed]
6. Leal, B.B.J.; Wakabayashi, N.; Oyama, K.; Kamiya, H.; Braghirolli, D.I.; Pranke, P. Vascular Tissue Engineering: Polymers and Methodologies for Small Caliber Vascular Grafts. *Front. Cardiovasc. Med.* **2021**, *7*, 592361. [CrossRef] [PubMed]
7. Caliskan, E.; de Souza, D.R.; Böning, A.; Liakopoulos, O.J.; Choi, Y.H.; Pepper, J.; Gibson, C.M.; Perrault, L.P.; Wolf, R.K.; Kim, K.B.; et al. Saphenous Vein Grafts in Contemporary Coronary Artery Bypass Graft Surgery. *Nat. Rev. Cardiol.* **2020**, *17*, 155–169. [CrossRef] [PubMed]
8. Zierler, R.E.; Jordan, W.D.; Lal, B.K.; Mussa, F.; Leers, S.; Fulton, J.; Pevec, W.; Hill, A.; Murad, M.H. The Society for Vascular Surgery Practice Guidelines on Follow-up after Vascular Surgery Arterial Procedures. *J. Vasc. Surg.* **2018**, *68*, 256–284. [CrossRef] [PubMed]
9. L'Heureux, N.; Dusserre, N.; Marini, A.; Garrido, S.; de la Fuente, L.; McAllister, T. Technology Insight: The Evolution of Tissue-Engineered Vascular Grafts-From Research to Clinical Practice. *Nat. Clin. Pract. Cardiovasc. Med.* **2007**, *4*, 389–395. [CrossRef] [PubMed]
10. Harskamp, R.E.; Lopes, R.D.; Baisden, C.E.; De Winter, R.J.; Alexander, J.H. Saphenous Vein Graft Failure after Coronary Artery Bypass Surgery: Pathophysiology, Management, and Future Directions. *Ann. Surg.* **2013**, *257*, 824–833. [CrossRef]

11. Mi, H.Y.; Jing, X.; Yu, E.; Wang, X.; Li, Q.; Turng, L.S. Manipulating the Structure and Mechanical Properties of Thermoplastic Polyurethane/Polycaprolactone Hybrid Small Diameter Vascular Scaffolds Fabricated via Electrospinning Using an Assembled Rotating Collector. *J. Mech. Behav. Biomed. Mater.* **2018**, *78*, 433–441. [CrossRef]
12. Voorhees, A.B.; Jaretzki, A.; Blakemore, A.H. The Use of Tubes Constructed from Vinyon "N" Cloth in Bridging Arterial Defects. *Ann. Surg.* **1952**, *135*, 332–336. [CrossRef] [PubMed]
13. DeBakey, M.E.; Crawford, E.S.; Garrett, H.E.; Beall, A.C.; Howell, J.F. Surgical Considerations in the Treatment of Aneurysms of the Thoraco-Abdominal Aorta. *Ann. Surg.* **1965**, *162*, 650–662. [CrossRef] [PubMed]
14. Sarkar, S.; Sales, K.M.; Hamilton, G.; Seifalian, A.M. Addressing Thrombogenicity in Vascular Graft Construction. *J. Biomed. Mater. Res. B Appl. Biomater.* **2007**, *82*, 100–108. [CrossRef]
15. Berger, K.; Sauvage, L.R. Late Fiber Deterioration in Dacron® Arterial Grafts. *Ann. Surg.* **1981**, *193*, 477–491.
16. Weber, C.; Reinhardt, S.; Eghbalzadeh, K.; Wacker, M.; Guschlbauer, M.; Maul, A.; Sterner-Kock, A.; Wahlers, T.; Wippermann, J.; Scherner, M. Patency and in Vivo Compatibility of Bacterial Nanocellulose Grafts as Small-Diameter Vascular Substitute. *J. Vasc. Surg.* **2018**, *68*, 177S–187S.e1. [CrossRef]
17. Fang, S.; Ellman, D.G.; Andersen, D.C. Review: Tissue Engineering of Small-Diameter Vascular Grafts and Their In Vivo Evaluation in Large Animals and Humans. *Cells* **2021**, *10*, 713. [CrossRef]
18. Baguneid, M.S.; Seifalian, A.M.; Salacinski, H.J.; Murray, D.; Hamilton, G.; Walker, M.G. Tissue Engineering of Blood Vessels. *Br. J. Surg.* **2006**, *93*, 282–290. [CrossRef]
19. Joseph, J.; Patel, R.M.; Wenham, A.; Smith, J.R. Biomedical Applications of Polyurethane Materials and Coatings. *Trans. Inst. Met. Finish.* **2018**, *96*, 121–129. [CrossRef]
20. Wendels, S.; Avérous, L. Biobased Polyurethanes for Biomedical Applications. *Bioact. Mater.* **2021**, *6*, 1083–1106. [CrossRef]
21. Szycher, M. Polyurethanes: Medical Applications. In *Encyclopedia of Biomedical Polymers and Polymeric Biomaterials*; CRC Press: Boca Raton, FL, USA, 2016; pp. 6647–6670.
22. Padsalgikar, A.D. Applications of Polyurethanes in Medical Devices. In *Applications of Polyurethanes in Medical Devices*; Elsevier Inc.: Amsterdam, The Netherlands, 2022; pp. 1–269. [CrossRef]
23. Hsu, S.H.; Xu, J.; Lin, S.H.; Wu, S.D.; Cheng, Q.P.; Wong, C.W. Creative Transformation of Biomedical Polyurethanes: From Biostable Tubing to Biodegradable Smart Materials. *J. Polym. Res.* **2022**, *29*, 70. [CrossRef]
24. Naureen, B.; Haseeb, A.S.M.A.; Basirun, W.J.; Muhamad, F. Recent Advances in Tissue Engineering Scaffolds Based on Polyurethane and Modified Polyurethane. *Mater. Sci. Eng. C* **2021**, *118*, 111228. [CrossRef]
25. Adipurnama, I.; Yang, M.C.; Ciach, T.; Butruk-Raszeja, B. Surface Modification and Endothelialization of Polyurethane for Vascular Tissue Engineering Applications: A Review. *Biomater. Sci.* **2017**, *5*, 22–37. [CrossRef] [PubMed]
26. Szczepańczyk, P.; Szlachta, M.; Złocista-Szewczyk, N.; Chłopek, J.; Pielichowska, K. Recent Developments in Polyurethane-Based Materials for Bone Tissue Engineering. *Polymers* **2021**, *13*, 946. [CrossRef] [PubMed]
27. Herman, I.P. *Physics of the Human Body*; Johns Hopkins University Press: Baltimore, MD, USA, 2010; Volume 48, ISBN 9783319239309.
28. Esmaeili, S.; Shahali, M.; Kordjamshidi, A.; Torkpoor, Z.; Namdari, F.; Samandari, S.S.-; Ghadiri Nejad, M.; Khandan, A. An Artificial Blood Vessel Fabricated by 3D Printing for Pharmaceutical Application. *Nanomed. J.* **2019**, *6*, 183–194. [CrossRef]
29. Zhou, X.; Nowicki, M.; Sun, H.; Hann, S.Y.; Cui, H.; Esworthy, T.; Lee, J.D.; Plesniak, M.; Zhang, L.G. 3D Bioprinting-Tunable Small-Diameter Blood Vessels with Biomimetic Biphasic Cell Layers. *ACS Appl. Mater. Interfaces* **2020**, *12*, 45904–45915. [CrossRef]
30. Zumbardo-Bacelis, G.A.; Meza-Villegas, L.A.; Pérez-Aranda, C.A.; Vargas-Coronado, R.; Castillo-Cruz, O.; Montaño-Machado, V.; Mantovani, D.; Cauich-Rodríguez, J.V. On Arginine-based Polyurethane-blends Specific to Vascular Prostheses. *J. Appl. Polym. Sci.* **2021**, *138*, 51247. [CrossRef]
31. Asadpour, S.; Yeganeh, H.; Ai, J.; Ghanbari, H. A Novel Polyurethane Modified with Biomacromolecules for Small-Diameter Vascular Graft Applications. *J. Mater. Sci.* **2018**, *53*, 9913–9927. [CrossRef]
32. Yuan, X.; Li, W.; Yao, B.; Li, Z.; Kong, D.; Huang, S.; Zhu, M. Tri-Layered Vascular Grafts Guide Vascular Cells' Native-like Arrangement. *Polymers* **2022**, *14*, 1370. [CrossRef]
33. Wu, C.; Wang, H.; Cao, J. Tween-80 Improves Single/Coaxial Electrospinning of Three-Layered Bioartificial Blood Vessel. *J. Mater. Sci. Mater. Med.* **2023**, *34*, 6. [CrossRef]
34. Cetina-Diaz, S.M.; Chan-Chan, L.H.; Vargas-Coronado, R.F.; Cervantes-Uc, J.M.; Quintana-Owen, P.; Paakinaho, K.; Kellomaki, M.; Di Silvio, L.; Deb, S.; Cauich-Rodríguez, J.V. Physicochemical Characterization of Segmented Polyurethanes Prepared with Glutamine or Ascorbic Acid as Chain Extenders and Their Hydroxyapatite Composites. *J. Mater. Chem. B* **2014**, *2*, 1966–1976. [CrossRef]
35. Fitzgerald, P.J.; St. Goar, F.G.; Connolly, A.J.; Pinto, F.J.; Billingham, M.E.; Popp, R.L.; Yock, P.G. Intravascular Ultrasound Imaging of Coronary Arteries: Is Three Layers the Norm? *Circulation* **1992**, *86*, 154–158. [CrossRef] [PubMed]
36. Potkin, B.J.; Bartorelli, A.L.; Gessert, J.M.; Neville, R.F.; Almagor, Y.; Roberts, W.C.; Leon, M.B. Coronary Artery Imaging with Intravascular High-Frequency Ultrasound. *Circulation* **1990**, *81*, 1575–1585. [CrossRef] [PubMed]
37. Mintz, G.S.; Nissen, S.E.; Anderson, W.D.; Bailey, S.R.; Erbel, R.; Fitzgerald, P.J.; Pinto, F.J.; Rosenfield, K.; Siegel, R.J.; Murat Tuzcu, E.; et al. American College of Cardiology Clinical Expert Consensus Document on Standards for Acquisition, Measurement and Reporting of Intravascular Ultrasound Studies (IVUS). *J. Am. Coll. Cardiol.* **2001**, *37*, 1478–1492. [CrossRef] [PubMed]

38. Wong, M.; Edelstein, J.; Wollman, J.; Bond, M.G. Ultrasonic-Pathological Comparison of the Human Arterial Wall: Verification of Intima-Media Thickness. *Arterioscler. Thromb. Vasc. Biol.* **1993**, *13*, 482–486. [CrossRef]
39. *ASTM D882*; Standard Test Method for Tensile Properties of Thin Plastic Sheeting. ASTM: West Conshohocken, PA, USA, 2018.
40. *ISO 7198*; Cardiovascular Implants—Tubular Vascular Prostheses. International Organization for Standardization: Geneva, Switzerland, 1998.
41. Pérez-Aranda, C.; Gamboa, F.; Castillo-Cruz, O.; Cauich-Rodríguez, J.V.; Avilés, F. Design and Analysis of a Burst Strength Device for Testing Vascular Grafts. *Rev. Sci. Instrum.* **2019**, *90*, 014301. [CrossRef] [PubMed]
42. Voorneveld, J.; Oosthuysen, A.; Franz, T.; Zilla, P.; Bezuidenhout, D. Dual Electrospinning with Sacrificial Fibers for Engineered Porosity and Enhancement of Tissue Ingrowth. *J. Biomed. Mater. Res. B Appl. Biomater.* **2017**, *105*, 1559–1572. [CrossRef]
43. Laterreur, V.; Ruel, J.; Auger, F.A.; Valliè, K.; Tremblay, C.; Lacroix, D.; Tondreau, M.; Bourget, J.-M.; Germain, L. Comparison of the Direct Burst Pressure and the Ring Tensile Test Methods for Mechanical Characterization of Tissue-Engineered Vascular Substitutes. *J. Mech. Behav. Biomed. Mater.* **2014**, *34*, 253–263. [CrossRef] [PubMed]
44. Stankus, J.J.; Soletti, L.; Fujimoto, K.; Hong, Y.; Vorp, D.A.; Wagner, W.R. Fabrication of Cell Microintegrated Blood Vessel Constructs through Electrohydrodynamic Atomization. *Biomaterials* **2007**, *28*, 2738–2746. [CrossRef]
45. *ISO 10993-5*; Biological Evaluation of Medical Devices—Part 5: Tests for in Vitro Cytotoxicity. International Organization for Standardization: Geneva, Switzerland, 2009.
46. Huang, K.T.; Chen, Y.H.; Walker, A.M. Inaccuracies in MTS Assays: Major Distorting Effects of Medium, Serum Albumin, and Fatty Acids. *Biotechniques* **2004**, *37*, 406–412. [CrossRef]
47. Ross, A.M.; Jiang, Z.; Bastmeyer, M.; Lahann, J. Physical Aspects of Cell Culture Substrates: Topography, Roughness, and Elasticity. *Small* **2012**, *8*, 336–355. [CrossRef]
48. Karakaş, D.; Ari, F.; Ulukaya, E. The MTT Viability Assay Yields Strikingly False-Positive Viabilities Although the Cells Are Killed by Some Plant Extracts. *Turk. J. Biol.* **2017**, *41*, 919–925. [CrossRef] [PubMed]
49. Sotolongo-García, R.; Rodríguez-Velázquez, E.; Alatorre-Meda, M.; Oropeza-Guzmán, M.T.; Tirado-Guízar, A.; Pina-Luis, G. Optimizing the Efficiency of a Cytocompatible Carbon-Dots-Based Fret Platform and Its Application as a Riboflavin Sensor in Beverages. *Nanomaterials* **2021**, *11*, 1981. [CrossRef] [PubMed]
50. Montaño-Machado, V.; Chevallier, P.; Mantovani, D.; Pauthe, E. On the Potential for Fibronectin/Phosphorylcholine Coatings on PTFE Substrates to Jointly Modulate Endothelial Cell Adhesion and Hemocompatibility Properties. *Biomatter* **2015**, *5*, e979679. [CrossRef] [PubMed]
51. *ASTM F756-17*; Standard Practice for Assessment of Hemolytic Properties of Materials. ASTM: West Conshohocken, PA, USA, 2017.
52. Chan-Chan, L.H.; González-García, G.; Vargas-Coronado, R.F.; Cervantes-Uc, J.M.; Hernández-Sánchez, F.; Marcos-Fernandez, A.; Cauich-Rodríguez, J.V. Characterization of Model Compounds and Poly(Amide-Urea) Urethanes Based on Amino Acids by FTIR, NMR and Other Analytical Techniques. *Eur. Polym. J.* **2017**, *92*, 27–39. [CrossRef]
53. Shababdoust, A.; Ehsani, M.; Shokrollahi, P.; Zandi, M. Fabrication of Curcumin-Loaded Electrospun Nanofiberous Polyurethanes with Anti-Bacterial Activity. *Prog. Biomater.* **2018**, *7*, 23–33. [CrossRef] [PubMed]
54. Wang, F.C.; Feve, M.; Lam, T.M.; Pascault, J.P. FTIR Analysis of Hydrogen Bonding in Amorphous Linear Aromatic Polyurethanes. II. Influence of Styrene Solvent. *J. Polym. Sci. B Polym. Phys.* **1994**, *32*, 1315–1320. [CrossRef]
55. Bruckmoser, K.; Resch, K. Investigation of Ageing Mechanisms in Thermoplastic Polyurethanes by Means of IR and Raman Spectroscopy. *Macromol. Symp.* **2014**, *339*, 70–83. [CrossRef]
56. Weakley, A.T.; Warwick, T.; Bitterwolf, T.E.; Aston, D.E. Multivariate Analysis of Micro-Raman Spectra of Thermoplastic Polyurethane Blends Using Principal Component Analysis and Principal Component Regression. *Appl. Spectrosc.* **2012**, *66*, 1269–1278. [CrossRef] [PubMed]
57. Johnson, J.; Ohst, D.; Groehl, T.; Hetterscheidt, S.; Jones, M. Development of Novel, Bioresorbable, Small-Diameter Electrospun Vascular Grafts. *J. Tissue Sci. Eng.* **2015**, *6*, 2. [CrossRef]
58. Amabili, M.; Balasubramanian, P.; Ferrari, G.; Franchini, G.; Giovanniello, F.; Tubaldi, E. Identification of Viscoelastic Properties of Dacron Aortic Grafts Subjected to Physiological Pulsatile Flow. *J. Mech. Behav. Biomed. Mater.* **2020**, *110*, 103804. [CrossRef]
59. L'Heureux, N.; Dusserre, N.; Konig, G.; Victor, B.; Keire, P.; Wight, T.N.; Chronos, N.A.F.; Kyles, A.E.; Gregory, C.R.; Hoyt, G.; et al. Human Tissue-Engineered Blood Vessels for Adult Arterial Revascularization. *Nat. Med.* **2006**, *12*, 361–365. [CrossRef] [PubMed]
60. Salacinski, H.J.; Goldner, S.; Giudiceandrea, A.; Hamilton, G.; Seifalian, A.M.; Edwards, A.; Carson, R.J. The Mechanical Behavior of Vascular Grafts: A Review. *J. Biomater. Appl.* **2001**, *15*, 241–278. [CrossRef]
61. Rodríguez-Velázquez, E.; Taboada, P.; Alatorre-Meda, M. Biocompatible Hollow Polymeric Particles Produced by a Mild Solvent- and Template Free Strategy. *Colloids Surf. B Biointerfaces* **2017**, *160*, 732–740. [CrossRef]
62. Bischof, J.C.; Padanilam, J.; Holmes, W.H.; Ezzell, R.M.; Lee, R.C.; Tompkins, R.G.; Yarmush, M.L.; Toner, M. Dynamics of Cell Membrane Permeability Changes at Supraphysiological Temperatures. *Biophys. J.* **1995**, *68*, 2608–2614. [CrossRef] [PubMed]
63. Boccafoschi, F.; Habermehl, J.; Vesentini, S.; Mantovani, D. Biological Performances of Collagen-Based Scaffolds for Vascular Tissue Engineering. *Biomaterials* **2005**, *26*, 7410–7417. [CrossRef] [PubMed]

64. Shi, C.; Yuan, W.; Khan, M.; Li, Q.; Feng, Y.; Yao, F.; Zhang, W. Hydrophilic PCU Scaffolds Prepared by Grafting PEGMA and Immobilizing Gelatin to Enhance Cell Adhesion and Proliferation. *Mater. Sci. Eng. C* **2015**, *50*, 201–209. [CrossRef]
65. Haghjooy Javanmard, S.; Anari, J.; Zargar Kharazi, A.; Vatankhah, E. In Vitro Hemocompatibility and Cytocompatibility of a Three-Layered Vascular Scaffold Fabricated by Sequential Electrospinning of PCL, Collagen, and PLLA Nanofibers. *J. Biomater. Appl.* **2016**, *31*, 438–449. [CrossRef]

polymers

Article

Injection Molding Condition Effects on the Mechanical Properties of Coconut-Wood-Powder-Based Polymer Composite

Quach Van Thiem, Van-Thuc Nguyen, Dang Thu Thi Phan and Pham Son Minh *

Faculty of Mechanical Engineering, Ho Chi Minh City University of Technology and Education, Ho Chi Minh City 71307, Vietnam; thiemqv@hcmute.edu.vn (Q.V.T.); nvthuc@hcmute.edu.vn (V.-T.N.); thuptd@hcmute.edu.vn (D.T.T.P.)

* Correspondence: minhps@hcmute.edu.vn

Abstract: This study investigates the mechanical properties of coconut sawdust powder combined with polypropylene (PP). The effect of compatibility content, wood powder (WP) content, and injection molding parameters on the properties of coconut wood powder composite (WPC) is evaluated. The results could be used to figure out the optimal mechanical properties such as tensile strength, elongation, elastic modulus, and flexural strength by selecting suitable parameters and composition. The bonding between the WP particles and the PP matrix is good, and the WP is uniformly distributed across the composite matrix, as indicated in the scanning electron microscopy (SEM) results. Interestingly, with the presence of the compatibilizer oleamide, increasing the WP content from 20 wt.% to 40 wt.% did not result in WP accumulation in the composite matrix. Notably, at 20 wt.% WP, the elongation is the highest (at 7.40 wt.%), while at 30 wt.% WP, the elastic modulus reaches the highest value. The maximum ultimate tensile strength (UTS) value is obtained at 35 wt.% WP. Higher WP mostly results in greater flexural strength and shore D hardness. At 40 wt.% WP, the WPC achieves its peak shore D hardness of 77.6. The Taguchi results suggest that WP content is the most critical factor in the UTS value of coconut WPCs. The filling pressure ranks second, followed by the packing pressure. Finally, unlike the other characteristics, the melt temperature has a minimal impact on the UTS value.

Keywords: wood powder; tensile strength; flexural strength; optimization; injection molding

Citation: Thiem, Q.V.; Nguyen, V.-T.; Phan, D.T.T.; Minh, P.S. Injection Molding Condition Effects on the Mechanical Properties of Coconut-Wood-Powder-Based Polymer Composite. *Polymers* **2024**, *16*, 1225. https://doi.org/10.3390/polym16091225

Academic Editors: Jesús-María García-Martínez and Emilia P. Collar

Received: 29 March 2024
Revised: 24 April 2024
Accepted: 25 April 2024
Published: 27 April 2024

1. Introduction

Recently, exploring ways to reuse waste products has become an attractive method of reducing pollution, saving energy, and saving material resources [1,2]. For instance, wood-powder-based composites have become a trend in green policies to reduce the use of plastics [3–7]. The wood powder (WP) is the bypass product, collected from the process of wood production. Wood products can be burned to create heat sources. WP can also be combined with some polymers to create a wood-based composite [8–11]. WP can also be mixed with ceramics to generate ceramic composites [12–16].

Dikobe et al. [17] investigated the morphology, mechanical properties, and thermal properties of the WP composite with polypropylene (PP)/linear low-density polyethylene (LLDPE) and maleic anhydride-grafted polypropylene (MAPP)/LLDPE blend systems. The composite used 10, 20, and 30 wt.% WP. The MAPP, LLDPE, and WP composites have stronger interfacial bonding and better thermal stability than the PP, LLDPE, and WP composites. Nygård et al. [18] improved the feeding ability of the wood fiber during the extrusion process. They showed that pelletizing greatly reduced the length of wood fibers while allowing for a regulated feeding of the fibers into the extruder. Buddi et al. [19] reported the mechanical properties of a 5 wt.% WP and HDPE composite enhanced with nanoSiO_2. With 5 wt.% SiO_2, the composite reached the highest tensile and flexural strengths. The higher SiO_2 content led to the segregation of nanoparticles and a reduction in the ductility of the composite. The composite obtained the highest impact strength at 3 wt.% SiO_2. He et al. [20] mixed WP with polyvinyl alcohol (PVA) by applying solid-state shear milling and thermal processing, allowing for the high productivity of the composite. The composites achieved a good tensile strength of 22.5 MPa and elongation of 120.5 wt.%, which is greatly higher than the traditional mixing method.

Furthermore, wood-powder-based composite (WPC) could be processed by various technologies such as extrusion, injection, or compression molding [21]. WPC products can appear in automobiles, construction, and furniture due to the advantages of weather duration and biological impact. Notably, da Silva et al. [22] reported that polylactic acid (PLA)/poly ε-caprolactone (PCL) composites combined with WP. The initial mixture is mixed by a screw extruder, graded, and finally injected by an injection molding machine. The tensile strength and elastic modulus of the composites are reduced; however, the impact strength of the composite is higher than that of pure PLA. The composite material is good enough to generate disposable cups. Interestingly, the WPC could be used as a 3D printing material [23]. Huang et al. [24] mixed WP, acrylonitrile–butadiene–styrene (ABS) polymer, and maleic anhydride to create 3D printing materials. They found that the best percentage of Red Gum WP in the composites is 29 wt.%.

For a denser and stronger 3D-printed WPC product, it is suggested to use wood particles with more rounded shapes and smoother surfaces. Kajaks et al. [25] mixed coupling-agent-maleated polypropylene (MAPP) with WP. The use of a coupling agent significantly increases the tensile strength, flexural strength, and impact strength by up to 1.5–2.2 times. The mechanical properties of high-density polyethylene (HDPE) mixed with WP are also surveyed [26]. The WP percentage could reach 50 wt.%, while the mechanical properties are still good enough. The best water resistance is obtained at 5 wt.% MAPP. Scanning electron microscopy (SEM) results showed that MAPP increases the interfacial bonding between the WP and HDPE matrix. Salemane and Luyt [27] combined PP and WP with some compatibilizers to create WPC. The findings suggest that WP settles in the amorphous component of the matrix, forming new crystalline phases or zones. SEM micrographs reveal a relatively equal distribution of WP in the PP matrix, which contributes to reported improvements in material characteristics.

Most of the above studies did not mention the impact of the injection molding parameters on the characteristics of the WPC. Moreover, the coconut sawdust by-products are also rarely discussed when exploring the WPC. Coconut sawdust is a waste product which is popular in South and South East Asia. Finding the optimal ways to reuse it could minimize pollution, save energy, and save materials and resources [1–5]. This report concentrates on studying the mechanical properties of coconut sawdust powder mixed with PP. The WP is dried and then mixed with PP and additives, following an extrusion process to generate granular. The composite is then injected by an injection molding machine to create the test samples. The impact of compatibility content, WP content, and injection molding parameters on the characteristics of the coconut WPC are examined. The WPC samples are tested via tensile, flexural, and hardness tests. The samples are also tested with a water absorption test, a scanning electron microscope (SEM) test, a shore D hardness test, a water absorption test, and a differential scanning calorimetry (DSC) test. The optimal tensile strength, elongation, elastic modulus, tensile strength, and hardness are indicated. The results might be used to determine the best mechanical qualities, such as tensile strength, elongation, elastic modulus, and flexural strength, by choosing the appropriate parameters and composition.

2. Experimental Methods

Materials and Method

Figure 1a,b show the coconut sawdust powder before and after filtering by a fine mesh with a 60-mesh US standard. Then, it is dried at 90 °C for 8 h to remove humidity [28,29]. The WP is then mixed with PP polymer, 5 wt.% compatibilizer, and 2 wt.% lubrication additive for 30 min in a mixer. Oleamide O2136 lubrication additive is supplied by MERCK

Company, Darmstadt, Germany. Compatibilizer Scona TPPP 8112 GA is from BYK Kometra GmbH, Schkopau, Germany. A PP polymer named Advanced-PP 1100 N is supplied by Advanced Petrochemical Company in Al Jubail 35725, Saudi Arabia, and its characteristics are shown in Table 1. The coconut WP in the WPC varies from 20 wt.% to 40 wt.%. After that, the mixture is extruded at 160 °C and a screw rotation speed of 150 rpm and granulated to create composite granulation with the size of 3.2 mm × 5 mm, as shown in Figure 1c,d. The composition granulation is dried for 8 h at 90 °C to remove humidity before being injected into the injection molding machine Haitian-MA 1200III at different injection parameters. Each WPC composition has eight samples: four samples for the tensile test and four samples for the flexural test. In Sections 3.1 and 3.2, the WPC is injected at a filling pressure of 33 bar, a packing pressure of 40 bar, and a melt temperature of 220 °C.

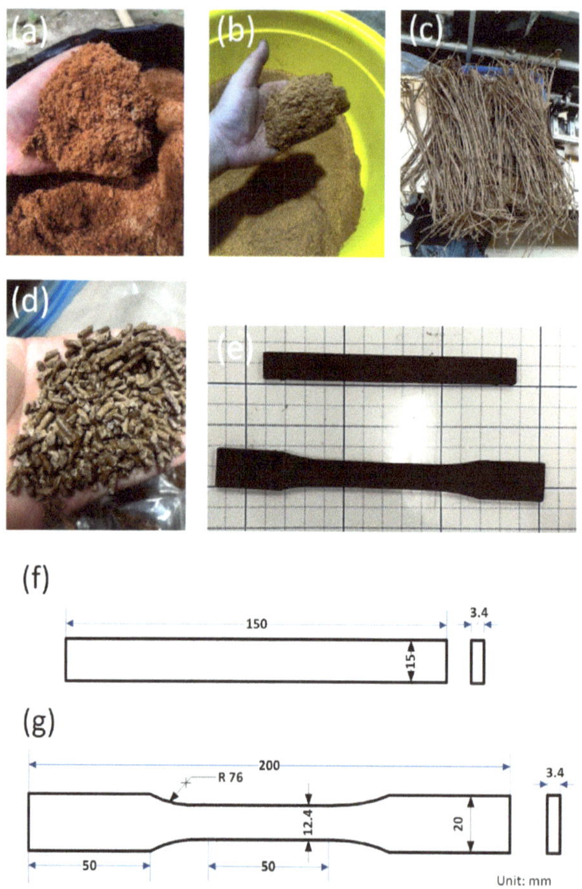

Figure 1. Sample preparation process: (**a**) coconut WP before filtering; (**b**) coconut WP after filtering; (**c**) coconut WP and PP after extrusion; (**d**) granulation; (**e**) tensile and flexural injection samples; (**f**) flexural sample dimension; (**g**) tensile sample dimension.

Table 1. Characteristics of Advanced-PP 1100 N.

Properties	Unit	Test Method	Value
Melt flow rate (230 °C/2.16 KG)	g/10 min	ISO 1133 [30]	12
Tensile modulus of elasticity	MPa	ISO 527-2 [31]	1550
Tensile stress at yield	MPa	ISO 527-2	35
Tensile strain at yield	%	ISO 527-2	8
Tensile strain at break	%	ISO 527-2	>50
Charpy impact strength notched	kJ/m^2	ISO 179/1eA [32]	3.0
Ball indentation hardness (H 358/30)	MPa	ISO 2039-1 [33]	78
Melting point, DSC	°C	ISO 3146 [34]	163
Heat deflection temperature—HDT/B (0.45 Mpa)	°C	ISO 75-2 [35]	85
Vicat softening temperature—VST/A50 (10 N)	°C	ISO 306 [36]	154
Density	g/cm^3	ISO 1183 [37]	0.91

The WPC samples are injected following ASTM D638 for the tensile test and ASTM D790 for the flexural test [38,39]. After being injected, the samples are subjected to tensile and flexural tests by a tensile test machine AG-X Plus 20 kN (Shimadzu, Kyoto, Japan) at a 5 mm/min speed. The fracture surfaces are examined via an SEM microscope TM4000 (Hitachi, Tokyo, Japan) with a voltage of 5 kV, magnification of ×100, working distance of 6.2–7.7 mm, and an emission current of 65,200–67,900 nA. The samples are also measured with DSC via DSC 214 Polyma (NETZSCH, Selb, Germany) to indicate the crystallinity percentage at a range of 40–250 °C, a speed of 10 K/min, and atmosphere of N2, 40.0–60 mL/min. In addition, the samples are also measured with shore D hardness tester LX-D (ETOPOO, Zhejiang, China). Finally, the water absorption is tested via standard ASTM D 570-88, using digital scale FA2204 (JOANLAB, Huzhou, China).

3. Results and Discussion

3.1. Tensile Test Results

Figure 2 shows the stress–strain diagram of the WPC at different WP percentages. The UTS values of the samples are 10.26 MPa, 10.42 MPa, 9.90 MPa, 10.57 MPa, and 10.47 MPa, corresponding to 20 wt.%, 25 wt.%, 30 wt.%, 35 wt.%, and 40 wt.% WP. The UTS value of the WPC varies from 9.90 MPa to 10.57 MPa. This result shows that changing the WP content does not strongly impact the UTS value of the WPC samples. Moreover, the highest UTS value is 10.57 MPa, which is achieved with 35 wt.% WP. With the WP from sanding dust birch wood veneer plywood in the PP matrix, Kajaks et al. [25] indicated that increasing from 20 wt.% to 40 wt.% WP does not strongly change the UTS value, which is consistent with this report.

The elongation at break comparison diagram of the WPC at different WP percentages is presented in Figure 3. The elongation values of the WPC samples are 7.40 wt.%, 5.53 wt.%, 4.97 wt.%, 4.31 wt.%, and 3.98 wt.%, corresponding to 20 wt.%, 25 wt.%, 30 wt.%, 35 wt.%, and 40 wt.% WP. The results indicate that increasing the WP content mostly leads to decreasing the elongation at break value. By limiting the mobility of the polymer chains, the presence of WP in the matrix reduces the sample's ability to deform [27]. WPC with 20 wt.% WP has the highest elongation value of 7.40 wt.%, while WPC with 35 wt.% WP obtains the lowest elongation value of 4.31 wt.%. Furthermore, the elongation at the break value of the WPC samples is smaller than the original PP polymer due to the low elongation of the WP. Kajaks et al. [25] also indicated the reduction of the elongation at break value when increasing the WP content from 20 wt.% to 40 wt.%. The reduction value is minor due to the existence of the compatibilizer in the WPC.

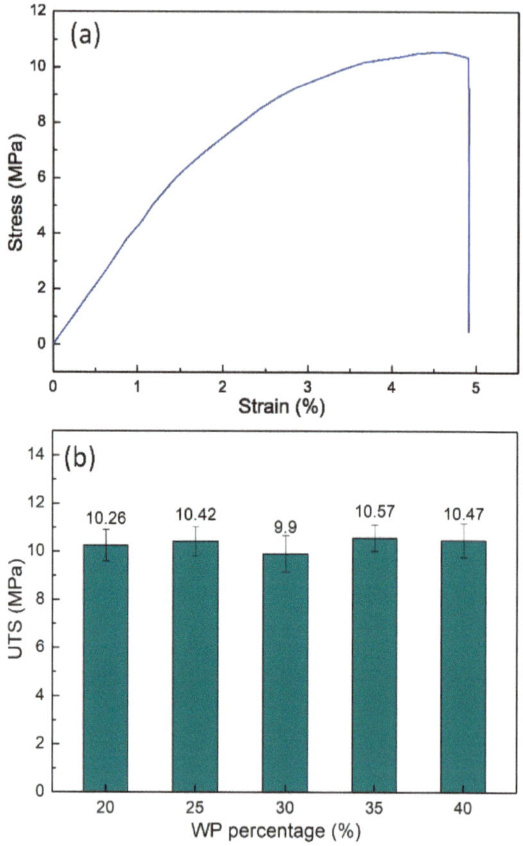

Figure 2. Stress–strain diagram and UTS comparison of WPC at different WP percentages: (**a**) stress–strain diagram of sample with 20 wt.% WP; (**b**) UTS comparison diagram.

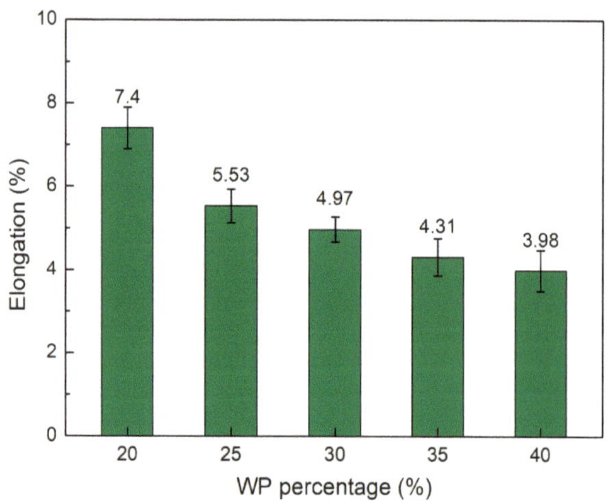

Figure 3. Elongation values comparison of WPC at different WP percentages.

Figure 4 shows the elastic modulus value comparison of WPC at different WP percentages. The elastic modulus values are 308.7 GPa, 364.7 MPa, 433.6 MPa, 422.7 MPa, and 320.8 MPa, corresponding to 20 wt.%, 25 wt.%, 30 wt.%, 35 wt.%, and 40 wt.% WP. Different from the elongation diagram, which goes down and then rises when increasing the WP content, the elastic modulus diagram goes up and then goes down. The reason is the reverse characteristic of the elongation and elastic modulus values. From 20 wt.% to 30 wt.% WP, with compatibilizer, there is a strong interaction between the PP and WP particles. This will reduce chain mobility and increase the elastic modulus with increasing WP content. On the other hand, the increase in WP could lead to the accumulation of WP, leading to a reduction in the elastic modulus due to the coarsening of WP. With 30 wt.% WP, the WPC reaches the highest value of elastic modulus at 433.6 MPa. On the contrary, with 20 wt.% WP, the WPC has the lowest elastic modulus at 308.7 MPa. Salemane et al. [27] also figured out that increasing the WP content will first lead to an increase in the elastic modulus, and then, further increases in the WP will result in a reduction in the elastic modulus of the WPC.

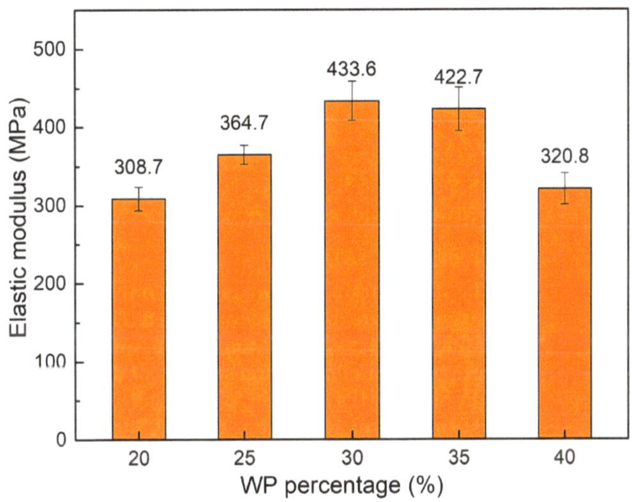

Figure 4. Elastic modulus values comparison of WPC at different WP percentages.

Figure 5 shows the stress–strain diagram of samples with different WP percentages. The area below the stress–strain curves reduces gradually when the WP percentage increases from 20 wt.% to 40 wt.%. Notably, from 20 wt.% to 30 wt.% WP, this area reduces rapidly. From 30 wt.% to 40 wt.%, the reduction speed is slower. These results are consistent with the UTS and elongation results, which are presented in Figures 2 and 3. The reduction of the area below the stress–strain when increasing the WP content also indicates the reduction in the toughness of the WPC when increasing the WP content.

In general, the UTS value of the WPC changes slightly around 10.4 MPa when changing the WP content from 20 wt.% to 40 wt.%. The elongation at break value reaches the highest value of 7.40 wt.% at 20 wt.% WP, while the elastic modulus obtains the highest value of 433.6 MPa at 30 wt.% WP. Notably, the WPC with 25 wt.% WP has good tensile characteristics with a UTS value of 10.42 MPa, an elongation value of 5.53 MPa, and an elastic modulus value of 364.7 MPa.

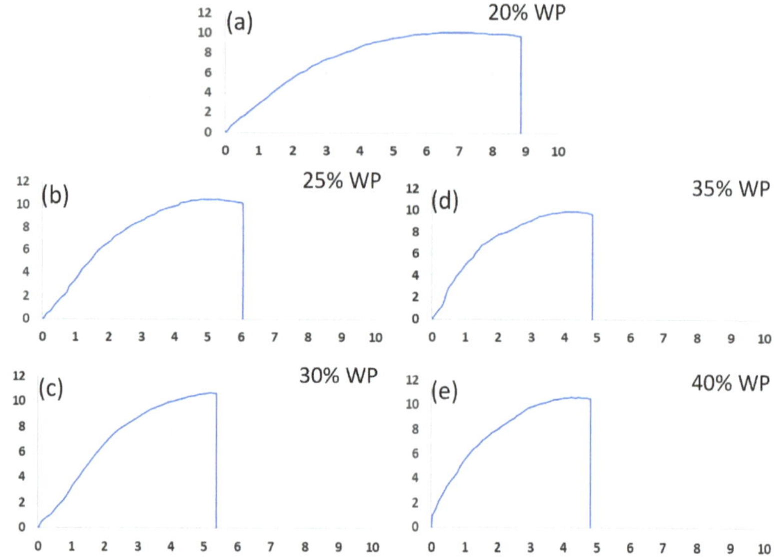

Figure 5. Stress–strain diagram of samples with different WP percentages: (**a**) 20 wt.% WP; (**b**) 25 wt.% WP; (**c**) 30 wt.% WP; (**d**) 35 wt.% WP; and (**e**) 40 wt.% WP.

3.2. Flexural Test Results

The stress–strain diagram and flexural strength comparison of WPC at different WP percentages are displayed in Figure 6. The flexural strength values are 14.06 MPa, 15.48 MPa, 14.64 MPa, 15.49 MPa, and 16.0 MPa. Generally, increasing the WP leads to an improvement in the flexural strength of the WC. The WPC has the highest flexural strength of 16.0 MPa at 40 wt.% WP, while at 20 wt.% WP, the WPC has the lowest flexural strength of 14.06 MPa. Overall, the flexural strength value oscillates gently around 15 MPa. This result is similar to the UTS value, which also indicates that the WP content has a slight impact on the UTS and flexural strength. However, the flexural strength value has a higher deviation value, indicating that WP content has a stronger impact on flexural strength than the UTS. Kaymakci et al. [40] mixed PP with 50 wt.% rice husk powder, achieving a flexural strength of 11.27 MPa, which is lower than this study's results. Valles-Rosales et al. [41] achieved a flexural strength of 16.2 MPa when mixing PP with 42.5 wt.% chili-stem waste particles.

The maximum flexural strain at break comparison diagram of the WPC at different WP percentages is presented in Figure 7. The elongation values of the WPC samples are 15.50 wt.%, 9.67 wt.%, 7.97 wt.%, 7.22 wt.%, and 5.06 wt.%, corresponding to 20 wt.%, 25 wt.%, 30 wt.%, 35 wt.%, and 40 wt.% WP. Interestingly, adding more WP to the WPC leads to a gradual decrease in the maximum flexural strain value. At 20 wt.% WP, the WPC has the highest maximum flexural strain value of 15.50 wt.%. It declines to the lowest value of 5.06 wt.% when the WP content increases to 40 wt.%. The reason is that the WP does not have as much ductility as the PP matrix; therefore, adding more WP results in a reduction in the maximum flexural strain at break value.

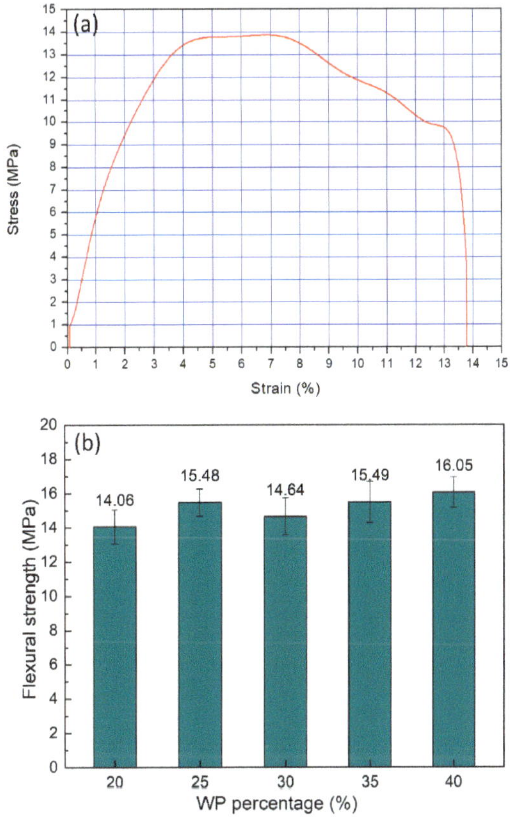

Figure 6. Stress–strain diagram and flexural strength comparison of WPC at different WP percentages: (**a**) stress–strain diagram of sample with 20 wt.% WP; (**b**) UTS comparison diagram.

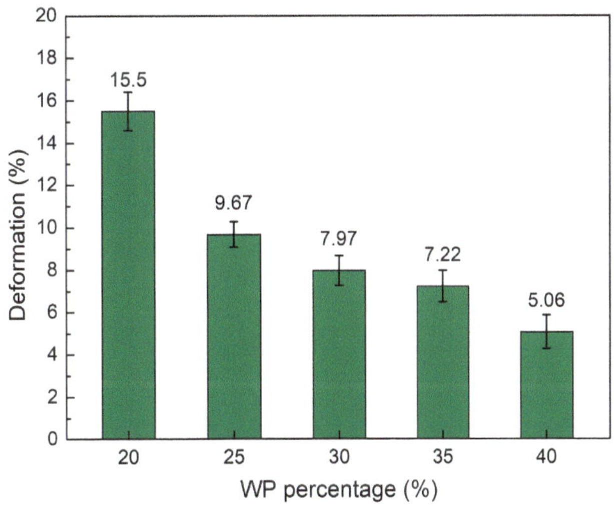

Figure 7. Maximum flexural strain at break values comparison of WPC at different WP percentages.

Figure 8 exhibits the flexural elastic modulus value comparison of WPC at different WP percentages. The elastic modulus values of the WPC are 511 MPa, 560.9 MPa, 565 MPa, 623.9 MPa, and 664.2 MPa, corresponding to 20 wt.%, 25 wt.%, 30 wt.%, 35 wt.%, and 40 wt.% WP. Increasing the WP content gives rise to the flexural elastic modulus value, which is reversed to the maximum flexural strain value. This result is consistent with Bessa et al.'s study [42], which also indicated that increasing the WP content leads to an improvement in the flexural elastic modulus of the WPC. However, increasing the WP content mostly leads to a reduction in the elongation at break and elastic modulus, as mentioned in the tensile test.

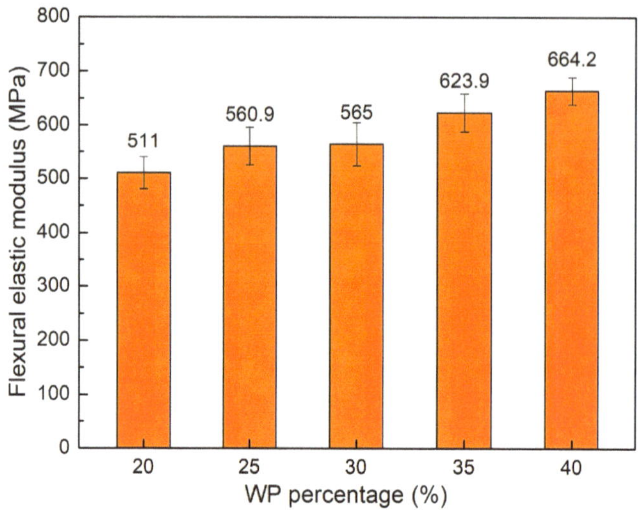

Figure 8. Flexural elastic modulus values comparison of WPC at different WP percentages.

Figure 9 shows the shore D hardness value comparison of WPC at different WP percentages. The shore D hardness values of the WPC are 70.8, 0.76 wt.%, 0.97 wt.%, 1.17 wt.%, and 1.58 wt.%, corresponding to 20 wt.%, 25 wt.%, 30 wt.%, 35 wt.%, and 40 wt.% WP. Increasing the WP content gives rise to the hardness of the WPC due to the higher hardness of the WP compared to the PP matrix. This is also the reason why increasing the WP content leads to a reduction in the maximum flexural strain and an increase in the flexural elastic modulus, as shown in Figures 7 and 8. This result is consistent with the Teymoorzadeh et al. [43] study, which also investigated the impact of WPC on hardness. In that study, increasing the WP from 0 wt.% to 40 wt.% led to a gradual increase in shore D hardness from 79 to 89 of the polylactic acid wood composite.

Overall, the flexural strength of the WPC varies around 15 MPa when the WP increases from 20 wt.% to 40 wt.%. The increase in the WP mainly leads to increased flexural strength. The flexural elastic modulus also increases when the WP content increases. On the contrary, the maximum flexural strain decreases when the WP content increases. The maximum flexural strain value achieves the highest number of 15.50 wt.% at 20 wt.% WP, while the flexural elastic modulus reaches the highest value of 664.2 MPa at 40 wt.% WP. Remarkably, if the maximum flexural strain is not critical, WPC samples with 40 wt.% WP are a good selection due to their good flexural strength of 16.0 MPa and flexural elastic modulus of 664.2 MPa. The highest shore D hardness of 77.6 of the WPC is obtained at 40 wt.% WP. Increasing the WP content will increase the hardness of the WPC.

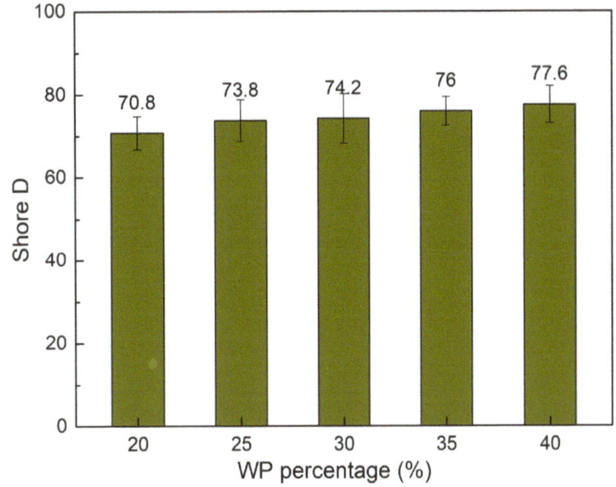

Figure 9. Shore D hardness value comparison of WPC at different WP percentages.

3.3. SEM, DSC and Water Absorption Tests Results

Figure 10 shows the SEM figures of WPC at different WP percentages. Before SEM testing, the samples are not coated with gold or platinum. With the compatibilizer Olenamide 5 wt.%, all WPC samples present good bonding between the WP and PP matrix. There are no separations between WP and PP matrix, indicating the high quality of the samples. The WP size varies around 90.2 μm–112.9 μm, which is a fine WP. Moreover, the WP was distributed evenly on the composite matrix, indicating the good effect of the compatibilizer Olenamide. Remarkably, with the presence of the compatibilizer Olenamide, increasing the WP content from 20 wt.% to 40 wt.% does not lead to the accumulation of the WP in the composite matrix. Ratanawilai et al. [44] indicated that reducing the average WP size from 315 μm to 180 μm led to a slight increase in the tensile and flexural strength. In this study, the WP size is smaller than that report, indicating that the WP size is small enough to generate a good WPC.

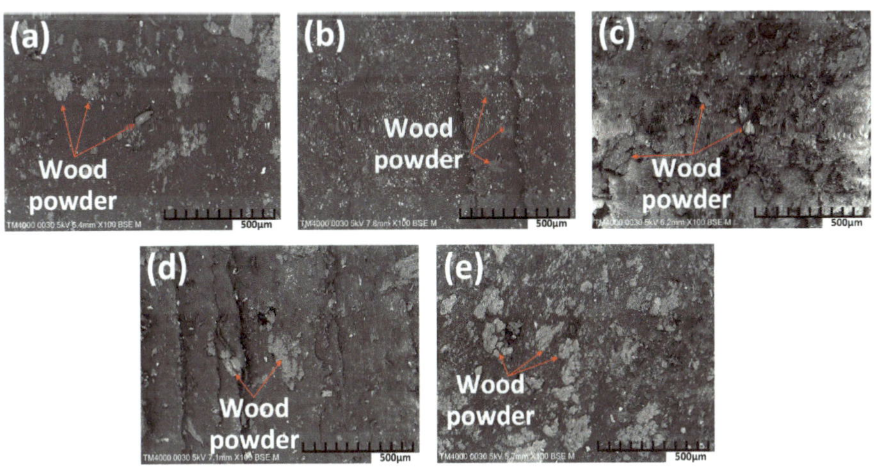

Figure 10. SEM figures of WPC at different WP percentages: (**a**) 20 wt.% WP; (**b**) 25 wt.% WP; (**c**) 30 wt.% WP; (**d**) 35 wt.% WP; and (**e**) 40 wt.% WP.

DSC results of WPC at different WP percentages are presented in Figure 11. Each sample composition uses about 15 mg to conduct the test. The crystallinity is calculated by the following equation:

$$X = \frac{\Delta H_m}{\Delta H_{m0}} \times 100 \text{ wt.\%}, \quad (1)$$

where ΔH_m is melting enthalpy; the ΔH_{m0} of PP is 207 J/g [45].

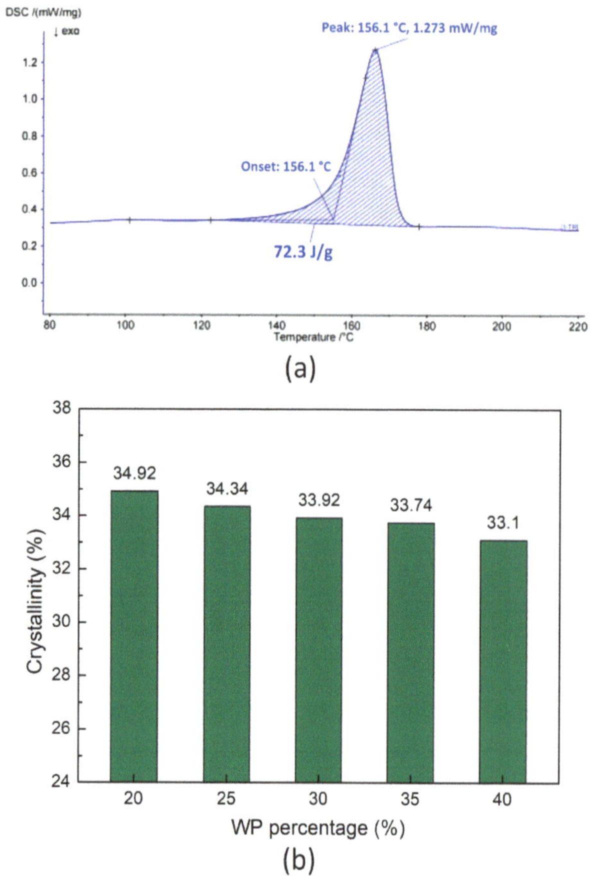

Figure 11. DSC results of WPC at different WP percentages: (**a**) DSC curve of sample with 20 wt.% WP; (**b**) crystallinity comparision.

The crystallinity values of the WPC are 34.92 wt.%, 34.34 wt.%, 33.92 wt.%, 33.74, and 33.1 wt.%, corresponding to 20 wt.%, 25 wt.%, 30 wt.%, 35 wt.%, and 40 wt.% WP. The crystallinity of the WPC reduces gradually and slightly when the WP content increases. The presence of coconut WP, which has an amorphous structure, leads to a lower crystallinity percentage of the WPC. However, if the true fraction of PP is considered during the crystallinity calculation, meaning the WP is not counted, the crystallinity values are higher. The crystallinity values of the PP matrix are 43.65 wt.%, 45.79 wt.%, 48.46 wt.%, 51.91, and 55.17 wt.%, corresponding to 20 wt.%, 25 wt.%, 30 wt.%, 35 wt.%, and 40 wt.% WP. It means that the PP matrix has a higher crystallinity when increasing the WP in the WPC.

Figure 12 shows the water absorption of WPC at different WP percentages. Water absorption measurements are tested via the standard ASTM D 570-88. Water absorption is calculated following equation:

$$X = \frac{W_1 - W_0}{W_0} \times 100\ \text{wt.\%} \quad (2)$$

where W_1 is the weight of the sample after water absorption and W_0 is the original weight of the samples.

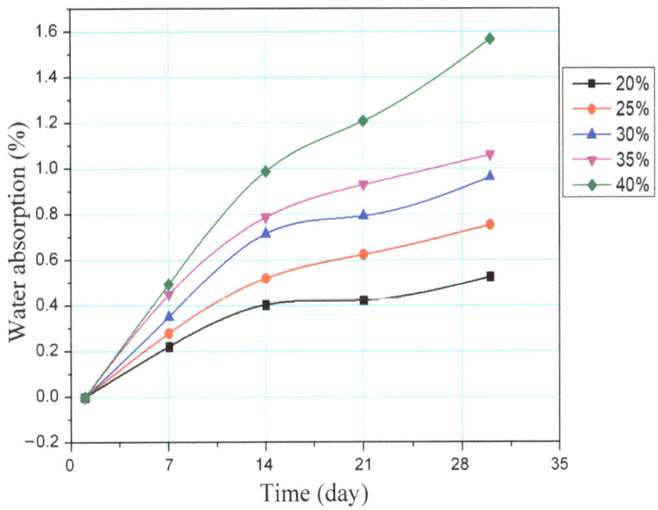

Figure 12. Water absorption of WPC at different WP percentages.

The maximum water absorption of WPC after 30 days is 0.52 wt.%, 0.76 wt.%, 0.97 wt.%, 1.17 wt.%, and 1.58 wt.%, corresponding to 20 wt.%, 25 wt.%, 30 wt.%, 35 wt.%, and 40 wt.% WP. Increasing the WP content leads to an improvement in the water absorption of the WPC, which is similar to Bessa et al.'s report [25]. The reason is the WP has a strong water absorption; therefore, increasing its content will result in a better water absorption of the WPC. Moreover, the water absorption process for 30 days could be divided into two stages. In the first stage, from 0 to the 14th day, the water absorption rises rapidly. Then, in the second stage, from the 14th day to the 30th day, the water absorption speed reduces as the WP becomes saturated. In general, the WPC samples have a low water absorption capacity.

3.4. Optimization by Taguchi Method

In this section, the study aims to optimize the tensile strength and flexural strength of the WPC using the Taguchi method. In Table 2, the injection parameters and WP are calculated following the Taguchi method via Minitab software version 20.3 with L25 orthogonal array, four factors, and five levels.

Table 3 presents the response table for means of coconut WPC for the UTS value. The results show that WP content plays the most important role in the UTS value of the coconut WPC. The filling pressure has the second rank, followed by the packing pressure. Finally, the surveyed melt temperature does not impact the UTS value as strongly as the other parameters. Therefore, controlling the WP content could lead to a suitable selection when using the WPC.

Table 2. Injection parameters and WP content of coconut WPC.

Sample No.	Filling Pressure (Bar)	Packing Pressure (Bar)	Melt Temperature (°C)	Powder Content (wt.%)
1	27	35	214	20
2	27	37.5	217	25
3	27	40	220	30
4	27	42.5	223	35
5	27	45	226	40
6	30	35	217	30
7	30	37.5	220	35
8	30	40	223	40
9	30	42.5	226	20
10	30	45	214	25
11	33	35	220	40
12	33	37.5	223	20
13	33	40	226	25
14	33	42.5	214	30
15	33	45	217	35
16	36	35	223	25
17	36	37.5	226	30
18	36	40	214	35
19	36	42.5	217	40
20	36	45	220	20
21	39	35	226	35
22	39	37.5	214	40
23	39	40	217	20
24	39	42.5	220	25
25	39	45	223	30

Table 3. Response table for means of coconut WPC for the UTS value.

Level	Filling Pressure (Bar)	Packing Pressure (Bar)	Melt Temperature (°C)	WP Content (wt.%)
1	10.352	10.326	10.374	10.276
2	10.302	10.386	10.324	10.446
3	10.366	10.344	10.334	9.994
4	10.382	10.352	10.338	10.556
5	10.332	10.326	10.364	10.462
Delta	0.080	0.060	0.050	0.562
Rank	2	3	4	1

Figure 13 presents the main effects plot for means of coconut WPC for the UTS value. The optimal parameters that could lead to the highest UTS value are a filling pressure of 36 bar, a packing pressure of 37.5 bar, a melt temperature of 214 °C, and a WP content of 35 wt.%. The predicted UTS value that applied these parameters is 10.658 MPa.

Table 4 represents the response table for means of coconut WPC for the flexural strength value. Similar to the UTS value, WP content also has the highest impact on the flexural strength value of the coconut WPC. The melt temperature has the second position on the impact range of the flexural strength value, followed by the packing pressure. Notably, the filling pressure has the weakest impact on the flexural strength value. This is different from the UTS value, in which the filling pressure ranks second place. Overall, the WP content is the most important parameter when considering both the UTS and the flexural strength values.

Figure 14 shows the main effects plot for means of coconut WPC for the flexural strength value. The optimal parameters are a filling pressure of 27 bar, a packing pressure of 45 bar, a melt temperature of 226 °C, and a WP content of 40 wt.%. The predicted flexural strength value that applied these parameters is 10.464 MPa.

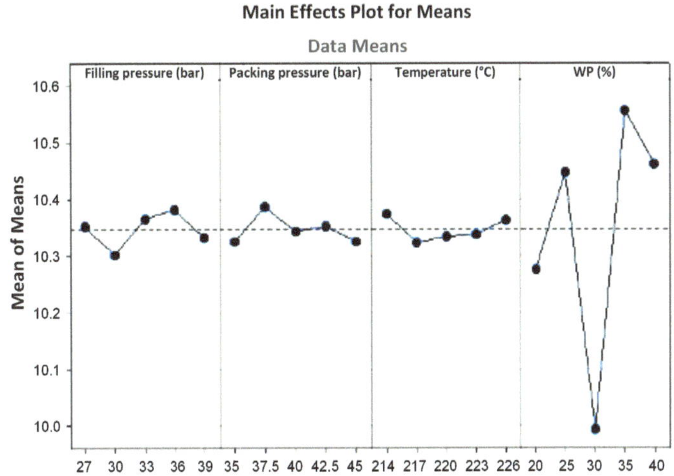

Figure 13. Main effects plot for means of coconut WPC for the UTS value.

Table 4. Response table for means of coconut WPC for the flexural strength value.

Level	Filling Pressure (Bar)	Packing Pressure (Bar)	Melt Temperature (°C)	WP Content (wt.%)
1	15.22	15.23	15.24	14.10
2	15.21	15.24	15.21	15.29
3	15.13	15.13	14.86	14.98
4	15.20	14.99	15.24	15.49
5	15.10	15.26	15.31	15.99
Delta	0.12	0.27	0.45	1.89
Rank	4	3	2	1

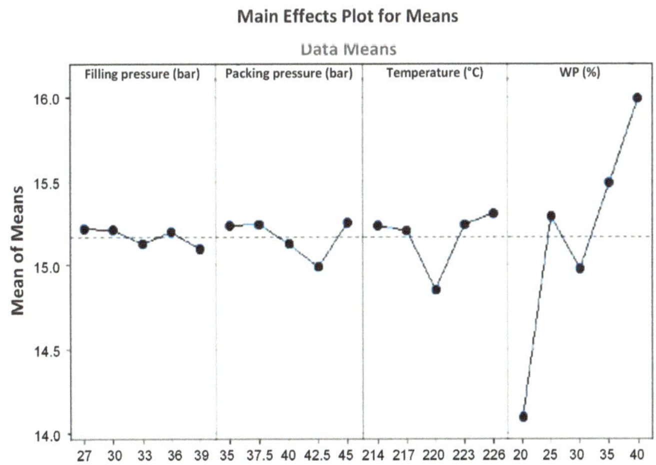

Figure 14. Main effects plot for means of coconut WPC for the flexural strength value.

4. Conclusions

This study studies the mechanical properties of coconut sawdust powder mixed with PP. The impact of compatibility content, WP content, and injection molding parameters on the characteristics of the coconut WPC is surveyed. Some important results are as follows:

- Depending on the coconut WP content and the desired properties, the mechanical properties of the WPC would achieve their best results at different WP contents. The highest UTS value is achieved at 35 wt.% WP. The elongation reaches the highest value of 7.40 wt.% at 20 wt.% WP, while the elastic modulus obtains the highest value of 433.6 MPa at 30 wt.% WP.
- The flexural strength, the flexural elastic modulus, and the shore D hardness have a linear relation to the WP content. An increase in WP content will increase these properties, reaching their best properties at 40% WP.
- SEM results show that all WPC samples present good bonding between the WP and the PP matrix. Moreover, the WP is distributed evenly on the composite matrix due to the presence of the compatibilizer.
- The crystallinity of the WPC reduces gradually and slightly when the WP content increases, indicating that the WP does not strongly impact the WPC crystallinity.
- Taguchi's results show that WP content plays the most important role in the UTS value of the coconut WPC. The filling pressure has the second rank, followed by the packing pressure. Finally, the surveyed melt temperature does not strongly impact the UTS value like the other parameters. Therefore, controlling the WP content could lead to a suitable selection when using the WPC. Further studies should focus on WPCs with a coconut WP percentage greater than 45%.

Author Contributions: P.S.M. and V.-T.N.: Conceptualization, funding acquisition; P.S.M. and V.-T.N.: writing original draft; D.T.T.P. and V.-T.N.: analysing, visualization; Q.V.T., P.S.M. and V.-T.N.: project administration; Q.V.T., P.S.M. and V.-T.N.: investigation; D.T.T.P., P.S.M. and V.-T.N.: writing, review, and editing. All authors have read and agreed to the published version of the manuscript.

Funding: Ho Chi Minh City University of Technology and Education, Vietnam, funded this research, and the Polymers Editorial Office supports APC.

Institutional Review Board Statement: Not applicable.

Data Availability Statement: The data used to support the findings of this study are available from the corresponding author upon request.

Acknowledgments: The authors acknowledge the support of HCMC University of Technology and Education.

Conflicts of Interest: The authors declare no conflicts of interest.

References

1. Sam-Daliri, O.; Ghabezi, P.; Flanagan, T.; Finnegan, W.; Mitchell, S.; Harrison, N. Recovery of Particle Reinforced Composite 3D Printing Filament from Recycled Industrial Polypropylene and Glass Fibre Waste. *Proc. World Congr. Mech. Chem. Mater. Eng.* **2022**, *177*, 3–4.
2. Shulga, E.; Karamov, R.; Sergeichev, I.S.; Konev, S.D.; Shurygina, L.I.; Akhatov, I.S.; Shandakov, S.D.; Nasibulin, A.G. Fused filament fabricated polypropylene composite reinforced by aligned glass fibers. *Materials* **2020**, *13*, 3442. [CrossRef]
3. Velmurugan, G.; Natrayan, L.; Chohan, J.S.; Vasanthi, P.; Angalaeswari, S.; Pravin, P.; Kaliappan, S.; Arunkumar, D. Investigation of mechanical and dynamic mechanical analysis of bamboo/olive tree leaves powder-based hybrid composites under cryogenic conditions. In *Biomass Conversion and Biorefinery*; Springer: Berlin/Heidelberg, Germany, 2023; pp. 1–13.
4. Raj, S.S.; Elangovan, E.; Arulmozhi, P.; Murugan, N. Selective micro wood reinforcements for bio-wood plastics deliberating mechanical and degradation properties. *Mater. Today Proc.* **2023**, *90*, 24–28. [CrossRef]
5. Cui, L.; Pan, M.; Zhou, Y.; Xu, H.; Ning, L.; Jia, S.; Wang, X.; Su, Z. A strong, biodegradable, and closed-loop recyclable bamboo-based plastic substitute enabled by polyimine covalent adaptable networks. *Chem. Eng. J.* **2023**, *477*, 146952. [CrossRef]
6. Burhani, D.; Sudarmanto; Wijayanto, A.; Andreansyah, I.; Widyawati, Y.; Nurhamiyah, Y.; Fransiska, D.; Agustina, S.; Kusumaningrum, W.B.; Fatriasari, W.; et al. Utilization of Indonesian seaweed in polyethylene-based composite with coconut husk powder as bio-compatibilizer. *Mater. Today Proc.* 2023. [CrossRef]

7. Khoaele, K.K.; Gbadeyan, O.J.; Chunilall, V.; Sithole, B. A review on waste wood reinforced polymer composites and their processing for construction materials. *Int. J. Sustain. Eng.* **2023**, *16*, 104–116. [CrossRef]
8. Velmurugan, G.; Karunakaran, P.; Sampath, P.S.; Varahamoorthi, R. Experimental Study of Sugarcane Bagasse Fiber with Rice Husk and Wood Powder Polymer Matrix Composite. *J. Nat. Fibers* **2023**, *20*, 2224977. [CrossRef]
9. Teacă, C.A.; Shahzad, A.; Duceac, I.A.; Tanasă, F. The Re-/Up-Cycling of Wood Waste in Wood–Polymer Composites (WPCs) for Common Applications. *Polymers* **2023**, *15*, 3467. [CrossRef]
10. Yang, Y.; Kang, X.; Yang, Y.; Ye, H.; Jiang, J.; Zheng, G.; Wei, K.; Ge, S.; Lam, S.S.; Ouyang, H.; et al. Research progress in green preparation of advanced wood-based composites. *Adv. Compos. Hybrid Mater.* **2023**, *6*, 202. [CrossRef]
11. Ribeiro, L.S.; Stolz, C.M.; Amario, M.; Silva, A.L.N.D.; Haddad, A.N. Use of Post-Consumer Plastics in the Production of Wood-Plastic Composites for Building Components: A Systematic Review. *Energies* **2023**, *16*, 6549. [CrossRef]
12. Pan, J.; Ren, J.; Xie, Y.; Wei, X.; Guan, Y.; Yan, X.; Tang, H.; Cheng, X. Porous SiOC composites fabricated from preceramic polymers and wood powders for efficient dye adsorption and removal. *Res. Chem. Intermed.* **2017**, *43*, 3813–3832. [CrossRef]
13. Guo, X.; Gao, Q.; Du, D.; Sun, C. Effects of Filling Rate and Resin Concentration on Pore Characteristics and Properties of Carbon Based Wood Ceramics. *Materials* **2021**, *14*, 2441. [CrossRef] [PubMed]
14. Amirthan, G.; Udayakumar, A.; Prasad, V.B.; Balasubramanian, M. Solid particle erosion studies on biomorphic Si/SiC ceramic composites. *Wear* **2010**, *268*, 145–152. [CrossRef]
15. Chen, H.; Bahmani, M.; Humar, M.; Cheng, D. Properties of wood ceramics prepared from thermo-modified Poplar. *Forests* **2020**, *11*, 1204. [CrossRef]
16. Yu, M.; Zhang, G.J.; Saunders, T. Wood-derived ultra-high temperature carbides and their composites: A review. *Ceram. Int.* **2020**, *46*, 5536–5547. [CrossRef]
17. Dikobe, D.G.; Luyt, A.S. Comparative study of the morphology and properties of PP/LLDPE/wood powder and MAPP/LLDPE/wood powder polymer blend composites. *Express Polym. Lett.* **2010**, *4*, 729–741. [CrossRef]
18. Nygård, P.; Tanem, B.S.; Karlsen, T.; Brachet, P.; Leinsvang, B. Extrusion-based wood fibre–PP composites: Wood powder and pelletized wood fibres—A comparative study. *Compos. Sci. Technol.* **2008**, *68*, 3418–3424. [CrossRef]
19. Buddi, T.; Rao, B.N.; Singh, S.K.; Purohit, R.; Rana, R.S. Development and analysis of high density poly ethylene (HDPE) nano SiO_2 and wood powder reinforced polymer matrix hybrid nano composites. *J. Exp. Nanosci.* **2018**, *13* (Suppl. S1), S24–S30. [CrossRef]
20. He, P.; Bai, S.; Wang, Q. Structure and performance of Poly (vinyl alcohol)/wood powder composite prepared by thermal processing and solid state shear milling technology. *Compos. Part B Eng.* **2016**, *99*, 373–380. [CrossRef]
21. Gardner, D.J.; Han, Y.; Wang, L. Wood–plastic composite technology. *Curr. For. Rep.* **2015**, *1*, 139–150. [CrossRef]
22. da Silva, W.A.; Luna, C.B.B.; de Melo, J.B.D.C.A.; Araujo, E.M.; Filho, E.A.D.S.; Duarte, R.N.C. Feasibility of manufacturing disposable cups using PLA/PCL composites reinforced with wood powder. *J. Polym. Environ.* **2021**, *29*, 2932–2951. [CrossRef]
23. Das, A.K.; Agar, D.A.; Rudolfsson, M.; Larsson, S.H. A review on wood powders in 3D printing: Processes, properties and potential applications. *J. Mater. Res. Technol.* **2021**, *15*, 241–255. [CrossRef]
24. Huang, Y.; Löschke, S.; Proust, G. In the mix: The effect of wood composition on the 3D printability and mechanical performance of wood-plastic composites. *Compos. Part C Open Access* **2021**, *5*, 100140. [CrossRef]
25. Kajaks, J.; Kalnins, K.; Uzulis, S.; Matvejs, J. Physical and mechanical properties of composites based on polypropylene and timber industry waste. *Cent. Eur. J. Eng.* **2014**, *4*, 385–390. [CrossRef]
26. Kajaks, J.; Kolbins, A.; Kalnins, K. Some exploitation poperties of wood plastic composites (WPC) based on high density polyethylene (HDPE) and plywood poduction waste. In *IOP Conference Series: Materials Science and Engineering*; IOP Publishing: Bristol, UK, 2016; Volume 111, p. 012003.
27. Salemane, M.G.; Luyt, A.S. Thermal and mechanical properties of polypropylene–wood powder composites. *J. Appl. Polym. Sci.* **2006**, *100*, 4173–4180. [CrossRef]
28. Yu, Y.; Yang, Y.; Murakami, M.; Nomura, M.; Hamada, H. Physical and mechanical properties of injection-molded wood powder thermoplastic composites. *Adv. Compos. Mater.* **2013**, *22*, 425–435. [CrossRef]
29. Lin, Z.; Xiang, Q.; Dongmei, X. Study on Water Resistance of Polypropylene Based Wood-Plastic Composites Used in Building. *Compos. Adv. Mater.* **2023**, *32*, 26349833231218017. [CrossRef]
30. *ISO 1133-1:2022*; Plastics—Determination of the Melt Mass-Flow Rate (MFR) and Melt Volume-Flow Rate (MVR) of Thermoplastics—Part 1: Standard Method. ISO: Geneva, Switzerland, 2022.
31. *ISO 527-2:2012*; Plastics—Determination of Tensile Properties—Part 2: Test Conditions for Moulding and Extrusion Plastics. ISO: Geneva, Switzerland, 2012.
32. *ISO 179-2:2020*; Plastics—Determination of Charpy Impact Properties—Part 2: Instrumented Impact Test. ISO: Geneva, Switzerland, 2020.
33. *ISO 2039-1:2001*; Plastics—Determination of Hardness—Part 1: Ball Indentation Method. ISO: Geneva, Switzerland, 2001.
34. *ISO 3146:2022*; Plastics—Determination of Melting Behaviour (Melting Temperature or Melting Range) of Semi-Crystalline Polymers by Capillary Tube and Polarizing-Microscope Methods. ISO: Geneva, Switzerland, 2022.
35. *ISO 75-2:2013*; Plastics—Determination of Temperature of Deflection under Load—Part 2: Plastics and Ebonite. ISO: Geneva, Switzerland, 2013.

36. *ISO 306:2022*; Plastics—Thermoplastic Materials—Determination of Vicat Softening Temperature (VST). ISO: Geneva, Switzerland, 2022.
37. *ISO 1183-1:2019*; Plastics—Methods for Determining the Density of Non-Cellular Plastics—Part 1: Immersion Method, Liquid Pycnometer Method and Titration Method. ISO: Geneva, Switzerland, 2019.
38. *ASTM D638*; Standard Test Method for Tensile Properties of Plastics. ASTM: West Conshehoken, PA, USA, 2022.
39. *ASTM D790*; Standard Test Methods for Flexural Properties of Unreinforced and Reinforced Plastics and Electrical Insulating Materials. ASTM: West Conshehoken, PA, USA, 2017.
40. Kaymakci, A. Effect of titanium dioxide on some mechanical, thermal, and surface properties of wood-plastic nanocomposites. *BioResources* **2019**, *14*, 1969–1979. [CrossRef]
41. Valles-Rosales, D.J.; Rodríguez-Picón, L.A.; Méndez-González, L.C.; del Valle-Carrasco, A.; Alodan, H. Analysis of the mechanical properties of wood-plastic composites based on agriculture Chili pepper waste. *Maderas Cienc. Y Tecnol.* **2016**, *18*, 43–54. [CrossRef]
42. Bessa, J.; Mota, C.; Cunha, F.; Fangueiro, R. Influence of different thermoplastic polymer/wood ratios on the mechanical and thermal properties of composite materials. *Procedia Eng.* **2017**, *200*, 480–486. [CrossRef]
43. Teymoorzadeh, H.; Rodrigue, D. Biocomposites of wood flour and polylactic acid: Processing and properties. *J. Biobased Mater. Bioenergy* **2015**, *9*, 252–257. [CrossRef]
44. Ratanawilai, T.; Nakawirot, K.; Deachsrijan, A.; Homkhiew, C. Influence of wood species and particle size on mechanical and thermal properties of wood polypropylene composites. *Fibers Polym.* **2014**, *15*, 2160–2168. [CrossRef]
45. Torrecillas, H.V.; Costa, L.C.; Souza, A.M.C. Influence of mixing protocol on the morphology and mechanical properties of PP/SEBS/MMT and PP/SEBS/PPgMA/MMT blends. *Polym. Test.* **2018**, *72*, 322–329. [CrossRef]

MDPI AG
Grosspeteranlage 5
4052 Basel
Switzerland
Tel.: +41 61 683 77 34

Polymers Editorial Office
E-mail: polymers@mdpi.com
www.mdpi.com/journal/polymers

www.ingramcontent.com/pod-product-compliance
Lightning Source LLC
Chambersburg PA
CBHW041407210926
51850CB00013B/3200